French Bean Diseases: Diagnosis, Epidemiology and Management

Other Related Books

Title	Author
Agricultural Microbiology	R.P. Pareek
Agrochemicals in Plant Disease Management	N.G. Ravichandran
Approaches and Trends in Plant Disease Management	Satish K. Gupta
Bacterial and Viral Disease and Their Management in Horticultural Crops	P.P. Reddy
Chemistry of Insecticides and Fungicides (3rd Revised Ed)	U.S. Sree Ramulu
Detection and Diagnosis of Plant Diseases	A K Meena
Disease Problems in Vegetable Production (2nd Ed)	Satish K. Gupta
Diseases of Horticultural Crops: Nematode Problems and Their Management	P.P. Reddy
Diseases of Vegetable Ornamental and Spice Crops	Satish K. Gupta
Fundamental of Plant Pathology	S. K. Tripathi
Fungal Diseases and Their Management in Horticultural Crops	P.P. Reddy
Instrumentation in Biosciences	S.M. Reddy
Integrated Disease Management and Plant Health	V. K. Gupta
Integrated Management of Plant Resources	M.K. Rai
Integrated Plant Disease Management	R.C. Sharma
Integrated Plant Pathology	S.M. Reddy
Laboratory Exercises in Plant Pathology (Part I-II) (Set)	A.B.A.M. Baudoin
Laboratory Manual of Microbiology	Vivek Kumar
Manual of Diseases and Insect-Pests of Important Vegetables	Satish K. Gupta
Methods in Fungal Biology: A Manual of Laboratory Protocols	A.K. Gautam
Objective Plant Pathology (2nd Ed)	Susanta Banik
One Hundred Years of Plant Pathology in India : An Overview	S.S. Chahal
Phytopathogenic Procaryotes and Plant Diseases	B.S. Thind
Plant Pathogens and Their Control in Horticulture	G.R. Dixon
Plant Pathology and Mycology: A Treatise	G. Bagyanarayana
Plant Pathology at A Glance (Encyclopedia of Plant Pathology) (2nd Ed)	D.P. Tripathi
Plant Pathology in India	S.S. Chahal
Recent Trends in Plant Pathology	B. Chakraborty
Research Methods in Plant Sciences: Allelopathy (Vol. 1-5) (Set)	S.S. Narwal
Soilborne Plant Pathogens Management of Diseases with Macro and Microelements	A.W. Engelhard
Vegetable Crops at A Glance	S.K. Tyagi

For more details visit website.

French Bean Diseases: Diagnosis, Epidemiology and Management

Dr. Satish K. Gupta
Professor (Plant Pathology)
MS Swaminathan School of Agriculture
Shoolini University of Biotechnology and Management Sciences, Solan-173 212

Dr. (Mrs.) Monica Sharma
Associate Professor (Plant Pathology)
Dept. of Plant Pathology
College of Horticulture and Forestry
Dr YSP University of Horticulture and Forestry
Neri, Hamirpur-177001

SCIENTIFIC PUBLISHERS

Published by
SCIENTIFIC PUBLISHERS (INDIA)
5 A, New Pali Road, P.O. Box 91
Jodhpur 342 001 (INDIA)
E-mail: info@scientificpub.com
Website: http://www.scientificpub.com

ISBN: 978-81-96153-02-1
eISBN: 978-81-96153-04-5

Printed in India

Shoolini University

Think Learning Think Success

FOREWORD

Agriculture has stood the test of time for India, both on the economy and food front. More than half of its population is engaged in production of various crops like cereals, pulses, fiber crops, fruits, vegetables, etc. The country has made phenomenal progress in the field of agriculture; and we are now a food-exporting nation from, not so long ago, a food-importing country. However, the nutritional status of a sizeable proportion of our population is still a challenge, and our policy makers are alive to this problem. Consequently, policies and programmes are being framed and implemented to address this issue. Addressing the goals of zero-hunger and malnutrition, emphasis is now on agricultural crops and production programmes that can be of help in this regard. Focus on millets, pulses, fruits and vegetables for research & development, steps being taken in this direction.

Protein-energy and micronutrient deficiencies can be addressed by increasing the consumption of pulses, which are a rich source of proteins, minerals, iron, and fiber. Although, India ranks first globally in terms of the area under pulses and the production of pulses, yet it is not self-sufficient and remains a net importer of pulses. Per capita availability of pulses in India today is 55 g/capita/day which is much less than the prescribed norm of 80 g/capita/day of World Health Organization.

Farmers are now patronizing pulses and other nutraceutical crops and French bean is one such crop. French beans (*Phaseolus vulgaris*) are rich source of proteins, minerals, and fiber. As a result, the area under this crop has reached 221 thousand hectares. Beans are grown both for green pods as well as dry beans (rajmash) used as pulse. However, successful cultivation of this crop is hindered due to various biotic and abiotic factors of which diseases of fungal, bacterial, viral and nematode nature play a major role. These diseases reduce the yield to the extent of about 25 %. To keep the ravages of these diseases in check, there is a need to have proper knowledge of diagnosis, etiology of the pathogen, disease cycle and epidemiology of the disease.

Conventional plant disease management based on chemical pesticides may not be acceptable in the present scenario, where protection of the environment and human health is a serious consideration. I am happy to know that a book entitled **"French-bean Diseases: Diagnosis, Epidemiology and Management"** is being published by the MS Swaminathan School of Agriculture of the University, wherein sincere efforts have been made to describe the fungal, bacterial viral and nematode diseases infecting the crop in a simplified manner. The symptomatology, pathogen, disease cycle, pre-disposing factors of each disease along with

Shoolini University
Post Box 9, Solan
173212, HP, India

T 01792-350000
M 78078-99712
W www.shooliniuniversity.com

management through cultural, host resistance, biological and chemical methods have been described in this book. I am sure that this book will not only serve the students and scientists but will also be helpful to extension workers and progressive farmers.

I appreciate the efforts made by the authors in publishing the book.

Dr. PK Khosla
Chancellor
Shoolini University, Solan- H.P. - 173 212

Preface

India is an Agricultural country and about 55% population is engaged in Agriculture. Contribution of Agriculture in India's Gross Domestic Product is immense i.e around 4%. We have already gained self sufficiency in food production but still we are importer of edible oils and pulses. Due to low production during 2021-22 crop season, India imported 9.44% more pulses compared to 2020-21 crop season. The value of the pulses imported in India has increased from 80.35 billion rupees in 2018-19 to 119.38 billion rupees in 2020-21. The major constraint in attaining self sufficiency in pulses is the non-availability of high yielding varieties and diseases caused by fungi, bacteria, virus (s) and nematodes which not only reduce the quantity but also impair the quality of the produce.

French bean (*Phaseolus vulgaris* L.) a leguminous vegetable grown for green pods and dry pulse (*Rajmah*) throughout the world and a rich source of proteins. During cultivation the crop is infected by various diseases which cause substantial yield losses. Due to climate change some diseases which were of minor importance earlier have become major. Moreover some diseases which were not earlier present were also introduced through import of seeds. In the present scenario, proper diagnosis and knowledge of epidemiological parameters of various diseases is very important for taking up suitable management practices by the farmers. Otherwise they will use pesticides indiscriminately to combat the ravages of these diseases which will not only cause environmental pollution but also leave harmful residues on the produce.

The present book entitled "**French Bean Diseases: Diagnosis, Epidemiology and Management**" is an effort to compile the available information on various diseases with regard to occurrence, symptomatology, etiology, disease cycle, favourable environmental conditions and management through cultural practices, host resistance, biological and chemical methods alone and their integration. The book contains 24 chapters dealing with the diseases caused by fungi, bacteria, virus and nematode of this crop. Beside these, one chapter on Scope and Cultivation of French bean is also included. The book will help the stakeholders in proper diagnosis of diseases and to decide management practice on the basis of prevailing weather conditions to be followed so that the quality of the produce meet international sanitary and phytosanitary standards. It is hoped that the information contained in this book will be useful for both under and post-graduate students, teachers, researchers, extension workers engaged in Agriculture and Horticulture as well as elite growers.

Authors are thankful to Dr. P.K. Khosla, Honb'le Chancellor, Professor Atul Khosla, Honb'le Vice Chancellor, Dr. Y.S. Negi, Dean, MS Swaminathan School of Agriculture, Shoolini University of Biotechnology and Management Sciences, Solan, Dr. R.S. Chandel, Honb'le Vice Chancellor, Dr. Som Dev, Dean, College of Horticulture and Forestry, Dr. YSP University of Horticulture and Forestry, Solan for their continuous encouragement and providing necessary facilities. Authors are also grateful to the scientists of the Department of Plant Pathology of Dr. YSP University of Horticulture and Forestry, Solan for their suggestions provided for improvement of the manuscript.

S.K. Gupta
Monica Sharma

Content

1

SCOPE AND CULTIVATION OF FRENCH BEAN

Various types of beans are grown throughout the world, all are belonging to legumes, but falling into different genera of the family leguminosae. The important members of broad group termed as beans are common bean (*Phaseolus vulgaris* L.), lima bean (*P. lunatus* L.), scarlet bean (*P. multiflorus* Wild), tepary bean (*Mucuna sp.*). Field bean or lablab (*Dolichos lablab*), sword bean (*Canavalia ensiformis* DC.) and cluster bean (*Cyamopsis tetragonoloba* (L.) DC.). These are grown as grain crops, as vegetables for green pods and as green manure crops for incorporating into the soil. The plants also form a good fodder for cattle but the pulses in general are grown for more than one purpose. The grains are rich source of protein constituting 20 per cent of the dry matter besides having certain medicinal properties. The green pods are known to have traces of glycosides which help in reducing blood sugar.

French bean (*Phaseolus vulgaris* L.) is one of the delicious culinary items being used throughout the world both as green pods and dry pulse. Besides French bean, other synonyms like Snap bean, String bean, Kidney bean, Haricot bean etc. are also popular. It has gained increasing popularity due to its quality proteins and nutritional balance besides certain medicinal properties, which results in higher biological efficiency. The dry seeds are rich in proteins, calcium and iron and contain appreciable amount of vitamin B_1 (Stiebeling and Clark, 1939). The green pods are known to have traces of glycosides which help in reducing blood sugar. The common bean (*Phaseolus vulgaris*) produces an alpha-amylase inhibitor, which has been demonstrated to have the ability to cause weight loss and hence could be helpful in the management of obesity in human beings (Barrett and Udani, 2011). In India, the crop is largely and traditionally grown in Himachal Pradesh, Jammu and Kashmir, Uttar Pradesh, North Eastern Hills, Darjeeling, South Plateau, South Indian Hills (Nilgiri and Palni hills) Mahabaleshwar, Ratnagiri (Maharastra) and Chickmangalore (Karnataka) having mild climate with humid environmental conditions. In India French bean covers an area of 2,21,000 ha with an annual production of 22,26,000 MT (NHB, 2020) whereas in Himachal Pradesh the area under French bean cultivation is 4,252 ha with an annual production of 63,735.50 MT (Anonymous, 2020) where it is grown as an off season crop during May-October. The green pods are transported to Northern plains where they are sold at a premium price.

FRENCH BEAN

French bean (*Phaseolus vulgaris* L.) has been in cultivation since ancient times. More than 400 commercial cultivars of this crop are known in the world. It is known by a large number of names. More popular are; French bean, Kidney bean, navy bean, haricot bean, pinto bean, field bean, garden bean, common bean and snap bean. *Phaseolus vulgaris* (common bean) is the second most important legume crop in the world after soybean. In India, the fresh pods for vegetable use are known as *faras bean* whereas the dried pods for pulse are known as *rajmash.* The tender pods, particularly round poded types, which are fleshier and have less string, are preferred as vegetable to the flat-podded types. The dried seed is prized as a highly nutritious pulse, which is cooked as a whole and also canned in sauce.

Origin and domestication

Radiocarbon dates suggested that French bean was domesticated in Central America about 7000 years ago (Kaplan, 1965). Vavilov (1951) reported its primary center of origin in South Mexico and Central America and secondary center of origin in Peruvian – Ecuadorian – Bolivian area. When the European reached America they observed the red Indian cultivating beans along with maize. It was introduced into Europe after the discovery of America by Columbus in 1542, perhaps by the Spanish explorers. From Europe, it spread to Africa, West Indies, India and other parts of the world. It reached India in 17th century (Simmonds, 1976).

Taxonomy

Family	:	Leguminoae
Sub-family	:	Lotordeae
Class	:	Papilionidae
Tribe	:	Phaseoleae
Genus	:	*Phaseolus*
Species	:	*vulgaris*

Phaseolus aborigineus is considered as the progenitor of French bean. There are around 30 species under *Phaseolus*, out of which 5 are domesticated, i.e. *Phaseolus vulgaris* (French bean), *P. coccineus* (Runner bean), *P. aculifolius* (Tepary bean), *P. lunatus* (Lima bean), *P. polyanthus* (year long bean). Out of which all are self-pollinated except *Phaseolus coccineus*, which is cross-pollinated.

Floral Biology

French bean is a self-pollinated crop. Anthesis or opening of lower starts from 5 A.M. and continues up to 7 A.M. Anther dehiscence continues up to 5.30 A.M. to 7.30 A.M. Pollen remains fertile on day of anthesis. Stigma receptivity is at peak between 7 to 8 A.M. So, any crossing work should be done at this time.

Genetics

French bean is diploid in nature with chromosome number $2n = 2x = 22$. The genetics and inheritance pattern of different characters of French bean have been studied and their description is given below:

Growth Characters	Description
Growth habit	It is a monogenic character and indeterminate growth habit is dominant over determinate growth habit.
Earliness	It is a monogenic character and early flowering is dominant over late flowering.
Colour of foliage	It is a digenic character and green leaf is dominant over variegated one
Flower colour	It is a monogenic character and dark colour is dominant over white colour
Pod shape	It is a polygenic character. Normal round is dominant over other shape
Pod surface	It is a monogenic character and non-shiny pod is dominant over shiny pod.
Pod stringiness	It is a monogenic character and stringiness is dominant over stringlessness
Groove of pods	It is a monogenic character and non-grooved pod is dominant over grooved pods.

Uses and nutritive value

Depending upon its usages, it may be divided into four classes i.e. vegetable bean, green-shelled bean, pulse, processed bean (canned bean, frozen bean, freeze dried bean). Beans have gained increasing popularity due to its quality proteins and nutritional balance (Rubatzky and Yamaguchi, 1997). The dry seeds are rich in proteins, calcium and iron and contain appreciable amount of vitamin B1 (Stiebeling and Clark, 1939). According to Gopalan *et al.* (1982), it contains, protein 1.7 gm, calcium 50 mg, phosphours 28 mg, iron 1.7 mg, carotene 132 mg, thiamine 0.08 mg, Riboflavin 0.06 mg, vitamin C 24 mg (per 100 gm of edible pods).Beans are said to be antidibetic (Barrett and Udani 2011) and good for bladder burns, cardiac carminative, depurative, diarrhea, diuretic, dropsy, dysentery, eczema, amollient, hiccups, kidney resolvent, rheumatism, sciatica, tenersus (Duke, 1981).

Changes due to domestication

The most obvious effect of domestication and subsequent plant breeding in *Phaseolus* has been the modification of growth habit. There has been reduction in the number of branches and leaves as a consequence of domestication and an increase in leaf size and stem diameter in all species except *P. aculifolius* (Simmonds, 1976). The most important changes which have taken place in the fruiting characteristics of these species are in pod and seed size, although there has also been an appreciable increase in flower size in *P. vulgaris* and *P. coccineus*. Seed number per pod has changed in *P. vulgaris* and *P. aculifoius*, where upto nine seeds are found in wild forms but rarely more than 5 in most of the cultivars. In all species, an increase in permeability of the seeds to water has taken place under domestication, which is an important property for uniform germination and easy cooking (Miranda, 1974).

A remarkable range of tests colours occurs in the seeds of *Phaseolus* species; five main colour groups have been identified, namely; white, black, red, ocre and brown. Superimposed on these ground colours may be various patterns of spots, flecks and strips.

Pod structure has also been altered in cultivars with a reduction in dehiscence and fibre content. Three distinct pod textures are found in the common bean: the parchmented which are very fibrous and dehisce strongly at maturity, the leathery types which are less dehiscent but split readily along the sutures and the fleshy or string less pods which are indehiscent and do not split readily. Varieties with parchmented pods are used only for dry seed production; leathery podded varieties can be used for green pod production when young or haricot production when fully matured. Fleshy podded or, string less types is usually grown for the production of green pods (Smartt, 1969).

A further change, which has taken place with the dispersal of Phaseolus bean to temperate regions of the world, is in response to photoperiod. Since the genus native to the tropics, many cultivars from this region are photoperiod sensitive and responsive to short days. There are also day neutral types and all the temperate cultivars have been selected for day neutrality or tolerance to long days (Simmond, 1976).

Present status of French bean in India

Status of production

(a) Area and production: French bean occupies an important position among various kharif pulses and vegetative crops grown in temperate hills of India at an altitude ranging from 900 m to 2500 m above mean sea level (Sharma and Sohi, 1989). The crop is largely and traditionally grown in Himachal Pradesh, Jammu and Kashmir, Uttar Pradesh, North Eastern hills, Darjeeling, South Plateau, South Indian hills (Nilgiri and Palni hills), Mahabaleswar, Ratnagri (Maharashtra) and Chikmagalure (Karnataka) having mild climate. Introduction of French bean in the non-traditional areas particularly in the plains of Uttar Pradesh, Bihar, Madhya Pradesh, Maharashtra and Karnataka as a rabi crop with appropriate production technology offer good possibilities for increasing the production of French bean in India. The area under French bean (green) cultivation in India was 145,000 ha with production of 3.8 lakh tonnes during 1994. The area under bean cultivation has increased to 30,000 ha with production of 23.0 lakhs during 2018 (NHB, 2018). The statistics of area and production of French bean (green) in India during different years are given below:

Year	Area ('000 ha)	Production ('000 tons)
1994	145	388
1995	145	390
1996	148	400
1997	148	400
1998	150	410
1999	150	410
2000	150	420
2001	150	420
2002	150	420
2003	150	420
2004	150	420
2014	218	2204
2015	232	2334
2016	198	2012
2017	228	2277
2018	236	2356
2019	215	2080

Standard Agro techniques for French bean

As French bean is grown in both plains and hills of India, the standard agro techniques also varies for these regions and is given below:

Agro techniques	Plains	Hills
Sowing time	July – September January – February	March – May (Bush) June – September (Pole)
Spacing	45-60 cm × 10-15 cm (Bush) 90-100 cm × 90-100 cm (Pole)	45-60 × 10-15 cm (Bush) 90-100 × 90-100 cm (Pole)
Seed rate	85-90 kg/ha (Bush) 25-30 kg/ha (Pole)	85-90 kg/ha (Bush) 25-30 kg/ha (Pole)
Suitable variety	Bountiful, Jampa, Wade, BKW-74, EC-30021, EC-24920, Premier, Master Piece. String less, Grey Pod	VL Boni-1, Contender, Arka Komal, Pusa Parvati, Pant Anupama, SVM-1, Lakshmi, Kentucky Wonder, Blue Queen
Nutrient requirement	FYM 25-30 t/ha N:P:K 50 : 75 : 75 kg/ha	FYM 25-30 t/ha N P K 50: 75: 75 kg/ha
Irrigation	Critical stage – flowering and pod formation Flooding/Sprinker system	Critical stage – flowering and pod formation Flooding/Sprinkler system

Weed control	Fluchloralin @ 2 lit/ha as presowing application Hand weeding	Basalin 45 EC ! 1-1.5 kg a.i./ ha as pre plant application Hand weeding
Harvesting	September – November March – April	May – July September – October/ November

Arya (2000) and Thamburaj (2004)

Advances in French bean production

After the domestication of French bean in India, a number of advances in production have been proposed by different workers.

(a) Green house cultivation: French bean can also be grown in Green houses, which gives round the year production of the crop. It can also be an alternative to grow French bean during off-season. Both pole and bush type of varieties can be grown in green houses. Greenhouse grown French bean produces good quality pods and also increases the soil condition of the green house.

(b) Container gardening: Container gardening of French bean has created a new option of growing French bean by city dwellers in terrace or roof top gardens as land is a limiting factor over there. The agro techniques for container gardening is given below:

Sowing time	**January – February and July - September**
Seed rate	2-3 seed/pot
Variety	Pusa Parvati, Contender
Nutrient management	5-10 gm Urea at 7-10 days interval from 3 weeks after sowing
Insect management	Sucking insects – Malathion or Endosulfan @ 2 ml/lit of water
Disease management	Damping off and wilt – Captan @ 2 gm/lt of water

(c) Use of growth regulator: Many workers have suggested that exogenous application of growth regulator has significant effect on growth and yield of French bean. Thamburaj (2004) suggested that PCPA @ 2 ppm, L-NAA % 5-25 ppm, B-NAA @ 5-25 ppm when sprayed during poor pod set, it increased the pod setting and thereby increases the yield. Mandal and Sanyal (2004) reported that a combination of NAA 50 ppm and CPPU @ 40 ppm has an effect on increase in yield.

Water stress is a major problem in growing French bean in hills as it is grown as a rain fed crop in these areas. Shinde *et al.* (2001) reported that application of potassium along with BA @ 10 ppm during water stress increases resistance to water stress and nullify the effect of water stress on yield. Rai and Yadav (2002) reported that paclobutrazol @ 150 ppm increases the yield of French bean and CCC @ 400 ppm reduces vegetative growth and promoter early flowering and yield of French bean.

(d) Use of mulches and biofertilizer: Application of mulch helps in the conservation of moisture and also create a barrier between soil and the host to avoid splashing of different soil borne pathogen to the foliage. Ashworth and Harrison (1983) reported that mulches like straw and black plastic have significant effect on yield of French bean and increased the yield by 25 and 55 q/ ha as compared to control. Seed treatment with *Rhizobium* also increased the yield by 14.1 per cent as compared to control.

Important varieties in India

Most of the varieties grown in India are of exotic origin and a few are also selection from indigenous germplasm. Some promising varieties with commercial importance in India are listed below.

(A) Pole Types

(i) **SVM-1:** It has been evolved by interspecific hybridization between *P. vulgaris* var. Contender and *P. multiflorus* var. PLB-257. It is resistant to angular leaf spot disease and recommended for mid hills. It is a climber with green pods and string less (13-14 m long pods). Average pod yield 95 q/ha and ready for picking in 65-76 days.

(ii) **Lakshmi:** It is a cross between Contender (bush) x local (Pole) varieties. It has been adapted in mid hills.13 – 14 cm long green string less round pods. Ready for picking within 65-70 days with an average green pod yield of 120-140 q/ha. It is tolerant to angular leaf spot disease.

(iii) **Kentucky Wonder:** It is an introduction from USA. Pods ready within 60-65 day for picking, 20 cm long green pods with average pod yield of 100-125 q/ha.

(iv) **Bountiful:** This is an introduction from USA and coming up well during September to February in South India. Bush type, average pod yield is 100-120 q/ha and the pods remain tender for 4-5 days when left unharvested.

(B) Bush Type

(i) **Premier:** This is an introduction from Sweden. This bushy variety is well adapted for late sowing and less susceptible to wilt and mosaic diseases, ready for pod picking in 55-60 days, average green pod yield is 75-90 q/ha.

(ii) **Jampa:** This is an introduction from Mexico. This is a dwarf variety. This has given better yield than Local Poona and Rajendranagar. Average green pod yield is 80-85 q/ha. It is highly resistant to wilt diseases and can withstand warmer conditions.

(iii) **Contender:** An introduction from USA, bush type, takes 50-55 days for fruit picking, pod string less and average green pod yield is 80-95 q/ha. It is tolerant to powdery mildew and mosaic diseases.

(iv) **Pusa Parvati:** Released by IARI and evolved by mutation breeding from EC 1906. Early bearer and pods become ready within 45-50 days. Pods are round, tender, string less green and 15-18 cm long. Pod yield ranges from 80-85 q/ha. It is resistant to mosaic.

(v) **Arka Komal:** Released by IIHR Bangalore. This variety is an improvement over Australian introduction IIHR-60. Pods are green, straight, flat and tender. Pod yield is about 90 q/ha. The cultivar is very good for cooking and transport quality.

(vi) **VL Boni-1:** Pods are long, round, fleshy and string less. Pods are ready for harvesting within 45-60 days; average pod yield is 150-115 q/ha. It is suitable for multi-cropping.

(vii) **Pant Anupama:** Pods are round, smooth, tender, string less green and are ready for picking in 55-65 days. Average pod yield is 89 q/ha. It is resistant to angular leaf spot and moderately resistant to common mosaic virus.

Recently some private sector companies have also introduced some pencil type varieties like Faguni and Payal in India for cultivation. These are very high yielder, string less and meaty.

Constraints in production

Despite of several advances in production, still there are some constraints, which reduces the production potential of French bean in India.

Some important production constraints of French bean is described below:

- Use of old local varieties since years by the farmers is the main reason for lower production.
- Many improved production technologies like use of mulch, use of biofertilizers etc. have been proposed but farmers avoid adopting these techniques due to ignorance.
- Diseases are the major production problem and lack of varieties with multiple disease resistant is another cause of reduced production.
- Physical disorders like flower and pod drop, which occurs due to high or low temperature, is another production constraint.
- In hills, pole type varieties are grown as a kharif crop and lack of cheap materials and techniques for staking purpose.
- For processing purpose, mechanical harvesting is best but due to indeterminate growth habit with non-synchronous pod maturity mechanical harvesting is restricted.
- Post harvest technological aspects i.e. drying, shelling, storage, have not been paid adequate attention which causes loss due to post harvest insect damage.

REFERENCES

Ali M A. 1950. Genetics of resistance to French bean mosaic virus (virus I) in the bean. *Phytopathology* **40** : 69-79

Anonymous. 2020. Area and production of French bean in Himachal Pradesh, Directorate of Agriculture, Shimla-5(HP).

Arya P S. 2000. Off season vegetable growing in hills. A.P.H. Publishing Corporation, New Delhi.

Ashworth S and Harrison H. 1983. Evaluation of mulches for use in the home garden. *Horticultural Science* **18**(2) : 180-183.

Barrett ML and Udani JK. 2011. A proprietary alpha-amylase inhibitor from white bean (*Phaseolus vulgaris*): a review of clinical studies on weight loss and glycemic control. *Nutrition Journal* 17:10-24.

Biswas A K and Bhattacharya N K. 1976. Induced polyploidy in legumes III. *Phaseolus vulgaris* L., *Cytologia* **41** : 105-110.

Duke J A. 1981. Handbook of legumes of world economic importance. USDA, Beltsville, Maryland. Plennum Press, New York and London.

Gopalan C, Shastri B V and Balasubranamian S C. 1982. Nutrtive value of Indian Food. ICMR, NIN, Hyderabad.

Joshi B D and Rana R S. 1995. French bean in India. NBPGR, Shimla, India.

Kaplan L. 1965. Archeology and domestication in American Phaseolus bean. *Eco. Botany* **19**: 358.

Lal T and D S Padda. 1974. Varietal variation in relation to root rot and mosaic diseases in French beans under natural field conditions. *Journal of Research* PAU, **10**(1): 53-57.

Mandal S and Sanyal D. 2004. Responses of French bean to foliar application of NAA and CPPU. *Horticultural Journal* **17**(3): 235-240

Miranda C S. 1974. Evolutionary genetics of wild and cultivated *P. vulgaris* and *P. coccineus.* Ph.D. Thesis, Univ. of Cambride.

Nagmani A and Bhalla J K. 1979. Improve quantitative characters induced by fast neutron in *P. vulgaris*, p. 69. In: *Symposium on the role of induced mutation in crop improvement*, Hyderabad, Dept. of Botany.

Narasimhan V et al. 1983. Evaluation of French bean cvs. to rust diseases. *South Indian Horticulture* **31**: 149.

NHB. 2018. National Horticulture Board. www.nhb.org (accessed on dated 04.04.22)

NHB. 2020. National Horticulture Board. www.nhb.org.

Pandey G K and Seth J N. 1975. Chemical induced mutation in dwarf French bean. *Punjab Horticultural Journal* **15**(3/4): 132-138.

Parthasarthy V A. 1986. French bean. In : *Vegetable Crops of India*, T K Bose and Suri M G (eds), Dept. of Horticulture, BCKV, Kalyani.

Patel P N and Walker J C. 1967. Inheritance of tolerance of late blight in bean. *Phytopathology* **56**: 681-682.

Patil M M and Chaudhari A N. 1986. Heterosis in French bean. *Journal of Maharashtra Agricultural University* **11**(1): 72-73.

Pawar D B et al. 1987. Reaction of some French bean varieties to stemfly impertation under field condition. *Journal of Maharashtra Agricultural University* **12**(2):

263.

Peter K V. 1998. Genetics and Breeding of Vegetables, ICAR, New Delhi.

Prasad P R and Prasad A B. 1981. Induced variability in methionine content in *P. vulgaris*. *Prospect Vegetable in Cytology and Genetics* **3**: 679-688.

Rai N, Yadav D S and and Yadav R K. 2002. Standardization of paclobutrazol concentrations in vegetable crops. *Vegetable Science* **29**(2) : 149-153.

Rajendran R and Satyanarayana. 1988. Grow New French Beans. *Indian Horticulture* **33**(2): 6-7.

Rawat P S. et al. 1984. Chemical composition of some cultivars of French bean under Kumaon hill conditions. *Progressive Horticulture*i **16**(3/4): 326-330.

Rubatzky, V.E. and Yamaguchi, M. 1997. *World Vegetables: Principles, Production and Nutritive values*. Chapman and Hall, New York, USA, 843 pp.

Sekhon B and Chopra A K. 1975. Some eventual amino acid contents of various varieties of kidney bean *P. vulgaris*. *Journal of Research*, PAU, **12**(2) : 210-211.

Sharma P P and Joshi A K. Advances in French bean. In : *Advances in Horticulture*, Vol. 5, K L Chadha (ed), Malhotra Publishing House, New Delhi.

Sheopuria R R and Tiwari D K. 1970. Introduction of polyploidy in French bean. *JNKVV Research Journal* **4**(12) : 106-107.

Shinde A K, Jamamadagni B M and Mhatra V V. 2001. Water relation of five grain lelgumes influenced by potassium and growth regulators under water stress. *Journal of Potassium Research* **17**(1/4) : 107-111.

Shyam K R and Chakraborty P. K. 1985. Seed transmission of *Colletotrichum lindemuthianum* in different cultivars of beans. *Seed Research* **13**(1) : 211-213.

Sibernagel M J. 1986. Snap bean Breeding, pp. 244-277. In : Breeding Vegetable Crops, M J Bassett (ed), AVI Pub, Comp.

Siingh A K and Saini S S. 1982. A note on combining ability on beans. *Haryana Journal of Horticultural Science* **11**(3/4) : 270-273.

Simmonds N W. 1976. Evolution of crop plants. Longman, London and New York, p. 171.

Singh A K and Saini 1986. Combining ability and heterosis in French bean. *Indian Journal of Horticulture* **43**(1-2) : 121-126.

Singh A K and Saini S S. 1983 a. Stuides on heterosis and inbreeding depression for primary components and yield components in French bean. *South Indian Horticulture* **31**(4/5) : 187-195.

Singh A K and Saini S S. 1983b. Heterosis and combining ability studies in French bean. *SABRO Journal* **15**(1) : 17-22.

Singh B P, Singh B M, Saxena R K and Brahmi P. 1986. Germplasm resources in vegetable crops for stress conditions, resistance to pests and diseases, p. 422. In : *National Symposium on New Horizons in Resistance Breeding of Vegetable Crops. Vegetable Sciences*.

Singh R N. 1976. A new strain of bean common mosaic virus on bean in India. *Indian Journal of Mycology and Plant Pathology* **6** : 156-159.

Smart J. 1970. Interspecific hybridization between cultivated American species of the genus *Phaseolus*. *Euphytica* **19** : 480-489.

Stiebeling, H.K. and Clark, F. 1939. Planning for good nutrition. *U.S. Dept. Agri. year Book* 1939. Pp 321-349.

Subramanium D. 1979. Gamma rays induced mutants in *P. vulgaris* L. and *P. limenis* Ma. In: *Symposium on the role of induced mutation in crop improvement*, Hyderabad, Sept., 1979.

Swarup V and Gill H S. 1968. X-ray induced mutation in French bean. *Indian Journal of Genetics* **28**(1): 44-58.

Thamburaj S and Singh N. 2004. Vegetable, tuber crops and spices. ICAR, New Delhi.

Thomas T A, Singh R and Prasad R. 1983. Genetic resources for improvememnt of vegetable crops. South Indian Hort., 30th Commemoration issue, pp. 59-73.

Vaid K, Singh R M and Gupta V P. 1985. Combining ability in dry beans. *Crop Improvement* **12**(2) : 115-158.

Vavilov N I. 1950. The origin, variation, immunity and breeding of cultivated plants. *Chron. Bot.* **13** : 1-364.

Yarnell S H. 1965. Cytogenetics of the vegetable crops. *IV Lelgumes Bot. Rev.* **31**: 247-330.

2

ROOT ROT AND WEB BLIGHT

Root rot and web blight is a very serious disease of this crop throughout the world. In warm and wet conditions, it may assume serious proportions and cause significant losses.

GEOGRAPHICAL DISTRIBUTION

Root rot of French bean was first recorded during the year 1924 from New Jersey (Cook, 1924) and Philippines (Nacion, 1924). Thereafter, it was reported from Egypt Briton-Jones, 1925), England (Bunting and Dade, 1925), Ceylon (Gadd, 1927), Uganda (Snowden, 1927; Hansford, 1929), Tanganyika (Wallace, 1932), Rhodesia (Hopkins, 1939), Brazil (Viegas, 1945), Manitoba (Machaer and Brown, 1948), Medellin Columbia (Cardona Alvarez, 1944), Iran (Ahmadinejad and Okhovat, 1976), Malaysia (Nik and Yap, 1980), Japan (Naiki and Ili, 1981), Venezuella (Diaz Pollanco*et al.*, 1975), Spain (Tello *et al.*, 1985), Cuba (Perez and Gonzalez, 1988), North Minas Gerais (Goulart, 1988), Turkey (Temizel and Ertunc, 1992) and France (Kouassi, 1993).

Web blight was first described by Matz (1917) as the cause of fig decline in Florida and later on a destructive wilt of beans and cowpeas in Puerto Rico (Matz, 1921). Subsequently, it was reported on beans from Sierre Leone (Deighton, 1928), Florida (Weber, 1935), Texas, Brazil, Burma, Ceylon, Japan, Philippines (Weber, 1939), Mexico (Crispin and Gallegos, 1963), Costa Rica (Echandi, 1965), Central America (Gallendo et al., 1982) and Trinidad (Rajnauth, 1987).

In India, root rot disease of French bean was recorded for the first time in 1980 from Bangalore (Sharma and Sohi, 1981). Later Srivastava and Gupta (1989) reported the aerial blight phase of this disease from Sikkim. Paul and Sharma (1990) reported the occurrence of root rot from the Solan area of Himachal Pradesh while; web blight was recorded during the 1994 crop season (Mathew and Gupta, 1996) from the same locality.

LOSSES

Losses from root rot vary greatly in different localities and from year to year. Five to ten per cent loss is not uncommon and more has been noted in situations especially favourable to the fungus (Zaumeyer and Thomas, 1957). Cardona-Alvarez (1954) observed that the percentage of natural infection increased from 68.6 to 100 per cent with the resultant crop reduction upto 93 per cent in Medellin Columbia.

Several workers have reported varying degree of losses due to web blight (Deighton, 1928; Weber, 1939). Rajnauth (1987) reported 25-100 per cent yield losses in Trinidad. In Brazil, under favourable environmental conditions the bean crop has been destroyed within 48-72 hours (Sartorato, 1988).

In India, Sharma and Sohi (1980) reported that yield losses of green pods at different stages of plant growth, varied from 8.45 to 64.68 per cent. They also observed up to 15 per cent incidence of root rot in fields around Bangalore (Sharma and Sohi, 1981).

SYMPTOMS

The disease was found to occur on roots, stem, leaves, pods and seed (Mathew and Gupta, 1996). Sanchez and Cardenaz (1988) reported that root rot affects plant populations from emergence to the first 30 days of crop growth. The distinguishing symptoms of the occur in two phases and are described below:

Root Rot

Symptoms appear on roots and stem above and below the surface of the soil as reddish brown cankers. The lesions enlarge rapidly and engirdle the stem at the collar region, extending longitudinally downward to the roots (Plate 1c). Roots are affected in advance stages of infection leading to partial to complete rotting of the root system.

Web blight

The symptoms on leaves appear as small, circular, water-soaked spots. These spots increase rapidly in size and cover extensive areas of the leaf blade. Initially, the spots are much lighther in colour than the surrounding unaffected area. At later stages, they became tan brown and surrounded by dark borders. The spots coalesced to form larger areas on the leaf blade (Plate 1a). Light tan hyphae develop on both surfaces of the infected leaf and at later stages are studded with minute sclerotia of the fungus. Defoliation occurs in advance stages of infection. The pods are attacked at all stages of growth. On young pods, circular slimy water soaked spots appear which are light tan in colour. On mature pods, the spots are dark brown, more or less circular, slightly zonate and definitely sunken. Spots coalesce to cover the entire pod. The distal end of the pods is the most affected (Plate 1b). In favourable weather, runner hyphae of the fungus are seen spreading from the point of infection to healthy parts, which later become studded with sclerotia measuring 1.0 to 2.0 mm in diameter. Tan brown to reddish brown discoloration is observed on infected seed (Plate 1e), located immediately below the spots on the affected pods.

THE PATHOGEN

De Candolle founded the genus *Rhizoctonia* in 1815 to include the species of *R. crocorum* (Pers.). Kuhn (1858) first reported this fungus on diseased potato tubers and named it *Rhizoctonia solani* Kuhn. E.A. Burt described it as *Corticiumvagum* var. *solani*. In Europe, this fungus is generally known as *Hyphonchussolani*Priti and Del or *C. solani*.

Later, Rogers proposed the name *Pelliculariafilamentosa* (Pat.) Rogers, to include the organism previously classified as *Corticiumvagum* var. *solani*. The causal organism of web blight was first described as *Rhizoctonia microsclerotia* Matz as the cause of the disease on fig in Florida (Matz, 1917) and later on beans (Weber, 1935). Weber (1939) reported the basidial stage on beans and figs from Texas and the mycelia and sclerotia from both sources differed from those obtained from basidiospores of *Corticiumsolani* and *C. stevensci* Burt. (Weber, 1939). The descriptions of the organism given by different workers lead to the conclusion that its name should be *C. microscler*otia (Matz.) Weber. Later, Rogers considered it as a synonym of *Pelliculariafilamentosa* (Pat.) Rogers and in 1953 the name *P. filamentosa* f. sp. *microsclerotia* (Matz.) Exner was proposed (Zaumeyer and Thomas, 1957). However, all these have been considered as synonyms of *Rhizoctonia solani* (Parmeter and Whitney, 1970 and Talbot, 1970). It has now become evident that *Rhizoctonia solani* is a collective species, consisting of atleast seven and possibly more, unrelated strains called anastomosis groups (Ogoshi, 1975) in which fusion occurs within the isolates.

(a) Symptoms on leaves

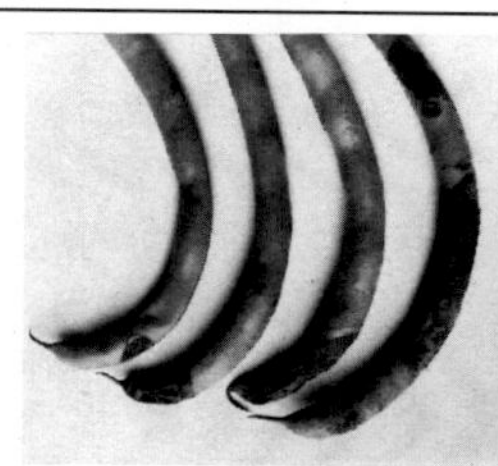

(b) Symptoms on pods

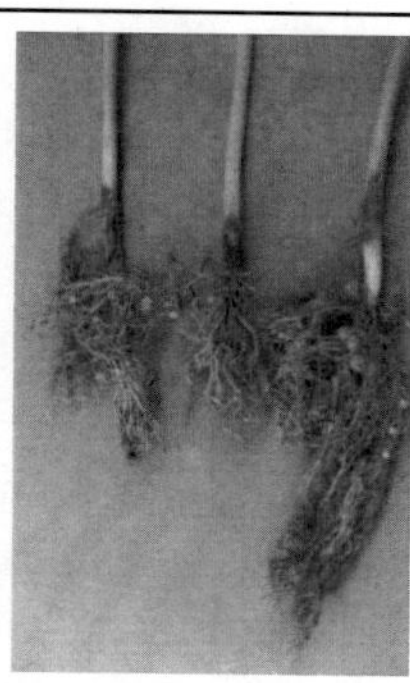

(c) Symptoms on stems and roots

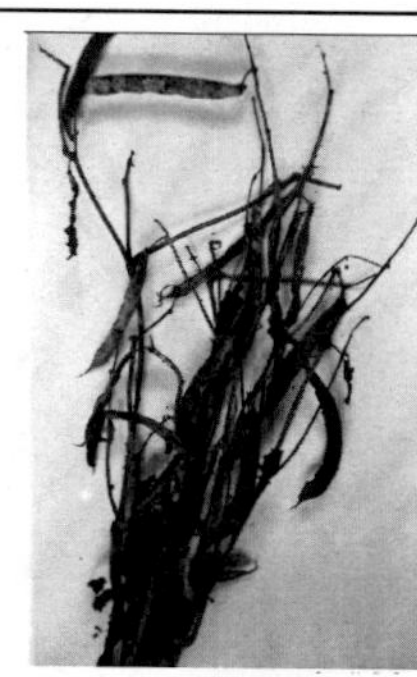

(b) Completely infected plant

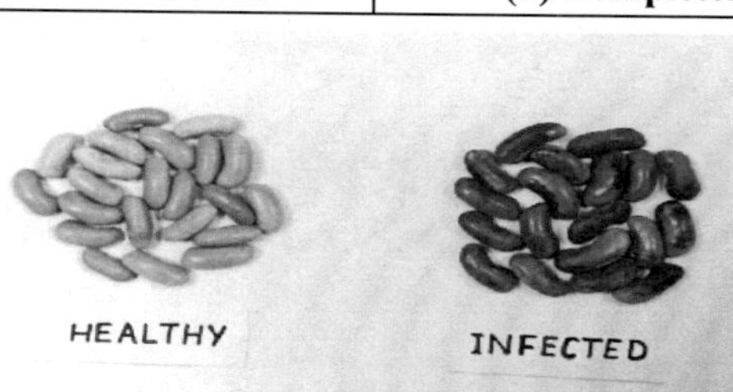

(e) Symptoms on seeds

Plate 1. Symptoms of root rot and web blight of French bean

The causal fungus (*Rhizoctonia solani* Kuhn) consists of septate mycelium which contains multinucleate cells in actively growing hyphae. The colour of the hyphae varies from hyaline when young to brown on maturity (Butler and Bracker, 1970). The mycelium is branched near the distal septum of the mother hyphal cells at right or acute angles in older hyphae and the branches are constricted at or near the point of origin with the septum (Plate 2b). The size of hyphae varies from 117.7 - 173.4 × 6.19 - 9.29 μm. In culture, sclerotia develop within 5-6 days (Plate 2a). The size of sclerotia ranges from 1 to 3 mm in diameter.

The sclerotia consist of brown barrel shaped monilioid cells measuring 20-35 x 10-15 μm (Plate 2c). The fungus could be seen growing as white crust on infected foliage and as white mycelial web at the distal ends of severely infected pods. In advanced stages, smooth sclerotia measuring 1.0 to 2.5 mm in diameter are seen adhering to infected plant parts. The basidial stage of the fungus (*Thanatephorus cucumeris*) also occurs but very rarely.

VARIABILITY

It is widely recognized that *R. solani* consists of many races, forms or groups of various isolates differing in pathogenicity, morphology in culture and ecology. There are number of reports on grouping of *R. solani* based on differences in pathogenicity, morphology in culture and/ or physiology (Ogoshi, 1975). In Himachal Pradesh, studies were carried out regarding physiological variation, anastamosis grouping and genetic variability among different isolates collected from north India (Upmanyu *et al.*, 2004, Sharma *et al.*, 2005, Upmanyu and Gupta, 2009). Different isolates exhibited slight variation to their physiological requirements. Most of the isolates preferred Czapeck's dox agar as solid medium while potato dextrose broth supported the maximum growth of the isolates (Upmanyu and Gupta, 2009). All the isolates grew rapidly within a temperature range of 25 to 30°C and RH above 95 per cent while pH ranging between 5.0 to 6.0 supported the maximum growth of the isolates. Glucose among carbon sources and ammonium tartrate as well as potassium nitrate among nitrogen sources yielded maximum mycelial growth of most of the pathogen isolates. Significant variation was recorded in aggressiveness of isolates within each cultivar having varied degree of horizontal resistance and their sensitivity towards Bavistin. 43 Isolates of *R. solani* differed in colony colour (from white to brown) as well as pattern of sclerotial formation and nine isolates did not produce sclerotia (Sharma, 2003) (Plate 2a). The anastomosis grouping based on hyphal fusion is considered as an easiest method of characterizing variability in *R. solani* (Richter and Schneider, 1953). Various isolates have been grouped in to AG-1, AG-2, AG-3 and AG-4 groups (Parmeter *et al.*, 1969) and at present at least 12 anstomosis groups (AG-1 to AG-11 and AG-BI) of distinctive physiology and genetic composition exist in *R. solani* (Ogoshi, 1987). On the basis of their behaviour, 18 different French bean isolates collected from different epidemiological areas of Himachal Pradesh were placed in AG-1-1A, AG-1-1B and AG-4 anastomosis groups (Upmanyu *et al.*, 2004).

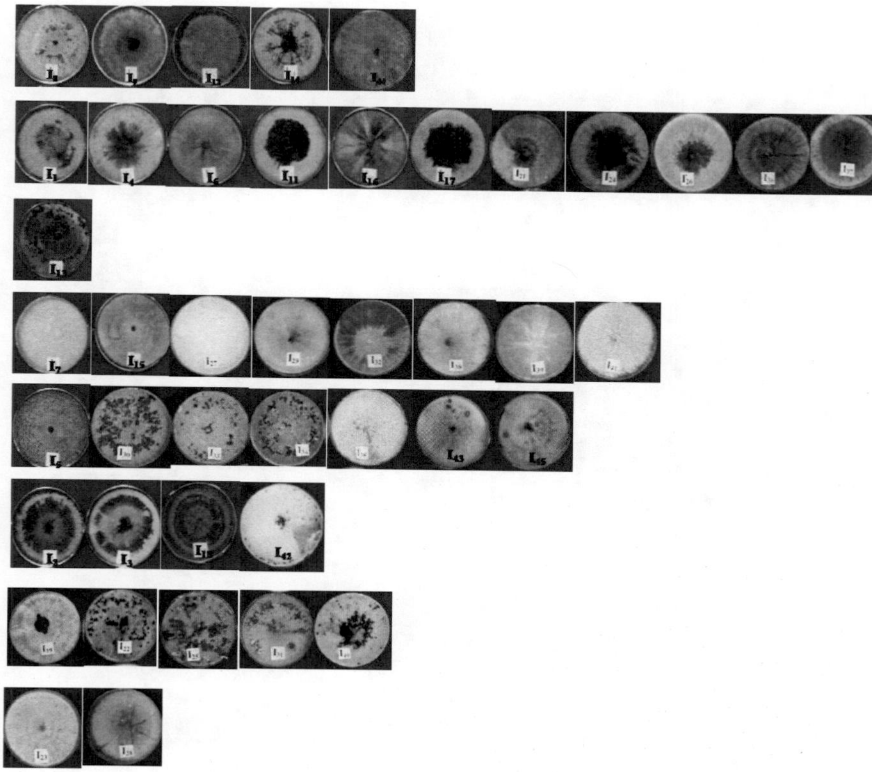

Plate 2a. Cultural and sclerotial variability of *R. solani*

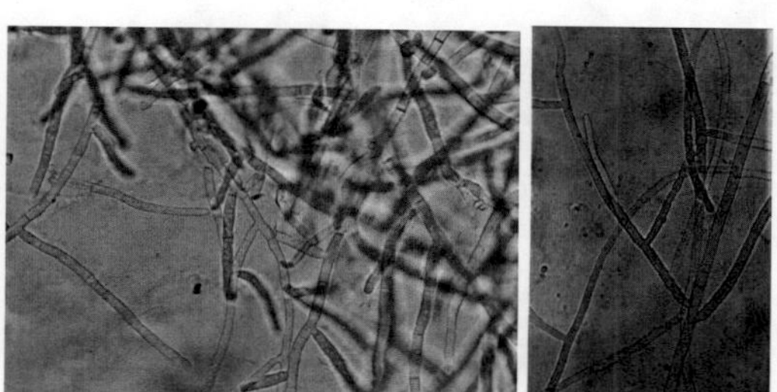

Plate 2b. Mycelium of *R. solani*

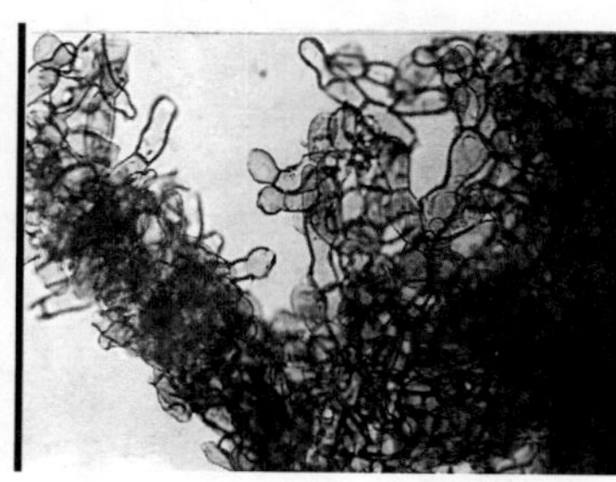

Plate 2c. Moniloid cells of Sclerotia of *R. solani*

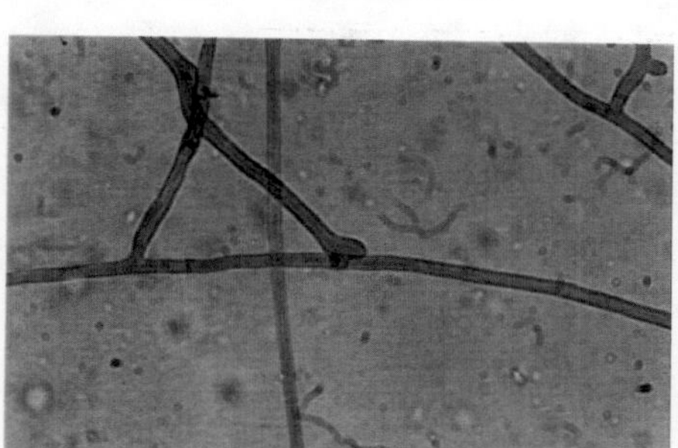

Plate 2d. Type 2 reaction (AG-4 × I15)

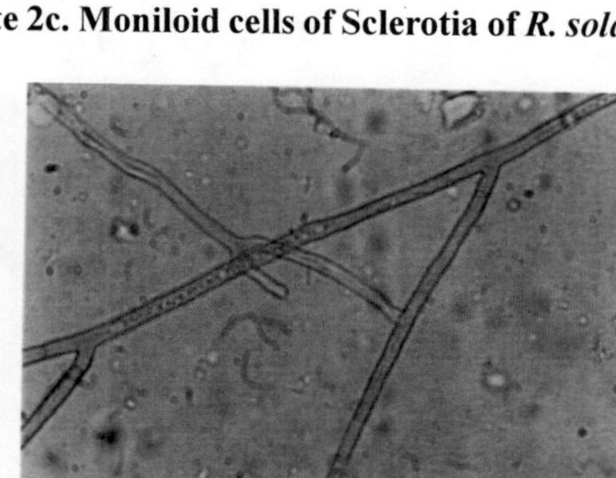

Plate 2e. Type 3 reaction (AG-1-IB × I4)

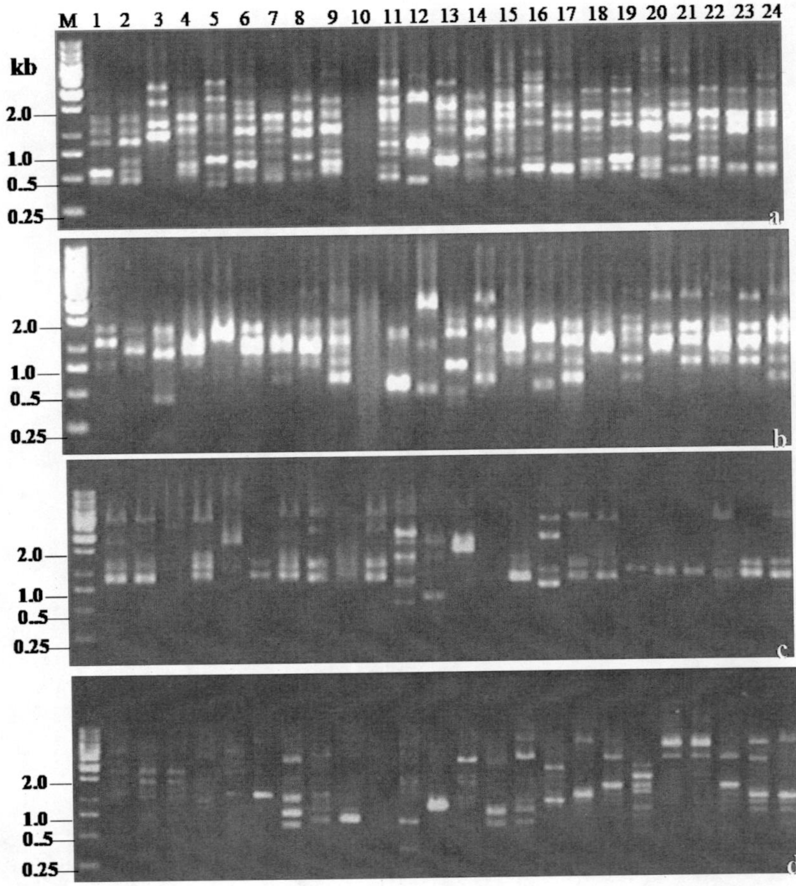

Plate 3a. DNA fingerprint patterns of isolates of *Rhizoctonia solani* generated by RAPD with primers (a) OPA-04 (b) OPA-10 (c) OPA-13 (d) OPA-18

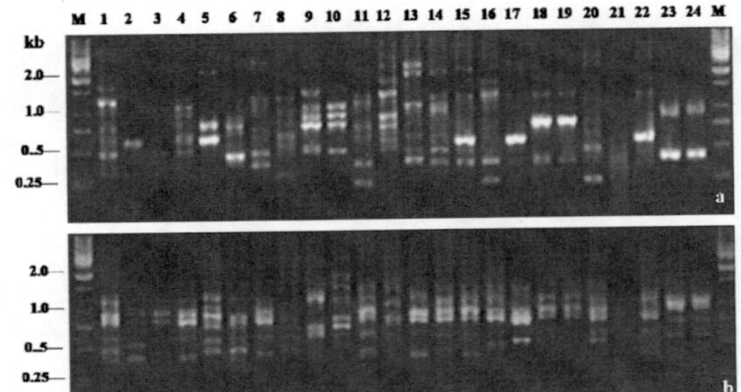

Plate 3b. DNA fingerprint patterns of isolates of *Rhizoctonia solani* generated by URP with primers (a) URP13R (b) URP17R

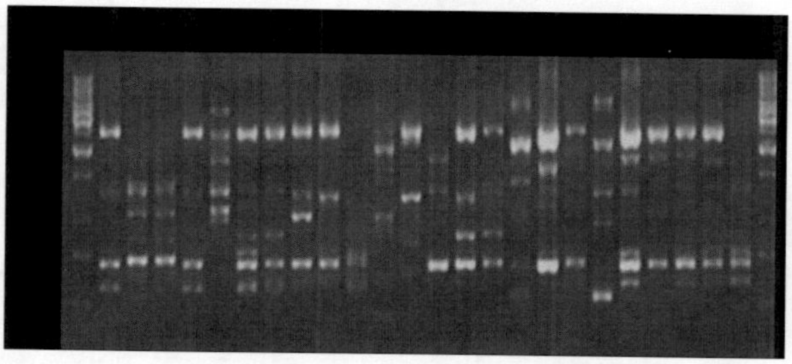

Plate 3c. DNA fingerprint patterns of isolates of *Rhizoctonia solani* generated by ISSR with primers SSR-02

Although many attempts were made to characterize *R. solani* in several AG types, reports of hyphal anastomosis involving more than one AG and the occasional loss of anastomosis ability complicates the identification and characterization of AG types (Kuninaga *et al.*, 1979; Neate and Warcup, 1985). Anastomosis grouping is a convenient but not the ideal method for classification of *R. solani* as misidentification is frequent because of the varying frequency of hyphal fusion (Neate and Warcup, 1985). This method also does not offer reliable informations on genetic variation or taxonomic relationships within and between AGs and sometimes isolates within an AG are found to be more similar than between AGs and even genetically distinct subgroups exist within AGs (Kuninaga and Yokosawa, 1985, Sneh *et al.*, 1991, Keijer *et al.*, 1996). In the absence of stable morphological and physiological characteristics (Mordue *et al.*, 1989) and anastomosis groups (Kuninaga *et al.*, 1979, Neate and Warcup, 1985), the identification and genetic characterization of *R. solani* isolates has proved to be very difficult and tedious.

Electrophoretic patterns of protein in polyacrylamide gels are distinctive among the intraspecific groups (ISGs) of *R. solani* (Ogoshi, 1978, Reynold *et al.*, 1983, Kuninaga, 1986). The patterns of isolates from the same group are similar or have common bands indicating that electrophoresis may be useful method of identifying groups of *R. solani*. Matsuyama *et al.* (1978) compared non-specific esterase isoenzme and showed that the electrophoretic pattern were different even among intraspecific groups of *R. solani*. Liu and Ge (1988) also observed significant differences in the soluble protein patterns among the anastomosis groups but only slight differences among isolates within the same group have been reported. The electrophoretic patterns of 163 isolates of *R. solani* AG-4 isolated from bean in Isfahan, Iran were grouped into seven zymogram groups (ZGs) according to the polygalacturonase (PG) and pectin estrase (PE) isozymes bands (Balali and Kowsari, 2004). French bean isolates were clustered into different groups on the basis of similarity coefficients for banding patterns for esterase as well as NADH dehydrogenase (Upmanyu *et al.*, 2004). Isolates also varied significantly in aggressiveness within each cultivar while cultivars showed significant variation in horizontal resistance. Two isolates (I_8 and I_9) were the least sensitive to Bavistin while I_{10}, I_{13} and I_{17} exhibited maximum sensitivity. The isolates were placed in AG-1-1A, AG-1-1B and AG-4 anastomosis groups on the basis of their behaviour. Isolates produced different banding patterns for esterase as well as NADH dehydrogenase and were clustered into different groups on the basis of similarity coefficients. Banding pattern of esterase clearly indicated the relationship with anastomosis groups of *R. solani* isolates. Isolates of AG-4 varied very little while AG-1-1B and AG-1-1A were slightly more variable (Plate 2d and 2e).

Molecular markers are used as important tools for the characterization of genetic diversity in the pathogens where morphological characteristics are either absent or not able to differentiate isolates properly (Sharma *et al.*, 1999). Moreover, morphological characters were also influenced by the environmental and cultural conditions. Therefore, problems associated with studying different levels of genetic diversity in *R. solani* have been suggested to be best addressed by the use of molecular markers (Todda *et al.*, 1999). Sharma *et al.* (2005) genetically differentiated and classified 24 isolates of *R. solani* collected from different hosts and epidemiological regions of north India into different groups by using RAPD, ISSR and URPs markers (Plate 3). French bean isolates collected from different geographical areas of Himachal Pradesh were grouped in to two separate clusters and 29.1 per cent isolate did not share any of the cluster indicating the high genetic variability in the pathogen population (Fig. 1). They further reported that URP markers were more efficient than other two i.e. RAPD and ISSR marker systems in their efficiency in detecting polymorphism among the isolates.

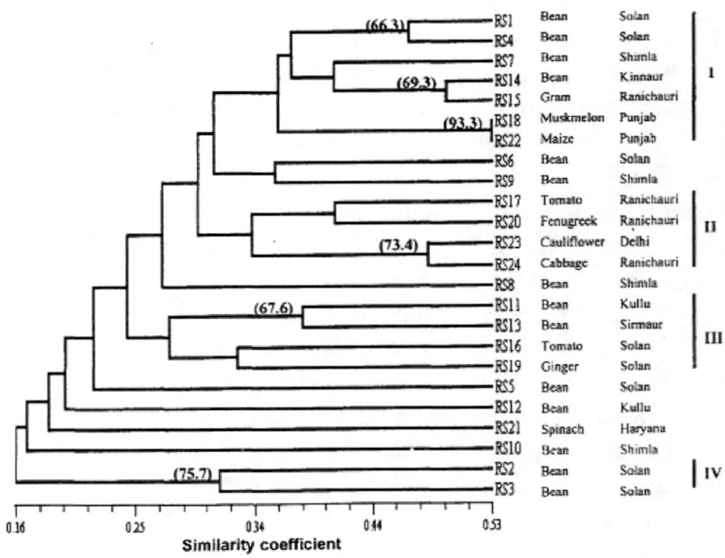

Fig 1. Dendrogram showing genetic relationships among *R. solani* isolates by using RAPD, ISSAR and URP primers

HOST RANGE

The fungus is polyphagous in nature and has been recorded on a number of hosts from various parts of the world. It is most widely distributed and attacks 287 genera belonging to dicotyledonous and 17 monocotyledonous families of Angiosperms and three each of Gymnosperms and Pteridophytes (Ahuja, 1976). The fungus was recorded for the first time in the year 1858 on potato tubers (Kuhn, 1858). Since then this fungus has gained the reputation of being a widespread and versatile plant pathogen.

DISEASE CYCLE

The pathogen perennates as mycelium or sclerotia in the soil on diseased plant debris. The fungus is present in most soils and once established in a field remains there indefinitely. The germination of sclerotia gives rise to infective propagules, which are sufficient to cause infection. The fungus spreads with rain, irrigation or floodwater, tools and with infected or contaminated propagation material (Fig. 2). The association of *R. solani* with bean seed has been reported (Tylkowska, 1984; Nitshe and Cafati, 1985). Seed transmission ranged from 36.66 to 39.66, 43.33 to 46.66 and 33.33 to 36.25 per cent in cultivars Kentucky Wonder and Contender by blotter, agar plate and growing on test methods, respectively (Gupta *et al.*, 2000). Shama and Shama (1989) indicated that seed penetration of *R. solani* occurs directly through intact seed coat or through the hilum tracheids and that the pathogen was transferred from infected seed to the soil. The infected seed after germination either rot in the soil itself or give rise to plant, which later damped off. The soil borne inoculum is splashed by rain to the foliage that causes web blight. On some plants or plant parts the pathogen can enter only through wounds or through openings in the epidermis (Chupp and Sherf, 1960) while in others, it thrives in the soil or in mud splashed on to the plants until it can enter through healthy tissue. In favourable weather, runner hyphae of the pathogen are seen spreading from the point of infection to healthy parts, which later become studded with sclerotia.

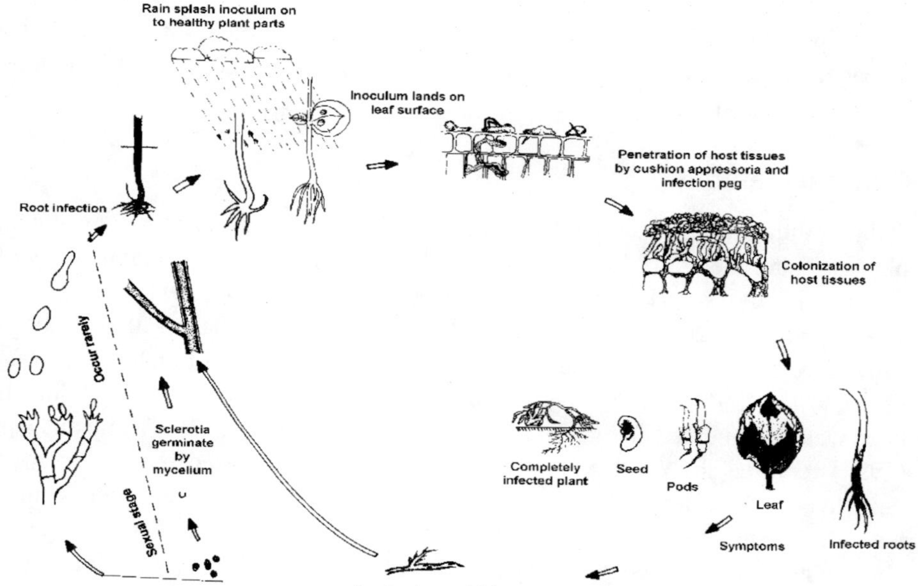

Fig. 2. Disease cycle of root rot and web blight of French bean

EPIDEMIOLOGY

Root rot is favoured by high soil temperature and soil moisture while, web blight is more prevalent in the wet season. Stan and Raichu (1973) observed that the disease was influenced to a greater extent by the temperature. *Rhizoctonia* diseases of roots are most severe under moist soil conditions (Tu *et al.*, 1992). Bruggen*et al.* (1986) reported positive correlation between lesion size and soil moisture. If the moisture is sufficient or there is frequent drizzle, pronounced blight symptoms will appear. During humid periods fan shaped whitish hyphae spread from the point of infection and produce the conspicuous and destructive phase of the disease (Weber, 1939; Schneider, 1953). Diaz Plaza *et al.* (1991) reported that disease incidence and severity increased after rain. Upmanyu and Gupta (2005) studied the role of abiotic factors on the development of this disease and reported that root rot development was favoured at high (80%) soil moisture and 25° C temperature while 100 per cent relative humidity coupled with 25° C temperature favoured web blight. Continuous leaf wetness for at least 6h was essential for disease initiation and increase in leaf wetness durations from 6 to 12 h showed a corresponding decrease in incubation period from 36 to 26 h whereas further increase in leaf wetness did not exert any effect on incubation period.High rainfall and soil moisture coupled with high relative humidity and soil temperature (23-25°C) were found to be favourable for web blight (Mathew and Gupta, 1996). A high soil moisture (80%) and a temperature of 25°C were the most favourable for root rot development, while web blight development was optimum at >85 per cent relative humidity coupled with 25°C temperature (Upmanyu and Gupta, 2005).

MANAGEMENT

The causal fungus persists in the soil, therefore, an integrated approach of disease management combining the use of cultural, biological and chemical methods is discussed in detail below: -

Cultural methods

Cultural practices viz., planting disease free seed, mulching, adjustment of planting date, medium (30-60 × 15 cm) plant spacing and crop rotations have been recommended to keep the disease under check (Sartorato *et al.*, 1988; Pieta, 1992; Tu, 1992; Upmanyu, 2002; Kumar and Kudada, 2018). Deep tillage (30 cm subsoiling) can reduce soil compaction and thereby, root rot severity in addition to increased yield (Tan and Tu, 1995). The suppression of root rot pathogens using soil amendments has been well documented (Lewis and Papavizas, 1971, Singh, 1983). Voland and Epstein (1994) reported the efficacy of fresh manure over composted manure in reducing bean hypocotyl lesions while soil amendment with cotton, mustard and neem cakes was found effective in reducing root rot incidence both under glass house and field conditions and increased yields (Upmanyu *et al.*, 2002). The effectiveness of these amendments may be due to enhanced populations of non-parasitic microorganisms exhibiting antagonistic effects on the pathogen thereby inhibiting the mycelial growth and sclerotia production of the pathogen. Soil amendments with low soil C: N ratio increased incidence of *R. solani* whereas high C: N ratio amendment did not interfere with the incidence while incidence was independent of soil moisture conditions (Fenille and Souzo 1999). Joshi *et al.* (2009) reported a high level of root rot suppression (>33%) in the poultry manure, composted *Lantana* sp. and composted *Urtica* treatments under field conditions.

Host resistance

Prasad and Weigle (1970; 1976) observed that dark seeded PI accessions of French bean viz. 109859, 163583, 174908, 226895 and 300665 were resistant to *R. solani* while most of the white seeded cultivars were susceptible to both seed infection and pre-emergence damping off. However, in later experiments Sharma and Sohi (1981) reported all the three French bean cultivars viz., "Contender", "Premier" and "Giant Stringless" to be equally susceptible irrespective of their colour. Bush type cultivars/ lines are more susceptible to the disease than pole type ones. The cultivars/lines viz., Wisconsin RRR 46, BAT 477, BAT 332, BAT 1753, RIZ 30, EMP 81, A 300, ICA Pijao and Jackson Wonder have been reported to be most resistant (Hagedorn and Rand, 1980; Turrialba, 1988; Tu and Park 1993; Muyolo *et al.*, 1993) while, PLB-43, EC-18834, IC-18593, PI-249554, Phagu Chitra, Pole wall-I, Pole wall-II, and K-13522 were found moderately susceptible to the disease (Gupta *et al.*, 1997). Upmanyu *et al.* (2004) evaluated 93 cvs./lines against web blight both under natural epiphytotic conditions and *in vitro* against 18 isolates of the pathogen, only line ET 8396 was found completely free and affected by only two isolates while 14 cvs./lines were moderately susceptible.Cultivar IPR 96-4 was also found resistant to this disease (Khati and Hooda, 2006). Naik *et al.* (2016) screened 60 germplasm lines of French bean against root rot in green house conditions and field conditions at the Agriculture Research Station, Naganahalli UAS, GKVK, Banglore, Karnataka. The six lines were showed resistant reaction, 13 lines were moderately resistant, 10 lines were susceptible, 28 lines were moderately susceptible, and 3 lines were highly susceptible reaction. The lines IC-272638, IC-258275, IIHR-909, VRF-3-2 and Arka Komal were found to be resistant in reaction to the disease in both natural and artificial inoculation.

Chemical methods

Different fungicides have been reported to be quite effective for the management of the disease. Seed treatment is effective in controlling *R. solani* infection of beans. Fungicides viz., benomyl (Segura and Diaz Polanco, 1975), captan (Sanchez, 1956), terrachlor (Quintozene) (Okhovat, 1977), thiram, thiophanate-methyl (Sesan and Dumitras, 1979), Benlate, Bavistin, Vitavax, Brassicol (Sharma and Sohi, 1981), captafol (Ito *et al.*, 1986), moncerine, quinolate (Raffat, 1992) and iprodione or iprodione + thiram (Maringoni *et al.*, 1992) gave maximum protection against pre and post emergence mortality and significantly increased the yield of green pods (Sharma and Sohi, 1981). However, seed treatment followed by foliar sprays with carbendazim, bitertanol, or benalaxyl + copper oxychloride gave best results against web blight (Thakur *et al.*,

1991). Gupta *et al.* (1999) reported the efficacy of Bavistin (0.2%), Celest (0.1%), Topsin-M (0.2%) and Raxil (0.2%) against pre-emergence damping off while Celest, Bavistin Indofil M-45 (0.25%), Topsin-M and Kavach (0.2%) were highly effective in reducing post-emergence damping off phase of this disease. Good control of web blight phase of this disease was achieved by seed treatment with Celest (0.2%) or Raxil (0.2%) or Bavistin (0.2%) followed by foliar sprays of Bavistin (0.05%) at fortnightly intervals (Mathew and Gupta, 1996). Efficacy of foliar sprays of Bavistin (0.1%) and tebuconazole (0.05%) against web blight have been reported (Upmanyu *et al.*, 2002).

Biological control agents

Exploitation of the populations of bioagents may prove useful to combat the ravages of soil borne diseases. Biological treatment of soil under glass house conditions with *Trichoderma viride* showed results similar to chemical control (Dumitras and Sesan, 1980). Elad *et al.* (1980) found that in naturally infested soils, wheat bran preparations of *Trichoderma harzianum* significantly decreased the *Rhizoctonia* disease of beans. Tu and Vaartaja (1981) reported that *Gliocladium virens*, a hyperparasite of *R. solani*, inhibited production of sclerotia. Management of this disease by seed/ soil treatment and foliar sprays with *T. harzianum, T. virens* and *T. viride* has also been reported (Mathew and Gupta, 1998; Hazarika and Das, 1998; Upmanyu *et al.*, 2002). Inhibition of mycelial growth and sclerotial germination of this pathogen by *T. harzianum* and *T. viride* culture filtrate has also been observed (Hazarika and Das, 1998). Roberti *et al.* (1993) indicated that *T. viride* produced non-volatile compounds while *T. harzianum* produced coils and hook pincer shaped structures, which prevented the growth of *R. solani*.

The water extracts obtained from green parts of potato, tomato and rape and *Trichoderma* PBG-1 showed marked antagonistic activity against *R. solani* (Smolinska and Kowalska, 2006). Out of sixty isolates of *Trichoderma*, 7 isolates significantly reduced the incidence of root rot compared to the control. *T. harzianum* isolates Tr-34 and Tr-14 were the most effective, resulting in only 20.8 and 23.4 per cent root-rot incidence, respectively (Joshi *et al.,* 2007). Dubey *et al.* (2017) reported that root rot and damping off in French bean could be managed by soil application and seed treatment with bio-formulations of *T. harzianum* alone and in combination with *P. fluorescens*.

Trichoderma species have reported to be as potent biocontrol agent against various phytopathogenic soil fungi including *R. solani* and *Pythium* spp. (Harmann *et al.*, 2004; Howell, 2006). The efficiency of *T. harzianum* to inhibit fungal growth may be through competition for space and nutrients, mycoparasitism and production of antibiotic compounds, it has found that the hyphae of *T. harzianum* coil around the hyphae of the pathogen and penetrate the host mycelium through degrading cell wall by secretion hydrolytic enzymes followed by assimilation of cell contents (Harmann *et al.,* 2004; Howell, 2006, Siameto *et al.*, 2011). *T. harzianum* exhibited high activity in controlling the root rot disease and increasing yield under glass house conditions (Matloob and Juber, 2013).

Beneficial bacteria have been intensively studied as biocontrol agents against soil borne diseases (Handelsman and Stabb, 1996; Weller, 1988). *Bacillus* spp. and *Pseudomonas* spp. are well reported to give very good management of many soil borne pathogens. In addition, few other bacterial antagonists are also reported to manage several plant pathogens. *Azotobacter chroococcum* reported to excrete beneficial effects on plant growth and yield of various crops by different mechanisms (Brenner *et al*, 2004; Saharan and Nehra, 2011). It was found to produce ammonia, vitamins and some growth regulators include indole acetic acid, gibberllins, and cytokines which enhance seed germination and promote plant growth (Mrkovacki and Milic, 2001; Karthikeyan and Sakthivel, 2011).

The efficacy of potential bacterial antagonists viz., fluorescent Pseudomonads and *Bacillus* spp. against root rot has also been reported (Sanchez *et al.*, 1994; Wolk and Sarkar, 1994, 1995). An endophytic *Pseudomonas florescens* carrying *Serratia marcescens* Chi A gene either on the plasmid or integrated into the chromosome controlled effectively the phytopathogenic fungus *R. solani* on bean seedlings under plant growth chamber conditions (Downing and Thomson, 2000). Co-inoculation of seed with *Bacillus subtilis* MB1600 (Epic) and *Rhizobium tropici* significantly reduced root rot severity and enhanced yield (Estevez-de-Jensen *et al.*, 2002). Efficacy of seed treatment and soil application of *B. subtilis* in the reduction of pre- and post-emergence root rot have also been reported (Sharma and Gupta, 2003). Sixty four per cent of 46 isolates of fluorescent pseudomonads (isolated from different area of Iran) and *Pseudomonas fluorescens* strain CHA0 showed antagonistic activity against *R. solani* (Afsharmanesh *et al.*, 2006).

Four bioagents, *Trichoderma viride, Trichoderma harzianum, Trichoderma virens* and *Aspergillus niger* significantly inhibited the mycelial growth as well as number of sclerotia of *R. solani* (Gupta *et al.,* 2014). The maximum mycelial growth inhibition (62.12 %) was observed by *T. harzianum* followed by *T. viride* (57.28 %), *T. virens* (51.38 %) and *A. niger* (46.72 %), respectively. The per cent reduction in the number of sclerotia was also maximum by *T. harzianum* (69.76 %) followed by *T. viride* (65.28 %), *T. virens* (60.05 %) and *A. niger* (52.82 %) after 168 h of incubation in lab. Kumar *et al* (2018) showed that seed treatment with *Trichoderma viride*@ 5.0 g/kg seed gave maximum seed germination (85.82%) and minimum pre- emergence (13.93%) and post- emergence (17.00%) root rot incidence. The maximum yield (63.44 q/ha) was also recorded by seed treatment with *T. viride* @ 5.0 g/10-6 spores/ml followed by *Aspergillus niger* @ 5.0 g/10^{-6} spores/ml (62.33 q/ha). Correlation coefficient studies exhibited negative and highly significant association of pre- and post-emergence root rot disease incidence with green pod yield during both the years rabi 2014-15 and 2015-16 cropping seasons. Korayem *et al.* (2020) reported that chitinolytic *Streptomyces* sp. for the biological control of *R. solani* and damping-off disease of green bean plants which could be used as an attractive alternative for pesticides in organic agriculture.

The biological control mechanism involving the use of bacterial anatagonists can be either induction of systemic resistance or antibiosis through production of antibiotic metabolites like phenazines and cyclic lipopeptides (Howell and Stipanovic, 1979; Nandakumar *et al.* 2001, Daes *et al.,* 2011).*Azotobacter* can also compete with pathogen for iron through production siderophores (Neilands, 1993; Hillel, 2005). Matloob and Juber (2013) reported that four isolates of *Azotobacterchroococcum*, isolated from wheat, bean and sesban, *Sesbania sesban* had high antagonistic activity against *R. solani* under *in vitro* conditions.

Arbuscular mycorrhizal fungi (*Glomus mosseae and Scutellospora erythropa)* treatment exhibited significant control of root-rot disease alone and in combination with *Pseudomonas fluorescens* and mustard oil cake amendment (Neeraj and Singh, 2009).

Integrated disease management

Integration of seed treatment with bioagents and fungicides has been found to be more effective to manage soil borne pathogens (Shrestha and Mukhopadhyay, 1991; Mukhopadhyay and Mukherjee, 1991; Mukhopadhyay *et al.*, 1992). Tu and Zheng (1993) studied the compatibility of seed treatment with *Bacillus subtilis*, *Pseudomonas fluorescence* or *Gliocladium virens* and subsequently dressed with diazinon + captan + thiophanate- methyl (DCT). They observed that fungicide treated seed showed inhibition of bioagents, which was transient, and the growth and sporulation of the bioagents resumed as the effect of the fungicides diminished. They further suggested that bioagents such as *B. subtilis* and *P. fluorescence* are more mobile in the rhizosphere of seedlings grown from treated seed and hence, their incorporation together with fungicides can extend the effect of the chemical to give protection against root rot. Upmanyu *et al.* (2002) also reported the tolerance in *T. viride* to carboxin, tebuconazole and carbendazim. The combined seed treatment of French bean with Contaf (0.025%) and *Gliocladium* (*Trichoderma*) *virens* (0.1%) resulted in higher seed germination, lower incidence of the disease and higher grain yield (Mukherjee *et al.*, 2001). Combined treatment of seed (soaking in *T. harzianum*, 1 x 10 conidia/ ml) and soil (20 and 100 kg/ feddan of calcium and minor sulfur, respectively) reduced the root rot incidence and increased green pod yield (Ziedan and Mahamoud, 2002). Soil amendment (mustard cake) + carbendazim (ST +FS) were found highly effective in reducing the pre-emergence and post-emergence root rot and web blight severity (Upmanyu *et al.*, 2002). A combination treatment including soil solarization (SS) for 50 days + soil amendment (mustard cake) + combination of *Allium sativum* ethanol extract and *Bacillus subtilis* was found effective in reducing pre- and post-emergence damping-off while web blight was best contained by SS + soil amendment + *A. sativum* (ST + FS) and increased yields (Sharma and Gupta, 2003). El-Mohamedy *et al.* (2015) managed the root rot of green bean by using seed treatment with *T. harzianum* and resistance inducer chemicals sorbic or benzoic acid.

REFERENCES

Ahmadinejad, A. and Okhovat, M. 1976. Pathogenicity test of some soil borne fungion some important field crops. *Iranian J. Pl. Path.* 12: 10-18, 13, 16.

Ahuja, S.C. 1976. Banded leaf and sheath blight of maize. Ph.D. Thesis, Post-graduate School, IARI, New Delhi. 153p.

Brenner, D.J., Krieg, N.R. and Staley, J.T. 2004. Bergey's manual of systematic bacteriology. Williams and Wilking. Baltimore. London .1136pp.

Briton-Jones, H.R. 1925. Mycological work in Egypt during the period 1920-22. *Egypt Min. Agri. Tech. Sci. Serv. (Bot. Sect.) Bull.* 49: 129 pp.

Bruggen, Van, A.H.O., Whalen, C.H. and Arneson, P.A. 1986. Emergence, growth and development of dry beans seedlings in respect to temperature, soil moisture and *Rhizoctonia solani. Phytopathology* 76: 568-572.

Bunting, R.H. and Dade, H.A. 1925. Gold Coast Plant Diseases. 124 pp. IIlus, London.

Butler, E.E. and Bracker, C.E., 1970. Morphology and cytology of *Rhizoctonia solani*. In: J.R. Parmeter Jr.(ed.) *Rhizoctonia solani*, *Biology and Pathology*. University of California Press, Berkley, pp. 32-51.

Cardona-Alvarez, C. 1954. Fungal root rot of the bean in the valley of Medellin-Colombia. *Rev. Fac. Agron. Medellin.* 15, 46: 137-209.

Chupp, C. and Sherf, A.F. 1960. Vegetable Diseases and Their Control. The Roland Press Company, New York, 693pp.

Cook, M.T. 1924. Common diseases of beans and peas. *Agri. Expt. Sta. Circ.* 142: 8.

Crispin, A. and Gallegos, C.C. 1963. Web blight a severe disease of bean and soyabean in Mexico. *Pl. Dis. Reptr.*47: 1010-1011.

Daes, J, Hua, G.K.H, Maeyer K.D, Pannecoucque J., Forrez I., Ongena M., Dietrich, L.E.P., Thomashow, L.S., Mavrodi, D.V. and Höfte, M. 2011. Biological control of Rhizoctonia root rot on bean by phenazine- and cyclic lipopeptide-producing *Pseudomonas* CMR12a. *Phytopathology* 101:996-1004.

Deighton, F.C. 1928. Mycological section. *Ann. Rep. Lands and Forests Dept. Sierre Leone for the year 1927*: 13-17.

Diaz, Polanco, C.; Diaz, S.D. and Saavedra, G. 1975. Effect of phytopathogenic fungi in reducing the bean population in plantations. *CIARCO*. 5: 27-30.

Diaz-Plaza, R., Teliz, O.D. and Munoz, D.A. 1991. Effect of diseases of bean grown under rainfed conditions at Mixtecapeblana. Revista-Mexicana-de-Fitopatologia. 9:21-30.

Downing, K.J. and Thomson, J.A. 2000. Introduction of the *Serratia marcescens* Chi A gene into an endophytic *Pseudomonas flourescens* for the biocontrol of phytopathogenic fungi. *Canadian J. Microbiol.* 46(4): 363-369.

Dubey, S.C., Tripathi, Aradhika, Tak, R and Devi, I. 2017. Management of root rot and damping-off complex in French bean by biocontrol agents. *Indian Phytopathology* 70(3): 386-390.

Dumitras, L. and Sesan, T. 1990. Feature of antagonism of *Trichoderma viride* Pers ex Fr. to *Rhizoctonia solani* Kuhn. *BiologieVegelata*. 32: 87-92.

Echandi, E. 1965. Basidiospore infection by *Pellicularia filamentosa* (*Corticium microsclerotia*), the incitant of web blight of common bean. *Phytopathology* 55: 698-699.

Elad, Y., Chet, I. and Katen, I. 1980. *Trichoderma harzianum* biocontrol agent effective against *Sclerotium rolfsii* and *Rhizoctonia solani*. *Phytopathology* 70: 119-121.

El-Mohamedy, Riad, R.S., Shafeek, M.R. and Rizk, F.A. 2015. Management of root rot diseases and improvement growth and yield of green bean plants using plant resistance inducers and biological seed treatments. *J. Agric. Tech.* 11: 1219-1234.

Estevez-de-Jensen, C., Percich, J.A., Graham, P.H. and de-Jensen-C-Estevez. 2002. Integrated management strategies of bean root rot with *Bacillus subtilis* and *Rhizobium* in Minnesota. *Field Crop Res*. 74(2-3): 107-115.

Fenille, R.C. and Souza, N.L. de. 1999. The role of the organic material amended and the soilmoisture on the pathogenecity of *Rhizoctonia solani* Kuhn AG-4 HGI in snap bean. *Pesquisa Agropecuaria Brasileira* 34(10): 1959-1967.

Gadd, C.H. 1927. Root diseases of economic crops. *Trop. Agri. (Ceylon*). 68: 363-370.

Gallendo, J.J., Abawi, G.S. and Thurston, H.D. 1982. Variability among isolates of *Rhizoctonia solani* associated with Snap bean hypocotyls and soils in New York. *Plant Dis*. 66: 390-394.

Goulart, A.C.P. 1988. Bean diseases in the Northern region of Minas Gerais. *FitopatologiaBrasileira* 13:230-242.

Gupta R., Yadav B., Singh S. and Singh S. 2014 Integrated management of web blight (*Rhizoctonia solani* Kühn) of French Bean. In: Kharwar R., Upadhyay R., Dubey N., Raghuwanshi R. (eds) *Microbial Diversity and Biotechnology in Food Security*. Springer, New Delhi. https://doi.org/10.1007/978-81-322-1801-2_24.

Gupta, S.K., Mathew, K.A and Shyam, K.R. 2000. Seed transmission of *Rhizoctonia solani* in French bean cultivars. *Pl. Dis. Res*. 15(1): 83-84.

Gupta, S.K., Mathew, K.A., Shyam, K.R. and Joshi, A.K. 1997. Screening of French bean germplasm against root rot and web blight. *Legume Res*. 20(1): 57-60.

Gupta, S.K., Mathew, K.A., Shyam, K.R. and Sharma, A. 1999. Fungicidal management of root rot (*Rhizoctonia solani*) of French bean. *Pl. Dis. Res*. 14(1): 20-24.

Hagedorn, D.J. and Rand, R.E. 1980. Wisconsin RRR 46 Snap bean breeding line. *Hort. Sci.* 15: 529-530.

Handelsman, J., and Stabb, E. V. 1996. Biocontrol of soilborne plant pathogens. *Plant Cell* 8:1855-1869.

Hansford, C.B. 1929. Annual report of the Government Mycologist. *Uganda Dept. Agri. Ann. Rept*. 1928. 46-47.

Harman, G. E., Howell, C. R.,Viterbo, A., Chet, I. and Lorito, M. 2004. *Trichoderma* species-opportunic, avirulent plant symbiosis. *Natural Rev. Microbiol*. 2:43-56.

Hazarika, D.K. and Das, K.K. 1998. Biological management of root rot of French bean (*Phaseolus vulgaris* L.) caused by *Rhizoctonia solani*. *Pl. Dis. Res.* 13(2): 101-105.

Hillel, D. 2005. *Plant Growth Promoting Bacteria*. Elsevier, Oxford, U. K.,103-115.

Hopkins, J.C.F. 1939. A descriptive list of plant diseases in southern Rhodesia and their control. *Rhodesia South Dept. Agri. Mem.* 2: 51.

Howell, C. R. 2006. Understanding the mechanisms employed by *Trichoderma virens* to effect biological control of cotton diseases. *Phytopathology* 96: 178- 180.

Howell, C. R. and Stipanovic, R. D. 1979. Control of *Rhizoctonia solani* on cotton seedlings with *Pseudomonas fluorescens* and with an antibiotic produced by the bacterium. *Phytopathology* 69:480-482.

Ito, M.F., Dudienas, C., Castrom, J.L.D.E., Soave, J. and Marda, J.A. 1986. Effect of fungicides applied to the aerial part on the quality of *Phaseolus vulgaris* seeds. *Fitopatologia Brasileira*. 11: 627-636.

Joshi, D., Hooda, K.S. and Bhatt, J. C. 2007. Biocontrol efficacy of indigenous *Trichoderma* isolates against root rot pathogens of French bean Uttrakhand hills. *J. Biol. Contrl.* 21: 119-126.

Joshi, D., Hooda, K.S., Bhatt, J. C., Mina, B. L. and Gupta, H. S. 2009. Suppressive effects of composts on soil-borne and foliar diseases of French bean in the field in the western Indian Himalayas. *Crop Protection* 28(7): 608-615.

Karthikeyan, A. and Sakthivel, K. M. 2011. Efficacy of *Azotobacterchroococcum* in rooting and growth of *Eucalyptus camaldulensis* stem cuttings. *Res. J. of Microbiol.* 1-7.

Keijer, J., Houterman, P.M., Dullemans, A.M. and Korsman, M.G. 1996. Heterogenicity in electrophoretic karyotype within and between anastomosis groups of *Rhizoctonia solani. Mycol. Res.* 100: 789-797.

Khati, Pinkey and Hooda, K.S. 2006. Multiple disease resistance in rajmash genotypes. *Indian Phytopath.* 59(4) : 521-522.

Korayem, A.S., Abdelhafez, A.A., Zaki, M.M. and Saleh, E.A. 2020. Biological control of green bean damping-off disease caused by *Rhizoctonia solani* by *Streptomyces parvulus* strain 10d. *Egyptian Journal of Microbiology* 55:87-94.

Kouassi, A.S. 1993. Tracing *Rhizotonia solani. UNILET-Informations* 81 : 18-19.

Kuhn, J.G. 1858. Die Kranbheiten der Kulturegenivachse ihre ursachen und ihre ver hutung. *Gustav Boselmann, Berlin*. 312p.

Kumar, M. and Kudada, N. 2018. Effect of sowing dates and weather conditions on Rhizoctonia root rot disease incidence and green pod yield of French bean. *Journal of Pharmacognosy and Phytochemistry* 7(3): 893-899.

Kumar, M., Kudada, N., Srivastava, J.N., Saurabh, A. and Kumari, A. 2018. Evaluation of bio-control agents against root rot disease of French bean caused by *Rhizoctonia solani* under field condition. *Journal of Pharmacognosy and Phytochemistry* SP1: 1581-1584.

Kuninaga, S. 1986. Differentiation of *Rhizoctonia solani* by physicochemical characteristics. *Plant Prot. Japan* 40: 137-142.

Kuninaga, S. and Yokosawa, R. 1985. DNA base sequence homology in *Rhizoctonia solani* Kuhn: IV. Genetic relatedness among seven anastomosis groups. *Ann. Phytopathol. Soc. Japan.* 51: 127-132.

Kuninaga, S., Yokosawa, R. and Ogoshi, A. 1979. Some properties of anastomosis group 6 and BI in *Rhizoctonia solani* Kuhn. *Ann. Phytopathol. Soc. Jpn.* 45: 207-214.

Lewis, J.A. and Papavizas, G.C. 1971. Contribution a I'etude de I'activitesystemique du benzimidazole 2-methylcarbamate (B.M.C.) du benomyl et du methylthiophonate. *Phytiatrie Phytopharmacie*. 21 : 237-254.

Liu, L. and Ge, Q.X. 1988. Studies on soluble protein patterns of *Rhizoctonia solani* by polyacrylamide gel electrophoresis. I. The influence of different source of isolates on protein patterns of anastomosis groups. *Acta MycologicaSinica* 7: 175-182.

Machacer, J.E. and Brown, A.M. 1948. Experiments on vegetable seed disinfection and observations on varietal resistance of beans, peas and sweet corn to some diseases in Manitoba. *Sci. Agric*. 28: 145-153.

Maringoni, A.C., Tofoli, J.G. and Fregonese, L.H. 1992. Fungicides action *in virto* on *Rhizotoniasolani* and on control of dry bean damping off. *Summa Phytopathologica* 18: 269-275.

Mathew, K. A. and Gupta, S.K. 1996. Studies on web blight of French bean caused by *Rhizoctonia solani* and its management. *Indian J. Mycol. Pl. Pathol*. 26: 171-176.

Mathew, K.A and Gupta, S.K. 1998. Biological control of root rot of French bean caused by *Rhizoctonia solani. J. Mycol. Pl. Pathol*. 28(2): 202-205.

Mathew, K.A. and Gupta, S.K. 1996. Root rot and web blight (*Rhizoctonia solani* Kuhn) of French bean. *Int. J. Trop. Pl. Dis*. 14: 135-146.

Matloob, A.A.H. and Juber, K.S. 2013. Biological control of bean root rot disease caused by *Rhizoctonia solan*i under green house and field conditions. *Agriculture and Biology Journal of North America* 4(5): 512-519.

Matsuyama, N., Moromizato, Z., Ogoshi, A. and Wakimoto, S. 1978. Grouping *Rhizoctonia solani* Kuhn with non-specific esterase zymogram. *Ann. Phytopathol. Soc. Japan* 44: 652-658.

Matz, J. 1921. Una enfermendaddamina de la haleichuela. *Porto Rico Insular Sta. Cir.* 57: 8.

Matz. J. 1917. A Rhizoctonia of the fig. *Phytopathology* 7: 110-118.

Mordue, J.E.M., Currah, R.S. and Bridge, P.D. 1989. An integrated approach to *Rhizoctonia* taxonomy, cultural, biochemical and numerical techniques. *Mycol. Res*. 92: 78-90.

Mrkovacki, N. and V. Milic. 2001. Use of *Azotobacter chroococcum* as potentially useful in agricultural application. *Annals of Microbiol.*51: 145-158.

Mukherjee, S., Tripathi, H.S. and Rathi, Y.P.S. 2001. Integrated management of wilt complex in French bean (*Phaseolus vulgaris* L.). *J.Mycol. Pl. Pathol.* 31(2): 213-215.

Mukhopadhyay, A.N. and Mukherjee, P.K. 1991. Innovative approaches in biological control of soil borne diseases in chickpea. Proc. Fourth Int. *Trichoderma* and *Gliocladium* Workshop, Belgirate, Italy, 17-20 July 1991. Petria 1: 146.

Mukhopadhyay, A.N., Shrestha, S.M. and Mukherjee, P.K. 1992. Biological seed treatment for control of soil borne plant pathogens. *FAO Plant Prot. Bull.* 40: 21-30.

Muyolo, N.G., Lipps, P.E. and Schmitthenner, A.F. 1993. Reactions of dry bean, lima bean, and soyabean cultivars to Rhizotonia root and hypocotyl rot and web blight. *Plant. Dis*. 77: 234-238.

Nacion, C.C. 1924. Study of Rhizoctonia blight of beans. *Phillipine Agri*. 12: 315-321.

Naik, T.S., Somasekhara, Y.M., Manu, T. G., Raja, Amaresh, Y. S. and Uma. 2016. Evaluation of French bean (*Phaseolus vulgaris* L.) germplasm against root rot caused by *Rhizoctonia solani*. *Plant Archives*. 16 (1): 83-87.

Naiki, T. and Iii, T. 1981. *R. solani* Kuhn root rot of bean, soyabean and adzukibean seedlings. *Memoirs of the Research Faculty of Agriculture*, Hokkaido University 12: 262-269.

Nandakumar, R., Babu, S., Viswanathan, R., Raguchander, T., and Samiyappan, R. 2001. Induction of systemic resistance in rice against sheath blight disease by *Pseudomonas fluorescens*. *Soil Biol. Biochem*. 33:603-612.

Neate, S.M. and Warcup, J.H. 1985. Anastomosis grouping of some isolates of *Thanatephorus cucumeris* from agricultural soils in south Australia. *Trans. Br. Mycol. Soc*. 85: 615-62

Neeraj and Singh, K. 2009. Impact of AM fungi along with *Pseudomonas fluorescens* and mustard oil cake in controlling root-rot of French bean. *Mycorrhiza News* 21: 12-14.

Neilands, J. B. 1993. Prespectives in biochemistry & biophysics sidero-phores. *Archives of Biochem. Biophys*. 302:1-3.

Nik, W.Z.W. and Yap, M.Y. 1980. *Rhizoctonia solani* a seed borne pathogen of French bean in Malaysia. *Pertanika* 2: 11-15.

Nitshe, M.J. and Cafati, K.C. 1985. Identification of internal fungi present in bean seeds. Agricultura *Tecnica* 45: 227-234.

Ogoshi, A. 1975. Studies on the anastomosis groups of *Rhizoctonia solani* Kuhn. *Jpn Agric. Res*. 9: 198-203.

Ogoshi, A. 1978. Intraspecific groups of *Rhizoctonia solani* and their significance. *9th Symp. Soil Borne Plant Dis.* Pp. 43-50. *Morioka: Phytopathol. Soc. Jpn.*

Ogoshi, A. 1987. Ecology and pathogenicity of anastomosis and intraspecific groups of *Rhizoctonia solani* Kuhn. *Ann. Rev. Phytopathol*. 25: 125-143.

Okhovat, M. 1977. Effect of few fungicides on *R. solani* Kuhn. causing root rot and damping off of bean. *Iranian J. Pl. Path*. 13: 1-8.

Parmeter, J.R. Jr., Sherwood, R.T. and Platt, W.D. 1969. Anastomosis grouping among isolates of *Thanatephorus cucumeris*. *Phytopathology* 59: 1270-1278.

Parmeter, J.R., Whitney, H.S. Jr. 1970. Taxonomy and nomenclature of the perfect state. In: In: *Rhizoctonia solani Biology and Pathology* (Parmeter, J.R. Jr. ed.). University of California Press, Berkely, pp. 20-31.

Paul, Y.S. and Sharma, R.C. 1990. Additions to the host range of *Rhizoctonia solani*. *Pl. Dis. Res*. 5: 234.

Perez, S.M.A. and Gonzales, B.J. 1988. Incidence of fungi affecting germination of bean seeds in Cuba. *Ciencias de la Agricultura* No. 34, 35: 5-9.

Prasad, K. and Weigle, J.L. 1970. Screening for resistance to *Rhizoctonia solani* in *Phaseolus vulgaris*. *Pl. Dis. Reptr.* 54: 40-44.

Prasad, K. and Weigle, J.L. 1976. Association of seed coats factors with resistance to *R. solani* in *Phaseolus vulgaris*. *Phytopathology* 66: 342-345.

Raffat, F.M. 1992. Efficiency of certain fugicides in controlling *Rhizoctonia* damping off of bean. *Assuit J. Agri. Sci.* 23: 37-47.

Rajnauth, G. 1987. Web blight, an important disease of bean and pak-choi in Trinidad. *Trop. Agri.* 84: 356-358.

Reynolds, M., Weinhold, A.R. and Morris, T.J. 1983. Comparison of anastomosis groups of *Rhizoctonia solani*by polyacrylamide gel electrophoresis of soluble proteins. *Phytopathology* 73: 903-906.

Richter, H. and Schneider, R. 1953. Untersuchungen zur morphologischen und biologischen differenzierung von *Rhizoctonia solani* Kuhn. *Phytopathol. Z.* 20: 167-226.

Roberti, R., Ghisellini, L., Pisi, A., Flori, P. and Filippini, G. 1993. Efficiency of two species of *Trichoderma* as a biological control against *Rhizoctonia solani* Kuhn. isolated from string bean root rot in Italy. *Adv. Hort. Sci.* 7: 19-25.

Saharan, B.S. and Nehra, V. 2011. Plant growth promoting rhizobacteria: A critical review. *Life Sciences and Medicine Research, LSMR-21*, 30 pp.

Sanchez, A., Echavez, B.R. and Schroder, E.C. 1994. *Psedudomonas cepacia*, a potential biofungicides for root rot pathogens of beans. *J. Agric. Univ. Puerto Rico* 78: 55-57.

Sanchez, A.J.H. and Cardenas, A.M. 1988. Etiology and damage of root rot disease in beans *Phaseolus vulgaris* L. in the state of Durango. *RevistaChapingo* 12: 43-49.

Sanchez, P.A. 1956. Effectiveness of various fungicides used alone or in combination in the control of damping off and seed decay in peas and beans. *Acta Agron. Palmira*. 6: 1-35.

Sartarato, A., Zimmerman, M.J., Rocha, M. and Yamada, T. 1988. Web blight. In *Cultura do Feijoero Factores que afetam a Productividade*, 503-520.

Sartorato, A. 1988. Web blight. In Cultura do Feijoero Factores que afetam a Productividade, 503-520.

Schneider, R. 1953. Untersuchungen Uber Feuchtigkeitsanspruche parasitischer Pilze. *Phytopath. Z.* 21: 63-78.

Segura, G. and Diaz Polanco, C. 1975. Treatment of seeds and seedlings of *Phaseolus vulgaris* with benomyl (benlate) to prevent infection by *Rhizoctonia solani*. CIARCO. 5: 37-40.

Sesan, T. and Dumitras, L. 1979. The effect of some fungicides and some seed borne fungi and nodule formation in kidney bean. *Analele Institututui de Cereatan Pentru Protectia Plantelov*. 15: 95-101.

Shama, Saad and Shama, S. 1989. Transmission of *Rhizoctonia solani* Kuhn in seeds of bean (*Phaseolus vulgaris* L.). *Curr. Sci*. 58: 972-974.

Sharma, M. 2003. Management of root rot and web blight (*Rhizoctonia solani*Kuhn) of French bean through eco-friendly methods. M.Sc. Thesis. Dr YS Parmar Univerisity of Horticulture & Forestry, Nauni, Solan. 83 p.

Sharma, M. and Gupta, S.K. 2003. Ecofriendly methods for the management of root rot and web blight (*Rhizocrtonia solani*) of French bean. *J. Mycol. Pl. Pathol.* 33: 345-361.

Sharma, M., Gupta, S.K. and Sharma, T.R. 2005. Characterization of variability in *Rhizoctonia solani* by using morphological and molecular markers. *J. Phytopathol*. 153: 449-456.

Sharma, S.R. and Sohi, H.S. 1980. Assessment of losses in French bean caused by *Rhizoctonia solani* Kuhn. *Indian Phytopath*. 33: 366-369.

Sharma, S.R. and Sohi, H.S. 1981. Effect of different fungicides against Rhizoctonia root rot of French beans (*Phaseolus vulgaris* L.). *Indian J. Mycol. Pl. Pathol.* 11: 216-220.

Sharma, T.R., Prachi and Singh, B.M. 1999. Application of polymerase chain reaction in phytopathogenic microbes. *Indian J. Microbiol.* 39: 79-91.

Shrestha, S.M. and Mukhopadhyay, A.N. 1991. Integrated approach - a successful method for controlling lentil wilt complex. Paper presented at Zonal Meeting of Indian Phytopathological Society, December 12-13, 1991. Pantnagar U.P. India (Abstr.).

Siameto, E. N., Okoth, S., Amugune, N. O. and Chege, N. C. 2011. Molecular characterization and identification of biocontrol isolates of *Trichoderma harzianum* from Embu District, Kenya. *Tropical and Subtropical Agroecosystems*. 13:81-90.

Singh, R.S. 1983. Organic amendments for root disease control through management of soil microbiota and the host. *Indian J. Mycol. Pl. Pathol.* 13: I-XVI.

Smolinska, U. and Kowalska, B. 2006. The effectivity of plant extracts and antagonistic microorganisms on the growth inhibition of French bean pathogenic fungi. *Veg. Crops Res. Bull.*64: 67-76.

Sneh, B., Burpee, I. and Ogoshi, A. 1991. Identification of *Rhizoctonia* species. APS Press, St. Paul, MN. USA.

Snowden, J.D. 1927. Report of the acting Mycologist. *Uganda Dept. Agri. Ann. Rept.* 1926. 30-32.

Srivastava, L.S. and Gupta, D.K. 1989. Aerial blight of French bean, groundnut, soyabean, black, green and horse gram - a new record from India. *Pl. Dis. Res.* 4: 163-164.

Stan, G.I. and Raicu, C. 1973. *Rhizoctonia solani* Kuhn. on bean. *Revista de Horticulturasiviticultura.* 12: 80-86.

Talbot, P.H.B. 1970. Taxonomy and nomenclature of the perfect state. In: *Rhizoctonia solani Biology and Pathology* (Parmeter, J.R. Jr. ed.). University of California Press, Berkely, pp. 20-31.

Tan, C.S. and Tu, J.C. 1995. Tillage effect of root rot severity, growth and yield of beans. *Canadian J. Pl. Sci.* 75:183-186.

Tello, J.C., Lacasa, A. and Molina, R. 1985. A plant pathological noteon the parasite complex of bean roots. *Complejo parasitario ITEA.* No. 61: 57-69.

Temizel, M.M. and Ertunc, F. 1992. Investigations on the detection of bean diseases of Van province. *J. Turkish Phytopath.* 21: 20-31.

Thakur, R.S., Sugha, S.K. and Singh, B.M. 1991. Evaluation of systemic fungicides for control of *Rhizoctoniasolani* under glass house conditions. *Indian J. Agri. Sci.* 61: 230-232.

Todda, T. Hyakumachi, M. and Arora, D.K. 1999. Genetic relatedness among and within different *Rhizoctonia solani* anastomosis groups as assessed by RAPD, ERIC and REP-PCR. *Microbiol. Res.* 154: 247-258.

Tu, J.C. 1992. Management of root rot diseases of peas, beans and tomatoes. *Canadian J. Pl. Path.* 14: 92-99.

Tu, J.C. and Park, S.J. 1993. Root rot resistance in common bean. *Canadian J. Pl. Sci.* 73: 365-367.

Tu, J.C. and Varrtaja, O. 1981. The effect of the hyperparasite (*Gliocladium virens*) on *Rhizoctonia solani* and *Rhizoctonia* root rot of the white beans. *Canadian J. Bot.* 59: 22-27.

Tu, J.C. and Zheng, J. 1993. Compatability of biocontrol agents with DCT. *Mededelingen van de FaculteitLandbouwwetenschappenUnivrsiteit Gent.* 58:1359-1364.

Tu, J.C., Tan, C.S. and Park, S.J. 1992. Effect of soil moisture in root rot soil on plant growth and root rot severity of susceptible and resistant bean cultivars. *Mededelingen van de FaculteitLandbouwwetenschappenUnivrsiteit Gent.* 57: 381-386.

Tylkowska, K. 1984. Occurrence of fungi on bean seeds reproduced in different regions of Poland. *BulletynInstytutuHodowlii. AklinetyZacji* No. 153: 185-202.

Upmanyu, S, Gupta, S.K. and Shyam, K.R. 2002. Innovative approaches for the management of root rot and web blight (*Rhizoctonia solani*) of French bean. *J. Mycol. Pl. Pathol.* 32(3): 317-331).

Upmanyu, S. and Gupta, S.K. 2005. Influence of environmental factors on the progress of root rot and web blight (*Rhizoctonia solani*) of French bean. *Indian Phytopath.* 58(1): 79-83.

Upmanyu, S. and Gupta, S.K. 2009. Physiological variation among French bean isolates of *Rhizoctonia solani. J. Pl. Dis. Sci.* 4(2): 160-163.

Upmanyu, S., Gupta, S.K. and Kaur, R. 2005. Variation in French bean isolates of *Rhizoctonia solani* Kuhn. *J. Mycol. Pl. Pathol.* 35: 168- 173.

Upmanyu, S., Gupta, S.K. and Sharma, M.K. 2004. Evaluation of French bean (*Phaseolus vulgaris*) germplasm against web blight (*Rhizoctonia solani*). *Indian J. Agric. Sci.* 74(8): 448-450.

Viegas, A.P. 1945. Algunsfungos do Brasil. VII-VIII Cyphellaceace E. Thelephoraceae. *Braganita Brazil* 5: 717-779.

Voland, R.P. and Epstein, A.H. 1994. Development of suppressiveness to diseases caused by *Rhizoctonia solani* in soils amended with composted and non-composted manure. *Plant Dis.* 78: 461-466.

Wallace, G.B. 1932. Preliminary list of fungi or diseases of economic plants in Tanganyika Territory. *Kew Roy. Bot. Gard. Bull. Misc. Inform.* 1932: 28-40.

Weber, G.F. 1935. An aerial Rhizoctonia on bean. *Phytopathology* 25: 38.

Weber, G.F. 1939. Web blight, a disease of beans caused by *Corticium microsclerotia. Phytopathology* 29: 559-575.

Weller, D. M. 1988. Biological control of soilborne plant pathogens in the rhizosphere with bacteria. *Annu. Rev. Phytopathol.* 26:379-407.

Wolk, M. and Sarkar, S. 1994. Antagonism *in vivo* of *Bacillus* spp. against *Rhizoctonia solani* and *Pythium* spp.*Anzeiger fur Schadlingskunde pflanzenschutz Umweltschutz* 67: 1-5.

Wolk, M. and Sarkar, S. 1995. *In vivo* antagonism of fluorescent pseudomonads against *Rhizoctonia solani* and *Pythium aphanidermatum* on cucumbers and beans. *Anzeiger fur Schadlingskunde pflanzenschutz Umweltschutz* 68: 3-8.

Zaumeyer, W.J. and Thomas, H.R.. 1957. A monographic study of bean diseases and methods for their control. *United States Department of Agriculture. Tech. Bull.* No. 868: 255.

Ziedan, E.H. and Mahamoud, S.Y.M. 2002. Calcium and sulfur soil treatment to improve biological control with *Trichoderma harzianum* for root rot disease control. *Assiut J. Agric. Sci.* 33(3): 149-160.

3

ANTHRACNOSE

This is one of the most devastating diseases of French bean (*Phaseolus vulgaris* L.) particularly in tropical and subtropical regions of the world (Pastor-Corrales *et al.*, 1994). Scribner (1888) first time used the name of this disease as Anthracnose which is derived from the Greek word –meaning ulcer which is appropriate as it causes sunken depressed spots on the pods.

GEOGRAPHICAL DISTRIBUTION

The pathogen was actually collected by Mycologists as early as 1843 in France (Desmazieres, 1843), but the first description of bean anthracnose in 1875, was based on the specimens collected by Lindemuth at Bonn, Germany (Zaumeyer and Thomas, 1957). Subsequently, it was reported from England, Italy, France, Germany, North America, Japan, Taiwan, East and West Indies, Belgian Congo, Uganda, Brazil, Guatemala, Venezuela, Canada, Cuba, Australia, New Zealand and erstwhile USSR (Zaumeyer and Thomas, 1957; Plopper, 1983).

In India, the occurrence of this disease was first time recorded in Nilgiri hills in 1915 (Hutchinson and Ram Ayyar, 1915). Since then it has been noticed in the entire bean growing tracts of the country having cool and moist weather during growing season.

LOSSES

Though its occurrence has been reported, the world over covering tropical and subtropical parts, but more losses due to this disease were observed in the temperate regions compared to tropics. In Italy, bean crop was destroyed completely due to this disease (Voglino, 1892). Fischer (1919) reported serious epidemics in 1915 and 1916 in Germany. During 1912-1920, this disease was accounted as one of the dreaded disease of this crop in the United States, east of Mississippi river (Zaumeyer and Thomas, 1957). Many infected seeds rotted in the ground without germination or, if they did germinate, the seedlings were destroyed before they emerged. Serious economic losses have also been reported from Latin American countries like Venezuela, Colombia (Echandi, 1976) and Brazil (Costa, 1972). Losses from this disease can approach to 100 per cent when highly contaminated seed is planted under favourable disease development conditions (Zaumeyer and Thomas, 1957). Up to 90-100 % yield losses has been observed in several countries including tropical America and Africa (Pastor-Corrales and Tu, 1989; Lenne, 1992). Such a high losses due to this disease not only reduce the availability of the food but also affect the nutrition, particularly in the Great Lakes region of Africa where beans are the major source of protein being a staple food of local population (Godderis, 1995). The incidence of this disease ranged from 5.0-65.0% in different localities in Himachal Pradesh, India, (Sharma *et al.*, 1994) however, the disease can cause huge yield losses in epidemic years (Padder and Sharma, 2010).

HOST RANGE

The host range of *Colletotrichum lindemuthianum* besides *Phaseolus vulgaris* includes *P. lunatus, P. limensis, P. acutifolius var. latifolius, P. coccineus, P. aureus,P. multiflorus, Macrotyloma uniflorum,Dolichoslablab, D. biflorus,Canavalia ensiformis, Vigna radiata, V. sinensis,V.* faba and *V. unguiculata* (Barrus, 1921; Patel, 1949; Walker, 1950; Zaumeyer and Thomas, 1957; Onesirosan and Barker, 1971).

SYMPTOMS

This malady may manifest on all above ground plant parts but its severity depends upon the time of infection and inoculums source, however they are more conspicuous on immature pods. The first symptoms are seen on the cotyledons as small dark brown to black lesions. Conidia developing on these lesions are splashed by rain to the developing leaves and young stem. On the leaves blackened dead areas appear on the veins. Similar spots may develop on the upper surface of leaves also. On stems and petioles, the lesions are sunken and rust in colour. On pods, black depressed spots with lighter or grey coloured central area (Plate-1) are noticed which in moist weather are covered by pinkish spore mass. Later the sides of these spots become raised and dark brown in colour. The seeds developing these spots in heavily infested pods show brown to light chocolate coloured sunken cankers on the seed coat. Yellowish to brown sunken lesions appear on the seeds when infection is mild and difficult to differentiate from the lesions produced by other organisms.

THE PATHOGEN

The disease is csaused by *Colletotrichum lindemuthianum* (Sacc and Magn.) Briosi and Cav. Perfect stage is *Glomerella cingulata* (Stonem) Spauld & Shrenk) which occur very infrequently (Kimati and Galli, 1970).Mycelium is septate and branched initially and become dark with age. On culture media, conidia may form singly at the ends of hyphal threads. When stromatoid masses form, acervuli are produced upon them (Plate 2) and on the host, they form stroma below the cuticle, rupturing the latter if it is still present on the lesion. Under favourable environmental conditions, conidia borne acrogenously on short conidiophores and appear on the surface as pinkish masses. Setae occur sparingly (Plate 2). Conidia are hyaline, cylindrical with rounded ends or somewhat pointed at one end measuring 4.4-5.3 x 13 -22 µm in size. A clear vacuole like body generally found near the centre of each conidium. One to four, usually two germ tubes arise from a conidium. Appresoria are formed at the tips of germ tubes, especially when they come into contact with host.

Originally, the perfect stage of the fungus was named as *Glomerella lindemuthianum* Shear (Shear and Wood, 1913), but now it is renamed as *G. cingulata* (Kimati and Galli, 1970). The fungus produces perithecia with a diameter of 120 to 210 μm and rostrum 30 to 80 um in length. Perithecium contains hyaline and filiform periphyses until 27 days of age and asci which measure 8 X 48- 68 μm disappear after 27- 30 days. Each ascus contains eight ascospores, which may be alantoid (605 X 20 μm) in shape. Ascospores are ejected from the tip of ascus.

Meyer et al. (2001) reported that during infection of this disease, CO_2 assimilation, stomatal conductance to water vapour, and photosynthetic electron transport rate decreased whereas in dark respiration increased.

Plate 1. Symptoms of anthracnose on pods of French bean

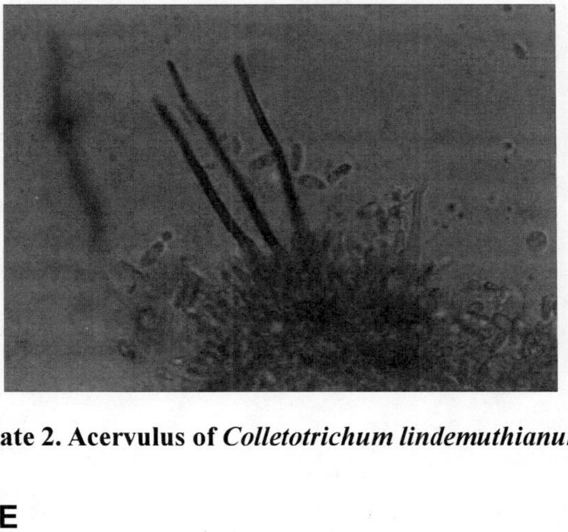

Plate 2. Acervulus of *Colletotrichum lindemuthianum*

DISEASE CYCLE

The pathogen survives in infected seeds and plant debris from one season to another (Zaumeyer and Thomas, 1957; Tu, 1983). In seed the pathogen was mostly present in seed coats and cotyledons and rarely in embryonic axes (Ravi *et al.*, 1999). In infected seed, conidia can remain viable for many years (Tu, 1983) and infect young seedlings. The conidium may germinate in 6 to 9 h under favourable environmental conditions and produces one to four germtubes, which in turn form appressorium and get attached to the host cuticle by a gelatinous layer (Dey, 1919). The later produces infective hyphae, which penetrates the cuticle and epidermis by mechanical pressure, applied by appressorium. Infective hyphae initially produce biotrophic primary hyphae (100-130 nm) that are wide and entirely intracellular, followed by necrotrophic secondary hyphae (25-40 nm) that are narrow and either intercellular or intracellular (Perfect *et al.*, 2001). The secondary hyphae were not surrounded by extracellular matrix. The hyphae enlarge and grow between the cell wall and protoplast for 2 to 4 days without apparent damage to the host cells. Several days later, the cell walls are degraded, probably by L-galactosidase (English and Albersheim, 1968), and the protoplast dies, leading to the appearance of water-soaked lesions. Mycellium then may aggregate within the lesion site and form an acervulus, which ruptures the host cuticle (Figure 1). Each acervulus contains a layer of 3 to 50 conidiophores depending on the lesion size (Zaumeyer and Thomas, 1957). The conidia are dispersed by rain splashes to nearby plants and the cycle can be repeated several times during the growing season. The perfect stage of the fungus occur rarely in nature. Perithecia are formed in the infected tissue and asci and ascospores are produced within the perithecia. The role of ascospores in disease cycle has not been investigated because their occurrence is very rare.

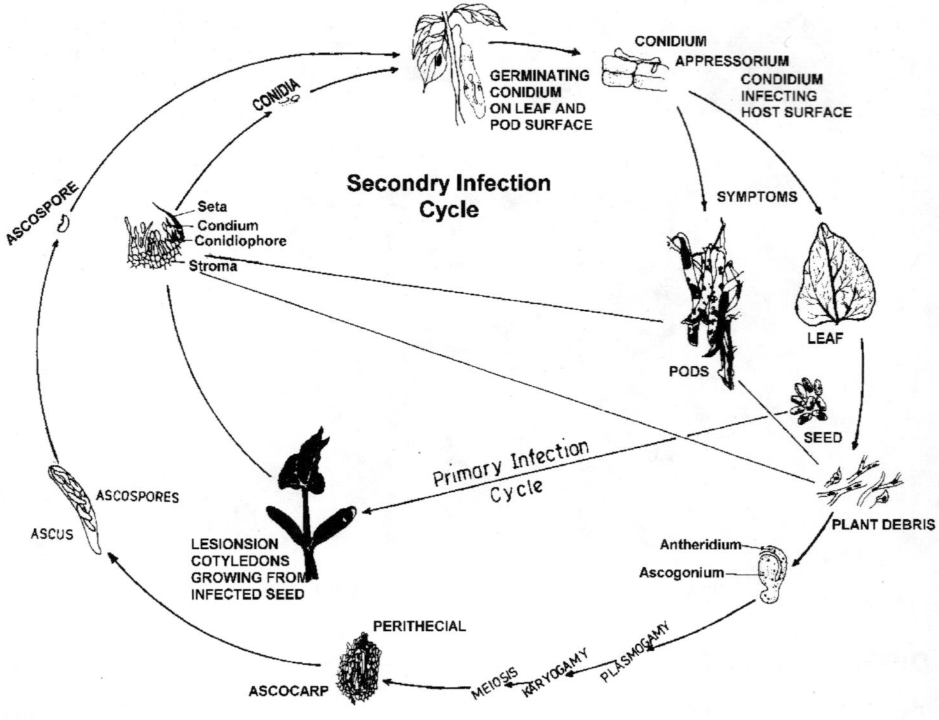

Figure 1. Disease cycle of anthracnose of French bean

EPIDEMIOLOGY

The pathogen overwinters in infected seed as well as in infected plant debris either as conidia or dormant mycelium. Either of these germinate or become active and cause infection of young seedlings. Tu (1983) observed the viability of the fungus in infected seed for many years. On germination of infected seed, conidia appearing on cotyledonay lesions act as the source of secondary inoculum and are splashed by rain or dew to primary leaves and hypocotyls. Primary leaves and hypocotyls are the centre of secondary infections. For the initiation of the disease, wet season in early growth period along with availability of inoculums source is very important (Borucka and Marcinkowska, 2001).For the growth and sporulation of the pathogen, cool and wet weather conditions are essential and may disappear in hot and dry weather. More than 92% relative humidity optimum being 100% is essential for infection..At least 10 mm of rain is required to establish initial infection (Tu, 1981). Optimum temperature for disease development ranged in between 18-27°C with maximum intensity at 21° C (Sindhan, 1963) which is distinctly reduced at 13°C (Sartorato, 1988). Horizontal and vertical spread of the disease is favoured by heavy and persistent rains along with moderate temperatures (19-25°C) and 70% relative humidity (Kumar *et al.*, 1999). For the appearance of severe disease epidemics, medium rainfalls at regular intervals are required for local dissemination of the conidia present in water soluble gelatinous matrix. Insects, animals and man also help in spreading of conidia by moving particularly in moist foliage. Meyer *et al.* (2001) reported that during attack of this disease, CO_2 assimilation, stomatal conductance to water vapour, and photosynthetic electron transport rate decreased whereas in dark respiration increased.

VARIABILITY

Barrus (1911) was the first to differentiate two distinct pathogenic races within a morphologic species of *C. lindemuthianum* on the basis of reaction of host varieties and designated them as Alpha and Beta. Subsequently, Burkholder (1923) found a bean field of cv. White Imperial (resistant to Alpha and Beta) to be severely affected this disease and he designated this isolate as 'Gamma' strain. Later strain 'Delta' from North Carolina in the USA was described by Andrus and Wade (1942) and it was distinct from all the 34 strains described by Schrieber (1932). Occurrence of strain 'Ebnet' was also reported from Germany which resulted due to breakdown of the resistance of bean cv. Cornell 42-242 (Hoffman *et al.*, 1974). Since then many pathotypes of this pathogen has been described in Europe (Charrier and Bannerot, 1970; Kruger *et al.*, 1977; Fouilloux, 1979), South and Central America (Menezes and Dianese, 1988), Canada (Tu *et al.*, 1984), Uganda (Pastor-Corrales and Tu, 1989) etc. Unfortunately, differential bean cultivars were not identical which made it difficult to compare results and to exchange data or resistant germplasm (Pastor-Corrales and Tu, 1989). To overcome this difficulty, a set of twelve bean differential (Table- 1) cultivars has been defined with binary values (Pasto-Corrales, 1991). A number designates a race, which is the sum of the binary values assigned to differentials susceptible to the isolate (Pastor-Corrales, 1991).

Table 1. Gene pool, growth habits and seed characteristics of differential bean cultivars for *Colletotrichum lindemuthianum* race characterization

S. No.	Differential	Binary Value	Gene pool[a]	Growth habit [b]	Seed colour[c]	Seed size[d]
1.	Michelite	1	MA	II	1	S
2.	M.D.R.K.	2	A	I	6	L
3.	Perry Marrow	4	A	II	1	L
4.	Cornell 49-242	8	MA	II	8	S
5.	Widusa	16	MA	I	1	M
6.	Kaboon	32	A	II	1	L
7.	Mexico 222	64	MA	I	1	M
8.	PI 207262	128	MA	III	2	S
9.	To	256	MA	I	2	M
10	Tu	512	MA	III	8	S
11.	AB 136	1024	MA	IV	6	S
12.	G 2333	2048	MA	IV	6	S

a. Gene pool: A= Andean; MA= Middle-American

b. Growth habit: 1= determinate upright or bush; II= indeterminate upright bush; III= indeterminate, week-stemmed, prostrate, non-climbing or semiclimbing; IV= indeterminate, strong climbing.

c. Seedcolour: 1 = White; 2 = cream-beige; 3 = yellow; 4 = brown maroon; 5 = pink; 6 = red; 7 = purple; 8 = black

d. Seed size: S = small, <25g/ 100 seeds; M= Medium, 25-40 g/100seeds; L= Large, >40g/ 100 seeds.

Using this system, atleast 130 different *C. lindemuthianum* races have been identified. Race characterization has been carried out on isolates from South America (Rava *et al*., 1994; Tamayo-Molano *et al*., 1995; Sicard *et al*., 1997a & b; Balardin, 1997), Central America (Balardin and Kelly, 1998; Gonzalez *et al*., 1998), USA (Kelly *et al*., 1994), Canada (Balardin *et al*., 1997), the Netherlands (Balardin *et al*., 1997), Brazil (Carneiro, 1999), India (Sharma *et al*., 1999) and Spain (Fernandez *et al*., 2000). In many African countries, even though anthracnose is severe, little or no information exists on diversity of *C. lindemuthianum*. Only 56 isolates have been characterized using the binary system in Africa, and a total of 36 races have been identified up till now (Beshir, 1992, Beshir, 1995; CIAT, 1996; Koch, 1996; Bigirimana *et al*., 2000). Races 401 and 485 have been recorded for the first time in Burundi, Central Africa (Bigirimana *et al*., 2000). Occurrence of Delta mutant (Race 7), a new race in North of Spain was recorded (Fernandez *et al*., 2000).

85 isolates of *C. lindemuthianum* were collected from different bean growing areas of Himachal Pradesh and were screened on International and CIAT differential sets (Drijfhout and Davis, 1989; Pastor-Corrales, 1991) and 12 and 19 races, respectively

were identified (Sharma, 1999). On International differentials, specifically Alpha-Brazil, Beta, Gamma and IND-I to IND- IX were characterized. The races were nominated as IND I to IND IX which were found distinct from those existing in Europe and USA, thus constituting a new race group from the Indian subcontinent. On CIAT differential set, only races 6 and 73 resembled the North American races. 12 races were resembled with American and European virulences, whereas 17 races were identified as new, and hence total number of races increased to more than 140. RAPD data analysis with nine random primers produced polymorphic bands and grouped 90 isolates into two major clusters with 36 and 54 isolates, respectively. However, there was no analogy between the RAPD pattern and virulence phenogram (Sharma *et al*., 2007). Five populations of *C. lindemuthianum* causing anthracnose in common bean were analyzed to know about genetic diversity, migration and the probable rate of spread of races capable of overcoming resistance present in elite cultivars (Padder *et al*., 2007). Diversity within each population (HS) was high with values 0.26 to 0.31. Average differentiation among populations (GST) was 0.12 and populations were isolated by distance between the two populations. The results demonstrate occurrence of less short distance gene flow in Himachal Pradesh. Ribosomal DNA analysis of fifteen isolates representing 15 races of *C. lindemuthianum* restricted with 2 restriction enzymes (*Msp I* and *Hea*III), five haplotypes were observed in the dendrogram thereby revealing low diversity (Padder *et al*., 2008). 50 physiological races of *C. lindemuthianum* with wide variability were identified in Brazil (Ribeiro *et al.,* 2016). Greater occurrences of races 65, 73, and 81 have been observed in Brazil along with the occurrence of pathogenic variability among isolates of a single race, destabilizing the resistance of commercial cultivars. Ribeiro *et al.* (2016) classified 51 isolates of race 65 in 10 different physiological races. Seventeen isolates of *C. lindemuthianum* from northern central Ecuador were characterized by the inoculation of a group of twelve standard differential bean varieties (Rodríguez-Ortega *et al.,* 2018). Among the analyzed samples, thirteen races were identified; five of those races had not been previously reported in the country. Thee differential G2333 (Co-42, Co-52 and Co-7) presented resistance to every characterized races in Ecuador. In addition, twenty - one improved varieties and elite bean lines were evaluated with sixteen of the seventeen isolates, three genotypes were identified (TB2, TB3 and INIAP 485 Urcuquí) with resistance to the mentioned isolates, which can be used as sources of resistance to anthracnose.

A draft of the complete genome sequences of two isolates of *C. lindemutuianum* has been reported (de Queiroz *et al*., 2017). The contig assembly using the CLC Genomics Workbench software resulted in 1,857 and 1,276 contigs for two isolates. Smallest contig contained 1.02-kbp band and the largest contig contained a 1.11 Mbp band. G+C content was 37.6% and 37.6% for two isolates.

MANAGEMENT

Various management strategies have been attempted to reduce yield losses due to bean anthracnose, which were based on fungicides (Tu, 1986b; Trutman *et al.,* 1992) resistant cultivars (Fouilloux, 1976; Alzate-Marin *et al.*, 1999), cultural practices (Pastor-Corrales and Tu, 1989) and biocontrol of seed borne infection (Ravi *et al.*, 1999). The most effective method of reducing the losses from this disease is to integrate as many of the available control measures as possible into a disease management programme. The measures include cultural practices, chemical control, host resistance and biological control.

Cultural practices

To keep the ravages of this disease in check, the use of following cultural practices like field sanitation, avoiding of sowing of the bean crop in the same field year after year and healthy seed have been advocated (Sharma and Sohi, 1989). The disease free seed can be produced in those regions where the prevailing temperature remains high while humidity is low along with surface irrigation which makes the conditions unfavourable for infection of the pathogen, or under a pedigreed seed programme, in which seed plots were isolated and subjected to strict inspection for disease free seed (Tu, 1986). Under hilly tracts, best sowing time of this crop for maximum yield and minimum disease incidence ranged from mid-April to mid-May (Sindhan and Bose, 1981). Hailu (2017) tested four management options for this disease and found mean disease severity and AUDPC significantly lowest compared to sole planting in which these were highest. Integration of intercropping with compost application was effective ecofriendly option for the of this disease

Host resistance

Use of resistant cultivars is the most efficient, cheap and environment friendly method to check the epidemic development of this disease under favourable environmental conditions (Lenne, 1992). However, the pathogen is highly variable and development of new races is continuous which is limiting factor in the effective management of this disease. The resistant genes should be identified and deployed in the commercially grown varieties but this gene can also be overcome by the pathogen new race(s) developed either by mutation or rearrangement of the genetic material (Agrios, 1997). It is possible that the new gene for virulence had already been present earlier at low level and only in a few pathogen individuals. Resistance in cultivar Charlivox against alpha and beta races of the pathogen is conferred by single independent dominant gene (Anderson, 1963). Cell wall of resistant beans absorbs higher quantities of endo-polygalacturonase secreted by the pathogen than susceptible ones. Sharma *et al.* (1986) reported the presence of a glycoprotein inhibitor in the cell wall of resistant cultivars. Since the pathogen is highly variable, non-race specific resistance has been used in breeding for this disease. Cultivar IAPAR-Rio Negro-8 were found resistant against all races under field conditions (Alberini*et al.* 1987) whereas varieties *viz.,* Aroana 80, Ayso, Carioca 80, Iguaco, Moruna 80, RiuoPiqiuri, Rio Vermelho, Mulatao, Olho-de-Pombo and Vermelhao were resistant to atleast 7 of the 9 races of the pathogen and no race was virulent to all the 70 cvs. screened in Brazil (Menezes and Dianese, 1988). Cultivars/ lines VL 60, VL 63, Jawala, HPR 33, B 4 B 6 and P 48 were found resistant against eight isolates of the pathogen (Chakrabarty and Shyam, 1990). However, later Gupta (1993) exhibited highly susceptible reaction in cultivar Jawala, which might be explained on the development of new pathotype of the pathogen. Kumar *et al.* (1997) screened sixty cultivars/ lines of kidney bean against ten races viz., beta, gamma, Ind I, Ind II, Ind II, Ind IV, Ind V, Ind VI (alpha-Brazil), Ind VII and Ind VIII under natural and artificial epiphytotic conditions and reported that two accessions (AB-136 and G 2333) were highly resistant to all prevalent races in Himachal Pradesh, India, whereas cultivars Cornell 49-242, EC 43036, EC 57080, KRC 5, PI 207-262 and Widusa showed resistance to more than five races. Rest genotypes were susceptible. Later Gonalves-Vidigal*et al.* (2001) reported that resistance of cv. AB 136 to both races 31 and 69 was controlled by a single dominant gene designated as Co-6. In addition, linkage analysis using RAPD marker indicated that Co-6 also controls the resistance of this cultivar to other races of the pathogen, or that different genes are present in the same linkage block.

Few cvs./ lines/ land races viz., Negro Otomi (Acosta-Gallegos *et al.*, 2001), SRC 74, SRC 89A, SRC 90, SRC 95, SRC 201, PDR 14 (Sud and Sood, 2002), Morden 003 (Mundel *et al.,* 2004) and HUR-15, HUR-137, IPR 96-4, VL-63, PDR-14, HUR -385 (Khati and Hooda, 2006) have been found resistant to bean anthracnose. Exotic accessions like G 2333, Cornell 49242, PI 207262, Mexique 222, TO, Perry Marrow, Kaboon and Widusa were found resistant to more than five Indian races, whereas two Indian accessions KRC-5 and Hans showed resistance against six and four races, respectively (Pathania *et al.*, 2006). Out of 65 indigenous and 34 exotic accessions evaluated under laboratory conditions against *C. lindemuthianum*, 47, 32 and 18 among indigenous and 16, 22, 18 and 11exotic accessions showed resistance to race 3, 515, 589 and 529, respectively (Sharma *et al.*, 2012). Seven indigenous collections viz., IC-328537, IC-328538, IC-448888, IC313294, IC-278723, IC339645, IC341862 and 3 exotic accessions *viz.*, EC-169813, EC-398530 and EC-500226 were found resistant to all evaluated races.

Gupta *et al.*, (2010) studied the role of oxidative enzymes and biochemical constituents in imparting resistance to French bean varieties against anthracnose and reported that higher amount of oxidative enzymes (catalase, polyphenol oxidase, peroxidase) was present in diseased leaves than healthy ones and resistant varieties (Kentucky Wonder and Pant Anupam) showed higher enzyme activity than their susceptible (Contender and Laxmi) ones. In case of biochemical constituents, higher amount of total, reducing and non-reducing sugars, phenol content were present in resistant cultivars than susceptible ones. Higher amount of total and ortho- dihydroxy phenols and protein content was observed in diseased pods as compared to healthy pods. Higher content of chlorophyll was observed in healthy leaves of resistant varieties as compared to diseased leaves.

Darben *et al.* (2017) evaluated 75 common bean accessions from Common Bean Germplasm Bank (CBGB) of Núcleo de PesquisaAplicada à Agricultura, Nupagri, Brazil against two races of races 73 and 2047 of *C. lindemuthianum* and found 14 accessions as resistant to both races of pathogen, 36 as resistant to race 2047, and 25 as incompatibility reaction to race 73.

Inheritance of resistance

Bean - *C. lindemuthianum* interaction is based on the gene-for gene model and incompatibility is characterized by an early hypersensitive response (Esquerre-Tugayee*et al.*, 1992). Until now, nine resistance genes have been described: A; Are, Mexique 1, Mexique 2, Mexique 3, also referred to as Co-1, Co-2, Co-3, Co-4, Co-5, respectively (Mastenbroek, 1960; Kelly and Young, 1996), Co-6, Co-7 and co-8 (Schwartz *et al.*, 1982; Pastor-Corrales *et al.*, 1994; Alzate-Marin *et al.*, 1997). Bean cultivar Michigan Dark Red Kidney contains Co-1 and has been used in breeding for resistance to the alpha race of pathogen in USA (Cardenas *et al.*, 1964). The Co-2 gene in cultivar Cornell 49-242 was used extensively in breeding programs in Europe (Fouilloux, 1979) and Ontario (Tu, 1992 a). Cultivars Mexico 222, To, Tu, AB 136 and G 2333 are other sources of resistance to *C. lindemuthianum* (Pastor-Corrales *et al.*, 1994) with resistance genes Co-3, Co-4, Co-5, C0-6/ co-8 and Co-7/Co-5/Co-4^2 respectively (Alzate-Marin *et al.*, 1997; Young and Kelly, 1996; Young *et al.*, 1997). Existence of gene Co-9 had been reported in Meso-American Cultivar BAT93 (Geffroy *et al.*, 1999).

Inheritance of resistance in common bean genotypes PI 207262 against one race (raĉe 69) of the pathogen was studied and it was found that 3 dominant genes were responsible for resistance, one gene was situated in PI 207262 whereas two genes were present in G 2333 (Poletine *et al.*, 1999). Xu *et al.* (1999) were also of the view that anthracnose resistance inheritance is boverned by single dominant gene however cross between Chinese cultivars demonstrated a more complicated and mixed inheritance pattern giving a segregation ratio of 9: 7 in F_2 generation. Inheritance of resistance in two land races KRC 5 and KRC 8 of kidney bean against race 903 (beta) of *C. lindemuthianum* showed segregation ratio of three resistant to one susceptible (3:1) and one resistant to three susceptible (1:3) in F_2 generations (Sharma *et al.*, 2000). This revealed the presence of single dominant gene in KRC 5 and one recessive gene in KRC 8 for resistance, respectively, which confirmed from segregation ratio of backcross generations.

A population of 185 recombinant inbred lines (RIL) derived from the cross PMB0225 × PHA1037 was evaluated for anthracnose resistance under natural and artificial photoperiod growth conditions to study the genetic basis of quantitative resistance to two races i.e., 23 and 1545 of *C. lindemuthianum* in stem, leaf and petiole different organs (González *et al.*, 2015). Using multi-environment QTL mapping approach, 10 QTLs for race 23 and 16 QTLs for race 1545 were identified. The homologous genomic regions corresponding to 17 of the 26 main identified QTLs showed the presence of resistance-associated gene cluster encoding nucleotide-binding and leucine-rich repeat (NL) proteins. Among them, the major QTLs detected on linkage group 05 for resistance to race 1545 in stem, petiole and leaf which were positioned within a 1.2 Mb region. The NL gene Phvul.005G117900 is located in this region, which can be considered an important candidate gene for the non-organ-specific QTL identified here. Furthermore, a total of 39 epistatic QTL (E-QTLs) (21 for resistance to race 23 and 18 for resistance to race 1545) involved in 20 epistatic interactions (eleven and nine interactions for resistance to races 23 and 1545, respectively) were identified. No significant effects of environment interaction were displayed with main and epistatic identified QTLs.

Tanscriptome analysis of two near isogenic lines (NILs) differing for the presence of the Co-1 gene on chromosome Pv01 during a time course following infection with race 73 identified 3,250 significantly differentially expressed genes (DEGs) within and between the NILs over the time course of infection (Padder *et al.*, 2016). Majority of DEGs were up regulated during the biotrophic phase in the susceptible NIL, whereas more DEGs were up regulated during the necrotrophic phase in the resistant NIL. Defense related genes *viz.,* PR proteins, peroxidases, lipoxygenases were up regulated in the resistant NIL. On the other hand, genes encoding sugar transporters were up regulated in the susceptible NIL during the later stages of infection. Additionally, numerous transcription factors and candidate genes within the vicinity of the Co-1 locus were differentially expressed, suggesting a global reprogramming of gene expression in and around the Co-1 locus. On the basis of this analysis, number of candidate genes at the Co-1 locus had been reduced to three than the previous reported eight genes.

Field screening of fifty-five common bean genotypes, including differential varieties, characterized resistance sources, elite lines, cultivars and controls, using artificial inoculation with a mixture of six virulent and prevalent races of *C. lindemuthianum*, showed that 28 genotypes were resistant to anthracnose (mean severity score ≤ 3.5) whereas 10 genotypes stood out because they presented a mean anthracnose severity score of 1.0 (Vieira *et al*., 2018). The genotypes were also screened with six SCAR markers linked to resistance genes and four SCAR markers were shown to be useful for the assisted selection of their respective target genes (SH18 and SAS13 for Co-4^2, SAB03 for Co-5, and SAZ20 for Co-6). Two carioca seeded elite lines were highlighted by the phenotypic and molecular screening: K10 (Co-3^4, Co-4^2, Co-5 and Co-6) and K13 (Co-4^2). The phenotypic and molecular characterization of candidate resistance sources against anthracnose on the basis of disease reaction in field inoculation trials and on the analysis with linked molecular markers to resistance genes is an effective strategy and will be useful in the selection of potential donor parents and resistant lines in breeding programs of common beans.

Chemical control

Different systemic and non-systemic fungicides (ferbam, Ziram, arasan, thiram, ceresan, Benlate, Bavistin, Vitavax, Agrosan, G.N., captafol) have been used to reduce the seed-borne infections (Sindhan and Bose, 1981; Trutman*et al.*, 1992). Benlate and Bavistin can persist in the plant up to 15 to 20 days after seed treatment (Sindhan and Bose, 1981). Efficacy of of Bavistin, Benlate, zineb, maneb and mancozeb as foliar sprays for the management of bean anthracnose has also been demonstrated (Silva and Loch, 1982; Chakrabarty and Shyam, 1988). Efficacy of EBI fungicides namely triadimefon and triforine against the mycelia growth of the pathogen has been demonstrated however when these chemicals were used as seed treatment, they showed phytotoxic effect to the seedlings of *P. vulgaris* (Chakrabarty and Shyam, 1988). A combination of seed treatment with thiram and followed by two or three sprays of mancozeb at 45, 60 and 75 days of sowing as Carbendazim alone or in effectively managed this malady (Sharma and Sugha, 1995). A combination of seed treatment with tebuconazole (0.1%) and foliar sprays of tricyclazole (0.03%) reduced the disease severity both on leaves and pods while maximum seed yield was recorded in case of seed treatment and foliar sprays of carbendazim (Gupta *et al.*, 2000).

Various strobilurin fungicides like azoxystrobin, trifloxystrobin, pyraclostrobin (Picinini and Fernandas, 2000; Rava, 2002; Oliveira and de Oliviera, 2003) significantly reduced the disease severity and increased yields. The combination of these fungicides with EBI's like epoxiconazole (Rava, 2002), tebuconazole and propiconazole (Oliveira and de Oliviera, 2003) significantly reduced the anthracnose severity. Triphenyltin hydroxide stannane was also found effective against this disease and provided >70 per cent control of this disease (Picinini and Fernandas, 2000).

Continuous use of systemic fungicides has been shown to be associated with the development of resistant biotypes (Tu and Jarvis, 1979; Tu and McNaughton, 1980). Maringoni *et al.* (2002) reported that all isolates of *C. lindemuthianum* collected from *P. vulgaris* fields located at Paranpanema Valley, Brazil were insensitive to benomyl, carbendazim and thiophanate methyl showing the occurrence of cross resistance to different benzimidazole fungicides while all isolates were sensitive to chlorothalonil. Carbendazim and fluquinconazole applied without rotation were found least effective in the management of this disease (Oliveira and de Oliviera, 2003). However, combination of systemic fungicides with non-systemics and rotation of fungicides can be an effective approach for management of this disease instead of constant use of one systemic fungicide.

Biological control agents

Although research in biocontrol has advanced considerably in many major plant diseases in recent years but this is not the case with bean anthracnose. Limited work has been done like *in vitro* evaluation of the antagonists against this pathogen (Gupta *et al.*, 1991) or the phenomenon of cross protection by inoculating a less virulent strain of the fungus to reduce the severity from severe strains (Sutton, 1979). Infected seed when soaked in 10 per cent culture filtrate or treated with 0.4 per cent talc formulation of *Trichoderma viride* recorded minimum seed infection and maximum seed germination (Ravi *et al.*, 1999). Treatment with *P. chlororaphis*PCL1391 resulted in best biocontrol of anthracnose, while *P. fluorescens* WCS365 showed no significant difference compared to the positive control (Bardas *et al.*, 2009). Padder*et al.* (2010) evaluated three bioagents (*Trichoderma viride, T. harzianum* and *Gliocladum virens*) and five biopesticides (Achook, Neemgold, Wannis, Spictaf and Neemazal) under *in vitro* and *in vivo* conditions against *Colletotrichum lindemuthianum*. All the three antagonistic fungi caused significant inhibition of mycelial growth, maximum being with *T. viride*(69.21%) followed by *T. harzianum*(64.20%). *T. viride*and biopesticide Wanis @ 1000 μl/ml when used in combination as seed treatment were found most effective in reducing the seed borne infection.

Efficacy of different bioagents, viz. *Trichoderma viride, T. harzianum, T. hamatum* and *Gliocladium virens* was evaluated both under *in vitro* and *in vivo* conditions against *C. lindemuthianum* and maximum inhibition of mycelial growth in dual culture (59.48%) and inverted plate (55.98%) was obtained with *T. viride*(Padder and Sharma, 2011). All the bioagents overgrew the pathogen and the principal mechanism of mycoparisitism observed was coiling, brusting and disintegration of pathogen hyphae. Culture filtrate from *T. viride* was found best as it completely inhibited radial growth at 25 and 50 per cent concentration and reduced the spore germination of test fungus significantly. However, lower concentrations of culture filtrate from all bioagents showed little effect on spore germination. Seed application of bioagents was found better as compared to soil application. A maximum increase in seed germination and inhibition of seed borne infection was observed with *T. viride* followed by *T. harzianum* under pot culture conditions. *T. viride* has the maximum potentiality to suppress the spore germination, mycelial growth, seed borne infection of *C. lindemuthianum* and increased seed germination when compared with the other biocontrol agents.

Biopesticides/ Plant origin pesticides

Seed treatment followed by spraying with biopesticides at 10% leaf extracts of *Adenocalymma alliaceae* (*Pseudocalymma alliaceum*), *Azadirachta indica* and *Lawsoniainermis* was evaluated under pot culture and field conditions against bean anthracnose (Ravi *et al.*, 2000). Seed treatment + foliar spray of aqueous leaf extract of *L. inermis*, as well as seed treatment with *L. inermis* extract alone, significantly improved the seedling emergence, reduced the disease incidence, increased the yield per plant and 100-seed weight under pot conditions. Extracts of *T. erecta* and *L. camara* have also been reported inhibitory to *C. lindemuthianum* (Gupta *et al* 2005). Padder*et al.* (2010) evaluated five biopesticides (Achook, Neemgold, Wannis, Spictaf and Neemazal) under *in vitro* and *in vivo* conditions against *Colletotrichum lindemuthianum* and found Wanis applied @ 1000 µl/ml caused maximum inhibition of 82.12 per cent followed by Spictaf (52.85 %). Under greenhouse conditions, the extracts of *M. argyrophylla* and *O. vulgare* caused maximum reductions (41.82% and 37.65%, respectively) in disease severity when a local effect assay was carried out (Pinto *et al.*, 2010). Antimicrobial activity of cascalote phenolics against phytopathogenic fungus *C. lindemuthianum* which causes anthracnose disease in common beans was evaluated. The results indicate that cascalote phenolics have *in vitro* fungistatic activity against *C. lindemuthianum* races R-0 and R-1472. Moreover, these compounds provoked on the whole, *in vitro* inhibition of spores germination and cellulase and polygalacturonase activities in these fungi. *In vivo* assays with cascalote phenolics under greenhouse conditions using susceptible cv. PI 206272 of common bean resulted in a good protection against anthracnose severity especially as a preventive treatment (Garcia *et al.*, 2010).

REFERENCES

Agrios, G.N. 1997. *Plant Pathology*. 4th Edition. Academic Press, San Diego, USA.

Alberini J.L., Mohan, S.K., Menezes, J.R. de, Silva, W.R. da, Oliari, L. and Meyer, R.C. 1987. IAPAR 8 - Rio Negro, a new French bean cultivar. *Pesquisa-Agropecuaria Brasileira 22:* (9-10): 995-998.

Alzate-Marin, A.L., Baia, G.S., de Paula Junior, T.J., de Carvalho, G.A., de Barros, E.G. and Moreira, M.A. 1997. Inheritance of anthracnose resistance in common bean differential cultivar AB 136. Plant Disease 81: 996-998.

Alzate-Marin, A.L., Menarim, H., de Carvalho, G.A., de Paula Junior, T.J., de Barros, E.G. and Moreira, M.A. 1999. Improved selection with newly identified RAPD markers linked to resistance gene to four pathotypes of *Colletotrichum lindemuthianum* in common bean. *Phytopathology* 89: 281-285.

Anderson, A.L., Adams, M.W. and Whitford, G. 1963. Charlevoix anthracnose resistant dark red kidney bean. *Mich. State Univ. Agr. Sta. Res. Rept.* 6.

Andrus, C.F., and Wade, B.L. 1942. The factorial interpretation of anthracnose resistance in beans. *U.S.Dept. Agric. Tech. Bull.* 810: 29.

Bardas, G.A., Lagopodi, A.L., Kadoglidou, K. and TzavellaKlonari, K. 2009. Biological control of three *Colletotrichum lindemuthianum* races using *Pseudomonas chlororaphis* PCL1391 and *Pseudomonas fluorescens* WCS365. *Biol. Cont.* 49(2): 139-145.

Balardin, R.S. and Kelly, J.D. 1996. Identification of race 65-Epsilon of bean anthracnose (*Colletotrichum lindemuthianum*) in Michigan. *Plant Dis.* 80(6): 712.

Balardin, R.S. 1997. Identificacao de racesfisiologcas de *Colletotrichum lindemuthianum* no Rio Grande do Sul-Brasil. *Fitopatologia Brasileira* 22(1): 50-53.

Balardin, R.S. and Kelly, J.D. 1998. Interaction between *Colletotrichum lindemuthianum* races and gene pool diversity in *Phaseolus vulgaris. J. Amer. Soc. Hort. Sci.* 123: 1038-1047.

Balardin, R.S., Jarosz, A.M. and Kelly, J.D. 1997. Virulence and molecular diversity in *Colletotrichum lindemuthianum* from South, Central and North America. *Phytopathology* 87: 1184-1191.

Barrus M. F. 1911. Variation of varieties of bean in their susceptibility to anthracnose. *Phytopathology* 1: 190-195.

Barrus, M.F. 1921. Bean anthracnose. *N.Y. (Cornell) Agric. Exp. Stn. Mem.* 42: 97-209.

Beshir, T. 1992. Research on anthracnose of haricot beans in Ethiopia. 34-38 in: *Proceedings of the Pan-Africa Bean Pathology Working Group Meeting. Thika, Kenya, May 26-30, 1992. CIAT African Workshop Series No.* 23.

Beshir, T. 1995. Studies on the importance, pathogenic variation and management of bean anthracnose in Ethiopia. 33--43 in: *Proceedings of the Second Meeting of the Bean Pathology Working Group Meeting. Kakamega, Kenya, June 5-8, 1995. CIAT African Workshop Series No.* 37.

Bigirimana, J., Fontaine, R. and Hofte, M. 2000. Bean anthracnose: virulence of *Colletotrichum lindemuthianum*isolates from Burundi, Central Africa. *Plant Dis*. 84: 491.

Borucka, K. and Marcinkowska, J.Z. 2001. Occurrence of *Colletotrichum lindemuthianum* on *Phaseolus* beans in Poland. *Veg. Crops Res. Bull.* 54(1): 147-151.

Burkholder, W.H. 1923. Gamma strain of *Colletotrichum lindemuthianum* Sacc. & Magn. *Phytopathology* 13: 316-323.

Cardenas, F., Adams, M.W. and Anderson, A. 1964. The genetic system for reaction of field beans (*Phaseolus vulgaris* L.) to infection by three physiologic races of *Colletotrichum lindemuthianum*. *Euphytica*13: 178-186.

Carneiro, S.M.de T.P.G. 1999. Physiological races of *Colletotrichum lindemuthianum* in Parana State – Brazil. *Summa Phytopathologica* 25(3): 275-278.

Chakrabarty, P.K. and Shyam, K.R. 1988. Evaluation of systemic fungitoxicants against *Colletorichum lindemuthianum*, the incitant of bean anthracnose. *Indian J. Plant Pathol.* 6(1): 67-70.

Chakrabarty, P.K. and Shyam, K.R. 1990. Reaction of some promising lines of French bean to different isolates of *Colletotrichum lindemuthianum* in Himachal Pradesh. *Indian J. Plant Pathol.* 8 (1&2):86-88.

Charrier, F. and Bannerot, H. 1970. Contribution a l'etude des races physiologiques de l'anthracnose du haricot. *Ann. Phytopathol*. 2: 489-506.

CIAT (Centro Internacional de Agricultura Tropical), 1996. Anthracnose. 46-48 in: *Bean Program, Annual Report 1993*. Working Document No. 161.

Costa, A.S. 1972.II. Investigacoes sobre molestias do feijoeiro no Brasil, pp. 305-311. Anals Do I Simposio Brasileiro de Feijao. Universidade Federal de Vicosa, Minas Gerais, Brazil.

Darben, L.M., Gonela, A., Elias, H. T., da Silva, C. R., Pastre, H. H., and Gonçalves-Vidigal, M. C. 2017. Common bean germplasm resistant to races 73 and 2047 of *Colletotrichum lindemuthianum*. *African Journal of Biotechnology* 16(19): 1142-1149.

Dey, P. K. 1919. Studies in the physiology of parasitism. V. Infection by *Colletotrichum lindemuthianum. Ann. Biol.* 33: 305-312.

de Queiroz, C.B., Correia, H.L.N., Menicucci, R.P., Vidigal, P.M.P. and de Queiroz, M.V. 2017. Draft genome sequences of two isolates of *Colletotrichum lindemuthianum*, the causal agent of anthracnose in common beans. *Genome Announc.* 5: e00214-17.

Drijfhout, E. and Davis, J.H.C. 1989. Selection of a new set of homogeneously reacting bean (*Phaseolus vulgaris*) differentials to differentiate races of *Colletotrichum lindemuthianum*. *Plant Path. 38:* 391-396.

Echandi, E. 1976. Principales enfermedades de hongo del frijol (*Phaseolus vulgaris*), en los tropicos americanos en diferentes zonas ecológicas. *Fitopatologia Brasileira* 1: 171-177.

English, P.D. and Albersheim, P. 1968. Host-pathogen interactions. I. A correlation between a- galactosidase production and virulence. *Plant Physiol.* 44 : 217-224.

Esquerre-Tugaye, M.T., Mazau, D., Barthe, J.P., Lafitte, C. and Touze, A. 1992. Mechanisms of resistance to *Colletotrichum* species. Pp. 121-133 in: *Colletotrichum*: *Biology, Pathology and Control.* J.A. Bailey and M.J. Jeger, eds. C.A.B. International, Wallingford, U.K.

Fernandez, M.T., Fernandez, M., Casares, A., Rodriguez, R. and Fueyo, M. 2000. Bean germplasm evaluation for anthracnose resistance and characterization of agronomic traits: a new physiological strain of *Colletotrichum lindemuthianum* infecting *Phaseolus vulgaris* L. in Spain. *Euphytica* 114(2): 143-149.

Fischer, W. 1919. Die Brennfleckenkrakheit der Bohnen. *Fiihling'sLandw. Ztg.* 68: 241.

Fouilloux, G. 1976. Bean anthracnose: new genes of resistance. *Bean Improv. Coop. Annual Report* 19: 36-37.

Fouilloux, G. 1979. New races of bean anthracnose and consequences on our breeding programmes. Pp. 221-235 in: Int. Symp. *Diseases of Tropical Food Crops.* H. Maraite, and J.A. Meyer, eds. UniversiteCatolique de Louvain-la-Neuve, Belgium.

Garcia, Veloz R., Marin, M.R., Veloz, R.R., RodriguezGuerra, R., Torres, P.I., Gonzalez, C.M.M., Anaya Lopez, J.L., Guevara Olvera, L., Feregrino Perez, A.A., Loarca Pina, G. and Guevara Gonzalez, R.G. 2010. Antimicrobial activities of cascalote (*Caesalpiniacacalaco*) phenolics-containing extract against fungus *Colletotrichum lindemuthianum. Indust. Crops Prod.* 31(1): 134-138.

Geffroy, V., Sicard, D., de Oliveira, J.C.F., Sevignac, M., Cohen, S., Gepts, P., Neema, C., Langin, T. and Dron, M. 1999. Identification of an ancestral resistance gene cluster involved in the coevolution process between Phaseolus vulgaris and its fungal pathogen *Colletotrichum lindemuthianum. Plant -Microbe Interactions* 12: 774-784.

Godderis, W. 1995. Ibiharage: la culture du haricot au Burundi. Administration Generale de la Cooperation au Developpement, Bruxelles, Belgium.

Gonalves-Vidigal, M.C., Sakiyama, N.S., Vidigal-Filho, P.S., Amaral-Junior, A.T., Poletine, J.P.and Oliveira, V.R. 2001. Resistance of common bean cultivar AB 136 to races 31 and 69 of *Colletotrichum lindemuthianum*: the Co-6 locus. *Crop Breeding App. Biotech.* 1(2): 99-104.

González, G.A., Yuste-Lisbona, Fernando J., Rodiño, A. P., De Ron, A. M., Capel, C., García-Alcázar, M., Lozano, R. and Santalla, M. 2015. Uncovering the genetic architecture of *Colletotrichum lindemuthianum* resistance through QTL mapping and epistatic interaction analysis in common bean. *Frontiers in Plant Science* 6:141. doi: 10.3389/fpls.2015.00141.

Gonzalez, M., Rodrigues, R., Zavala, M.E., Jacobo, J.L., Hernandez, F., Acosta, J., Martinez, O. and Simpson, J. 1998. Characterization of Mexican isolates of *Colletotrichum lindemuthianum* by using differential cultivars and molecular markers. *Phytopathology* 88: 292-299.

Gupta S, Kalha CS, Vaid A and Rizvi SEH 2005. Integrated management of anthracnose of French bean caused by *Colletotricum lindemuthianum*. *J. Mycol. Pl. Pathol.* 35(3): 432-436.

Gupta, S., Mallick, S.A and Gupta, Moni. 2010. Role of oxidative enzymes and biochemical constituents in imparting resistance to French bean varieties against bean anthracnose. *Indian Phytopath.* 63(10): 47-50.

Gupta SK. 1993. Annual Progress Report, Department of Plant Pathology Dr. YSPUHF, Solan 161 pp.

Gupta, S.K., Dohroo, N.P. and Shyam, K.R. 1991. Antagonistic studies on seed-borne mycoflora of French bean (*Phaseolus vulgaris* L.). *Indian J. Plant Pathol.* 9(1&2): 62-63.

Gupta, S.K., Mathew, K.A., Shyam, K.R. and Gupta, D.K. 2000. Role of fungicides in management of anthracnose of French bean. In: Proceedings of Indian Phytopathological Society-Golden Jubilee International Conference on "*Integrated Plant Disease Management for sustainable Agriculture*" Vol. II, Indian Phytopathological Society, IARI, New Delhi: 705-707.

Hailu, N. 2017. Management of Common bean anthracnose (*Colletotrichum lindemuthianum*Sacc and Magn.) through integration of intercropping and compost application at Haramaya, eastern Ethiopia. *Innovative Techniques in Agriculture* 1.2: 96-106.

Hoffmann, G.M., Schnock, M. and Kruger, J. 1974. On the occurrence of a new physiologic race of *Colletotrichum lindemuthianum* on *Phaseolus vulgaris*. *Zeitschrift fur pflanzen krankheitan und Pflanzenschutz* 81: 490-493.

Hutchinson, C. M. and Ram Ayyar, C.S. 1915. The Indian rice beer ferment. *Mem. Dept. Agri. Indian Bact. Ser.* 1: 137-168.

Issa, E., Sinigaglia, C. and Oliveira, D.A. 1983. Chemical control of anthracnose (*Colletotrichum lindemuthianum* (Sacc. & Magn.) Scrib.) and two other diseases of bean (*Phaseolus vulgaris* L.). *Biologico* 49(4): 89-95.

Kelly, J.D. and Young, R.A. 1996. Proposed symbols for anthracnose resistance genes. *Bean Improv. Coop. Annual Report* 39: 20-24.

Kelly, J.D., Afanador, L. and Cameron, L.S. 1994. New races of *Colletotrichum lindemuthianum* in Michigan and implication in dry bean resistance breeding. *Plant Dis.* 78: 892-894.

Khati, Pinkey and Hooda, K.S. 2006. Multiple disease resistance in rajmash genotypes. *Indian Phytopath.* 59(4): 521-522.

Kimati, H. and Galli, F. 1970. *Glomerella cingulata* (Stonem.) Spauld. *et* V. Schrenk. f. sp. *phaseolin*.f., faseascegena de agente causal de anthracnose do feijoeiro. *Annuals da Escola Superior* de Agricultura "Luisde Queiroz" 27: 411-437.

Koch, S.H. 1996. *Colletotrichum* spp. on dry beans and lupins in South Africa. Ph.D. Thesis, University of Pretoria, South Africa.

Kruger, J., Hoffman, G,M. and Hubbeling, N. 1977. The Kappa race of *Colletotrichum lindemuthianum* and sources of resistance to anthracnose in Phaseolus beans. *Euphytica*26: 23-25.

Kumar, A., Sharma, P.N., Sharma, O.P. and Tyagi, P.D. 1997. Resistance to *Colletotrichum lindemuthianum* in kidney bean accessions of diverse origin in Himachal Pradesh. *Indian Phytopath.* 50: 59-64.

Kumar, A., Sharma, P.N., Sharma, O.P. and Tyagi, P.D. 1999. Epidemiology of bean anthracnose *Colletotrichum lindemuthianum* under sub-humid mid-hills zone of Himachal Pradesh. *Indian Phytopath.* 52(4): 393-397.

Lenne, J.M. 1992. *Colletotrichum* diseases of legumes. In: *Colletotrichum*: Biology, Pathology and Control. J.A. Bailey and M.A. Jeger, eds. C.A.B. International, Wallingford, U.K. pp.134-166.

Maringoni, A.C., Barros, E.M. de and de Barros, E.M. 2002. Occurrence of isolates of *Colletotrichum lindemuthianum* resistant to benzimidazole fungicides. *Summa Phytopathologica* 28(2): 197-200.

Mastenbroek, C. 1960. A breeding programme for resistance to anthracnose in dry shell haricot beans, based on a new gene. *Euphytica* 9: 177-184.

Menezes, J.R. and Dianese, J.C. 1988. Race characterization of Brazilian isolates of *Colletotrichum lindemuthianum* and detection of resistance to anthracnose in *Phaseolus vulgaris. Phytopathology* 78: 650-655.

Meyer, S., Saccardy-Adji,K., Rizza, F. and Genty, B. 2001. Inhibition of photosynthesis by *Colletotrichum lindemuthianum* in bean leaves determined by chlorophyll fluorescence imaging. *Plant Cell Environ.* 24(9): 947-955.

Oliveira, S.H.F. de and de Oliveira, S.H.F. 2003. New fungicides and spraying schedules in the control of anthracnose and angular leaf spot of common bean. *Summa Phytopathol.* 29: 45-48.

Onesirosan, P.T. and Barker, L.N. 1971. Stem anthracnose of cowpeas in Nigeria. *Plant Dis. Reptr*. 55: 820-822.

Padder, B. A.,Sharma, P.N., Kapil, R., Pathania, A., Sharma, O.P. 2010. Evaluation of bioagents and biopesticides against *Colletotrichum lindemuthianum* and its integrated management in common bean. *Not Sci Biol*2 (3): xx-xx.

Padder, B. A. and Sharma, P.N. 2011. *In vitro and in vivo* antagonism of biocontrol agents against *Colletotrichum lindemuthianum* causing bean anthracnose. *Archiv.Phytopathol. Plant Prot.* 44 (10): 961-969.

Padder, B.A., Sharma, P.N. Sharma, O.P. and Kapoor, V. 2007. Genetic diversity and gene flow estimates among five populations of *Colletotrichum lindemuthianum* across Himachal Pradesh. *Physiol. Molec. Pl. Path.*70: 8–12.

Padder, B.A., Sharma, P.N., Sharma, O.P. and Kapoor, V. 2008. Ribosomal DNA analysis of *Colletotrichum lindemuthianum*virulences from Himachal Pradesh (India). *Appl. Biol. Res.*10:6-10.

Padder, B.A., Kamfwa, K., Awale, H.E. and Kelly, J.D. 2016. Transcriptome profiling of the *Phaseolus vulgaris - Colletotrichum lindemuthianum* pathosystem. PLoS ONE 11(11): e0165823.

Pathania, A. Sharma, P.N., Sharma, O.P., Chahota, R.K., Ahmad, B. and Sharma, P. 2006. Evaluation of resistance sources and inheritance of resistance in kidney bean to Indian virulences of *Colletotrichum lindemuthianum. Euphytica* 149: 97-103.

Pastor-Corrales, M.A. and Tu, J.C. 1989. Anthracnose. Pp. 77-104 in: *Bean Production Problems in Tropics*. 2nd edition, H.F. Schwartz and M.A. Pastor-Corrales eds. CIAT, Cali, Colombia.

Pastor-Corrales, M.A. and Tu, J.C. 1989. Estandarizacion de variedades differenciales y de designacion de razas de *Colletotrichum lindemuthianum* (Abstr.). *Phytopathology* 81: 694.

Pastor-Corrales, M.A., Erazo, O.A., Estrada, E.I. and Singh, S.P. 1994. Inheritance of anthracnose resistance in common bean accession G 2333. *Plant Dis*. 78: 959-962.

Patel, M.K., Kamat, M.N. and Bhide, V.P. 1949. Fungi of Bombay Supplement I. *Indian Phytopath.* 2: 142-155.

Perfect, S.E., Green, J.R. and O'-Connell, R.J. 2001. Surface characteristics of necrotrophic secondary hyphae produced by the bean anthracnose fungus, *Colletotrichum lindemuthianum*. *European J. Pl. Pathol.* 107(8): 813-819.

Picinini, E.C. and Fernandes, J.M. 2000. Chemical control of angular leaf spot and anthracnose of beans. *Fitopatologia Brasileira* 25: 92-94.

Pinto, J.M.A., Souza, E.A. and Oliveira, D.F. 2010. Use of plant extracts in the control of common bean anthracnose. *Crop Protection* 29(8): 838-842.

Ploper, L.D. 1983. Enfermedades del poroto en el Noroeste Argentino y su control, Publication Miscelanea No. 74, Tucuman, Argentina.

Poletine, J. P., Gonclaves-Vidigal, M. C., Vidigal Filho, P. S., Scapim, C. A. and Silveiro, L. 1999. Inheritance of anthracnose resistance to *Colletotrichum lindemuthianum* race 69 in common bean genotype PI 207262. *Acta Scientiarum* 21(3): 447-451.

Rava, C.A. 2002. Efficiency of fungicides to control anthracnose and angular leaf spot in common beans. *Summa Phytopathol*. 28: 65-69.

Rava, C.A., Molina, J., Kauffman, M., and Briones, I. 1994. Determination de razas fisiologicas de *Colletotrichum lindemuthianum* en Nicaragua. *Fitopatologia Brasileira* 18: 388-391.

Ravi, S., Doraiswamy, S., Valluvaparidasan, V. and Jeyalakshmi, C. 1999. Effect of biocontrol agents on seed- borne *Colletotrichum* in French bean. *Pl. Dis. Res*. 14(2): 146-151.

Ravi, S., Doraiswamy, S., Valluvaparidasan, V. and Jeyalakshmi, C. 1999. Impact of seed –borne *Colletotrichum lindemuthianum* on French bean seed quality. *Acta Phytopathologica et Entomologica Hungarica* 34(4): 283-291.

Ravi,S., Valluvaparidasan, V., Doraiswamy, S. and Jayalakshmi, C. 2000. Production of *Colletotrichum lindemuthianum* free French bean seeds. *Legume Research* 23(3): 170-173

Ribeiro, T., Esteves, J. A. de F., Silva, D. A., Gonçalves, J. G. R., Carbonell, S. A. M., and Chiorato, A. F. 2016. Classification of *Colletotrichum lindemuthianum* races in differential cultivars of common bean. *Acta Scientiarum. Agronomy*, 38(2): 179-184.

Rodríguez-Ortega, D., Vega-Jiménez, L., Murillo-Ilbay, A. and Peralta-Idrovo, E. 2018. Pathogenic variability of *Colletotrichum lindemuthianum* and Ecuadorian *Phaseolus vulgaris* L. germplasm resistance. *AgronomíaMesoamericana* 29(1): 1-11.

Saccardo, P.A. 1878. Fungi veneti novi vel critici, Vel mycologiae venetae addendi, *Isariopsis griseola* Sacc., *Michelia* 1: 273.

Sartorato, A. 1988. Anthracnose. In: *Cultura do Feijoeire: Fatores que Afetam a Productividade* (Zimmerman, M.J. de O., Rocha, M. and Yamada, T. Eds.) *Piracicaba Brazil, Associcao Brasileira para pesquisa de Potassa e do Fosfato*: 457-477.

Schrieber, F. 1932. Breeding for resistance in *Phaseolus vulgaris. Phytopath. Zeitschr.* 4: 415-454.

Schwartz, H.F., Pastor-Corrales, M. A. and Singh, S.P. 1982. New sources of resistance to anthracnose and angular leaf spot of bean (*Phaseolus vulgaris* L.). *Euphytica* 31: 741-754.

Scribner, F.L. 1888. Anthracnose of the bean *Gloesporium lindemuthianum*. *U.S. D.A. Rpt.* (1887): 361.

Sharma, P.N., Sharma, O.P. and Tyagi, P.D. 1994. Status and distribution of bean anthracnose in Himachal Pradesh. *Himachal J. Agril. Res. 20* (1&2): 91-96.

Sharma, P.N. and Sugha, S.K. 1995. Management of bean anthracnose through chemicals. *Indian Phytopath. 48*(3): 304-307.

Sharma, P.N., Kapila, R.K., Sharma, O.P. and Sud, D. 2000. Inheritance of resistance in two land races of *Phaseolus vulgaris* to *Colletotrichum lindemuthianum. Indian Phytopath.* 53(1): 83-86.

Sharma, P.N., Kumar, A., Sharma, O.P., Sud, D. and Tyagi, P.D. 1999. Pathogenic variability in *Colletotrichum lindemuthianum* and evaluation of resistance in *Phaseolus vulgaris* in the North-Western Himalyan region of India. *J. Phytopath.* 147(1): 41-45.

Sharma, P.N., Padder, B.A., Sharma, O.P., Pathania, A. and Sharma, P. 2007. Pathological and molecular diversity in *Colletotrichum lindemuthianum*(bean anthracnose) across Himachal Pradesh, a north-western Himalayan state of India. *Austral. Pl. Path.***36**: 191–197.

Sharma, P.N., Banyal, K., Rana, J.C., Nag, R., Sharma, S.K. and Pathania, A. 2012. Screening of common bean germplasm against *Colletotrichum lindemuthianum* causing bean anthracnose. *Indian Phytopath.* 65(1): 99-101.

Sharma, R.C., Barthe J.P., Lafitte, C. and Touze, A. 1986. *In vitro* absorption of endopolygalacturonase secreted by *Colletotrichum lindemuthianum* on cell walls monogenically resistant and susceptible cultivars to anthracnose. *Can. J. Bot.*64: 2903-2908.

Sharma, S.R. and Sohi, H.S. 1989. Diseases of French bean in India and their control. *Indian Rev. Life Sci. 9:* 153-168.

Shear, C.L. and Wood, A.K. 1913. Studies of fungus parasites belonging to the genus *Glomerella.U.S.D.A., Bureau of Plant Ind. Bull.* 252.

Sicard, D., Buchet, S., Michalakis, Y. and Neema, C. 1997a. Genetic variability of *Colletotrichum lindemuthianum* in wild population of common bean. *Plant Path.* 46: 355-365.

Sicard, D., Michalakis, Y., Dron, M. and Neema, C. 1997b. Genetic diversity and pathogenic variation of *Colletotrichum lindemuthianum* in the three centers of diversity of its host, *Phaseolus vulgaris. Phytopathology* 87: 807-813.

Silva SDVM and Loch LC. 1982. Some epidemiological aspects of bean (*Phaseolus vulgaris*) anthracnose and the effect of using benomyl on yield components. *Fitopatologia Brassileira* 7: 427-435.

Sindhan, G. S. and Bose, S.K. 1981. Evaluation of fungicides against anthracnose of French bean caused by *Colletotrichum lindemuthianum. Indian Phytopath.* 34(3): 325-329.

Sindhan, G.S. 1983. Effect of temperature and relative humidity on the development of anthracnose of French bean. *Prog. Hort. 15:* 132-135.

Sud, A.K. and Sood, V.K. 2002. Screening of indigenous and exotic collections of kidney bean against bean anthracnose under dry temperate conditions of Himachal Pradesh. *Crop Res Hisar* 24(1): 202-203.

Sutton, D.C. 1979. Systemic cross protection in bean against *Colletotrichum lindemuthianum. Austral. Pl. Pathol.* 8: 4-11.

Tamayo-Molano, P.J., Otoya, M.M. and Pastor-Corrales, M.A. 1995. Diversidad de razas de *Colletotrichum lindemuthianum* el patogeno de la anthracnosis del frijol, en Rionegro, Antioquia. *Fitopatologia Colombiana* 19: 1-6.

Trutmann, O., Paul. K.B. and Cishabayo. 1992. Seed treatments increase yield of farmer varietal field bean mixtures in the Central African highlands through multiple disease and beanfly control. *Crop Prot.* 11(5): 458-464.

Tu, J.C. and Jarvis, W.R. 1979. Ontogeny, organization, and longevity of sclerotium like structures produced by *Colletotrichum lindemuthianum* in the presence of benomyl. *Canadian J. Pl. Pathol.* 10: 17-22.

Tu, J.C. 1981. Anthracnose (*Colletotrichum lindemuthianum*) on white bean (*Phaseolus vulgaris*) in Southern Ontario: Spread of the disease from an infection focus. *Plant Dis.* 5: 477-480.

Tu, J.C. 1983. Epidemiology of anthracnose caused by *Colletotrichum lindemuthianum* on white bean (*Phaseolus vulgaris*) in Southern Ontario: Survival of the pathogen. *Plant Dis.* 67: 402-404.

Tu, J.C. 1986b. Control of anthracnose disease (*Colletotrichum lindemuthianum*) of navy bean (*Phaseolus vulgaris*). *Meded. Fac. Landbouw. Rijksuniv. Gent* 51: 645-651.

Tu, J.C. 1991. Comparison of the efficacy of *Gliocladiumvirens* and *Bacillus subtilis* in the control of seed rots and root rots of navy beans. *Rijksuniver-sitie-Gent. 56*:(2a): 229-234.

Tu, J.C. 1992. Management of root rot diseases of peas, beans, and tomatoes. *Canadian J. Pl. Path.* 14: 92-99.

Tu, J.C. 1992a. *Colletotrichum lindemuthianum* on bean: Population dynamics of the pathogen and breeding for resistance. in: *Colletotrichum: Biology, Pathology and Control*. J.A. Bailey and M.J. Jeger, eds. C.A.B. International, Wallingford, U.K. pp 203-224.

Tu, J.C. and McNaughton, M.E. 1980. Isolation and characterization of benomyl resistant biotypes of the delta race of *Colletotrichum lindemuthianum Canadian J. Pl. Sci.* 60: 585-592.

Tu, J.C., Sheppard, J.W. and Laidlaw, D.M. 1984. Occurrence and characterization of the epsilon race of bean anthracnose in Ontario. *Plant Dis*. 68: 69-70.

Vieira, A.F., Almeida, L.C.S., Rodrigues, L.A., Costa, J.G.C., Melo, L.C., Pereira, H.S., Sanglard, D.A and Souza, T.L.P.O. 2018. Selection of resistance sources to common bean anthracnose by field phenotyping and DNA marker-assisted screening. *Genetics and Molecular Research* 17 (3): gmr18066.

Voglino, P. 1892. *La ruggine perforatrice delle foglie e l' antracnosi dei fagiuoli,* Torino, Italy, (I Fungi piu Dannosi alle Piante Coltivate. 1891-95): 7.

Walker, J.C. 1950. *Plant Pathology*. McGraw - Hill Book Co. Inc. New York. 699p.

Xu, Z., Wang, S. and Wei, M. 1999. Inheritance of resistance to anthracnose (*Colletotrichum lindemuthianum*) in snap bean (*Phaseolus vulgaris*). *China Veg.* 4: 18-20.

Young, R.A. and Kelly, J.D. 1996. Characterization of the genetic resistance to *Colletotrichum lindemuthianum* in common bean differential cultivars. *Plant Dis*. 80: 650-654.

Young, R.A., Melotte, M., Nodari, R.O. and Kelly, J.D. 1997. Marker assisted dissection of the oligogenic anthracnose resistance in the common bean cultivar, 'G 2333'. *Theoretical and Applied Genetics* 96: 86-94.

Zaumeyer, W.J. and Thomas, H.R. 1957. A monographic study of bean diseases and methods for their control. *USDA Technical Bulletin* No. 868: 255 p.

4

POWDERY MILDEW

Powdery mildew is an important disease of snap and lima beans, which has a worldwide distribution. Moderate temperatures and humidity favour infection, however it can be prevalent within a wide range of environmental conditions (Zaumeyer and Thomas, 1957).

GEOGRAPHICAL DISTRIBUTION

The disease was first time reported from France by De Candolle in 1805 (Lamarck and Candolle, 1805), subsequently it was reported from Bermuda in 1889 (Galloway, 1889), Cuba (Cook and Horne, 1908), Virginia (Cook, 1931), Italian East Africa (Ciccarone, 1940), U.S.A. (Campbell and Hofman, 1952), Peru (De la Rocha, 1950), Nicaragua (Litzenberger and Stevenson, 1957), Saugar (Grover, 1953), Guadeloupe (Pauvert, 1991) and India (Bisht *et al*., 1997). In Solan, Himachal Pradesh, the occurrence of the disease was recorded under protected cultivation conditions and the disease was widespread in different polyhouses (Gupta and Bharat, 2010)

LOSSES

Powdery mildew can cause severe yield losses under favourable weather conditions in Spain (Trabanco *et al*., 2012).The losses from this disease vary from year to year, but in certain localities they may exceed half the crop (Cook, 1931). It reduces yields by arresting the development of the plants. Schwartz *et al.* (1981) reported that an epidemic of this disease reduced yields by 17-69 per cent depending on cultivars or breeding line. Arriel *et al.* (1991) recorded losses up to 40 and 50 per cent in cultivars Rio Vermelho and Eriparza, respectively. In Himachal Pradesh, India, the severity of this disease ranged from 50 to 75 per cent in different polyhouse (Gupta and Bharat, 2010).

SYMPTOMS

The disease affects all above ground parts of the plants. The characteristic symptoms include the appearance of faint, slightly dark areas on leaves that may be mistaken for mottling. These areas later develop into small, white, powdery spots, which enlarge rapidly and coalesce to include larger areas, finally covering the entire leaf with coating of a whitish powder (Plate-1). If infection occurs in early growth stages of the crop, affected leaves may become somewhat malformed, turn pale yellow and finally fall down pre-maturely. Similar symptoms also appear on petioles and stems. In advance stages of infection, pods are also infected which appeared to be covered with talcum-like powder.

Plate 1. Leaves of French bean showing the symptoms of powdery mildew, caused by *Erysiphe polygoni*

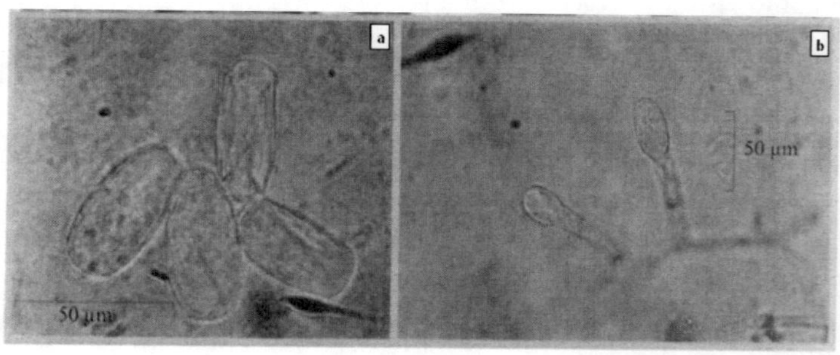

Plate 2. (a) Conidia and (b) conidiophores of *Erysiphe polygoni*

THE PATHOGEN

The disease is caused by *Erysiphe polygoni* DC. (Ferreira *et al*., 1999) but later studies suggest that it is closer to *Erysiphe diffusa* (Cooke & Peck) U. Braun & S. Takan, formerly *Microsphaera diffusa* Cke. & Pk. (Almeida *et al*., 2008). Conidiophores are upright, hyaline, bearing conidia either singly or in chains (Plate 2). The conidia are cylindrical, oval or nearly orbicular, simple, hyaline, one celled spore with rounded ends and successively cut off from the ends and measured 30-45 x 10 -20 µm in size (Plate 2a). The pathogen also produces cleistothecial stage for overwintering in dry temperate zones where favourable climatic conditions are prevalent. The cleistothecia are nearly spherical, sometimes flattened, dark-coloured bodies with hyphal appendages and no ostiole. The asci are released only after the thick cleistothecial wall disintegrates or is ruptured. The asci vary in number from 2-8. They are oblong or ovate and range from 45- 70µm in length and from 30- 40µm in width. Each ascus bears three to eight spores, 20-25 µm long and 9- 14 µm wide.

HOST RANGE

The host range of the fungus besides French bean (*Phaseolus vulgaris*) and other leguminous hosts like *P. aureus, P. coccineus, Pisum sativum, Glycin max, Vigna radiata, Vigna mungo*. Besides these, it also affects several genera of economic importance like *Brassica, Calendula, Catalpa, Dahlia, Delphinium, Lupinus, Lycopersicon, Medicago, Seabiosa, Trifolium, Valeriana* and *Vicia* (Stevens, 1913).

DISEASE CYCLE

The anamorphic stage of the powdery mildew fungus is mostly found in the cultivation areas except in the dry temperate regions where teleomorphic stage is produced due to the prevalence of favourable weather conditions. Wherever cleistothecia are produced, they explain the mode of perpetuation and serve as the source of primary inoculums. Ascospores formed in cleistothecia are released when cleistothecial wall is ruptured and are blown by wind to lower leaves where they cause primary infection (Fig. 1). In other areas where cleistothecia are not produced the pathogen overwinter in anamorphic stage on weed and other cultivated hosts as it is having very wide host range. Besides this, seed borne nature of the pathogen has also been reported (Zaumeyer and Thomas, 1957) when pods are infected, but the seed is probably not an important means of its distribution. In areas where cleistothecia are not produced, conidia surviving on weed or cultivated hosts or in protected structure and blown by wind and cause primary infections. Conidia produced on primary infections are further blown by wind and cause secondary infections.

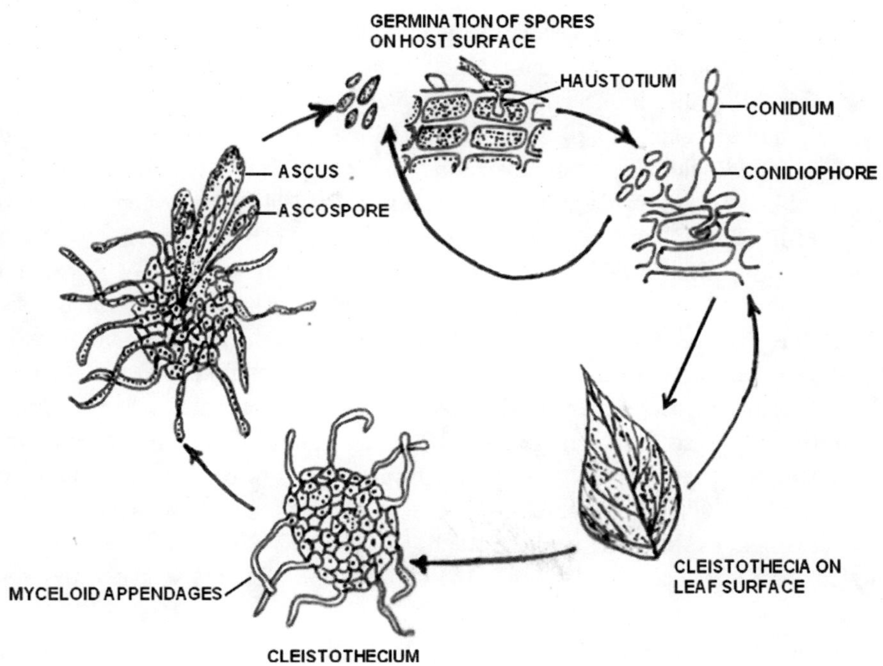

Fig. 1. Disease cycle of powdery mildew of French Bean

EPIDEMIOLOGY

The pathogen survives through conidia on annual and perennial leguminous and other hosts. Disease becomes more severe when plant growth is checked due to low temperature and insufficient soil moisture (Cook, 1931). Relatively dry weather also favours the development of the disease. Conidia of the fungus could germinate at RH 100 down to approximately 0 per cent but soon shriveled and killed at lower values (Yarwood, 1936). Moderate relative humidity is more favourable for production and germination of spores than high. Similarly, dry weather promotes the development of disease than rainy. The optimum temperature for this disease is reported to be 22°C. Singh *et al.* (2016) reported that 22 -27° C and leaf surface humidity (high relative humidity during night and low during day) are necessary for the occurrence of the powdery mildew. If such conditions are prevalent than there are all chances of powdery mildew infection and a greater number of infections/ leaf. If these conditions prevail for longer period than the disease will be more severe. Childs (1940) reported a diurnal cycle of conidiophore development and observed that the period of abstraction of conidia occurred between 10 a.m. and 2 p.m.

MANAGEMENT

The disease can be managed by integration of various available methods like host resistance, cultural practices, chemical and biological methods.

Host resistance

Use of resistant varieties is an economical and efficient method of plant disease control. Most common and serious resistant variety should be selected for plantation based on local adaptability and high yield potentials.Very few cultivars have been reported to have resistance against this disease, but resistance is complicated by the existence of different physiologic races (Zaumeyer and Meiners, 1975). Cultivars/ lines like Refujee bean (Cook, 1931), Florida White Wax (Zaumeyer, 1947), Fullgreen No. 1 (Andrus and Hoffman, 1952) and Seminde (Anonymous, 1953) have been reported to be resistant against this disease while Logan, Cooper Wax and Ashley Wax were tolerant (Zaumeyer, 1947). Contender has also been reported to have considerable resistance against this disease (Campbell and Hoffman, 1952). Hoffman and Kanapaux (1952) also developed one hybrid 'Wade' from Burpee's Tender Pod X B 1187-1-6, which was also found to be resistant to this disease. Also, Alabama No. 1, Tenderlong 15, Idaho Refugee, Sensation Refugee No. 1066 and 1067, US 5 Refugee and Top Crop have been reported to be resistant to one or more races of the fungus (Zaumeyer and Thomas, 1957). Out of 245 accessions/ cvs. evaluated against powdery mildew, only six accessions/ cvs. viz. Amanda, Belneb, Cornell 49242, Negro Sun Luis, PorilloS intetico and local accession BGE 003161 showed complete resistance without visible symptoms on leaves (Trabanco *et al.*, 2012).

Dundas (1936) reported that a single dominant gene governs the resistance. However, Rezende *et al.* (1999) have indicated that two genes are involved in genetic control due to a double recessive epistasis. In inheritance studies, cultivar Porrillo Sintetico carries two dominant and independent resistance genes (Trabanco *et al.*, 2012), one gene conferring complete resistance (IT0) and another gene conferring IT3, characterized by a moderate mycelial development on the leaves. Both genes showed a dominant epistatic relationship.

Two quantitative trait loci (QTLs) viz., $PWM2^{AS}$ and $PWM11^{AS}$ were detected on Pv2 and Pv11, respectively (Bassi *et al.,* 2017). Both QTLs on Pv11 were mapped on the same genomic region, suggesting that it is a pleiotropic region. The study identified new markers closely linked to powdery mildew resistance QTLs, which can be used for marker-assisted selection, fine mapping, and positional cloning.

Cultural Practices

Chemical pollution and health hazards make the use of cultural practices to be of immense value. Moreover, such practices also economies the crop production programme. Adjustment of date of planting is one such practice. Hsieh (1988) has reported that early autumn or late spring sowings were helpful in reducing the attack of powdery mildew. Yarwood (1939) obtained satisfactory control of the disease through water spray applied at a pressure of about 70lb to have a washing effect for few seconds. Mixed cropping of bean with maize has been reported to lower the incidence of powdery mildew in comparison to monocultures (Rheenen *et al.*, 1981). Careful timing and rate of application of Nitrogen fertilizer affect the progress of the disease. Wider plant spacing also helps in making the environment unfavaorable for disease development. Another way to manage the disease are crop rotation and elimination of the volunteer infected plants from in and around the field because it has very wide host range.

Chemical control

Finely divided sulfur dust is the cheapest and most effective method to control this disease (Janson, 1922; Ogilive, 1925). Cook (1931) suggested the application of sulfur in the form of spray. However, dusts were more effective, cheaper and simpler to handle than sprays (Moore, 1936). In New York, a single application of anisomycin (50ppm) or griseofulvin (200ppm) effectively eradicated the infection of this disease (Zaumeyer, 1957). On the other hand, Borders (1966) obtained excellent control by a 3: 4 mixture of captan 50WP and tribasic $CuSo_4$ (53% Cu), without any phytotoxicity. Lukens and Horsfall (1967) reported the efficacy of tetraidoethylene and hexachlore –2- propanol against this disease. In Colombia, Schwartz *et al* (1981) successfully managed the disease using benomyl. EBI fungicides like penconazole, hexaconazole, difenoconazole and fenarimol, recommended against pea powdery mildew (Gupta and Shyam, 1998), can also be used for the control of this disease.

Meena *et al.* (2016) evaluated different fungicides and *Trichoderma* spp. for the management of green gram powdery mildew (*E. polygoni*) and reported that fungicidal sprays were more effective than *Trichoderma* spp. Among fungicides, carbendazim sprays were more effective in reducing the disease severity and increased yield followed by wettable sulphur while mancozeb was least effective.

Bannet *et al.* (1984) used calcium silicate slag for treatment of powdery mildew disease of wheat that helps the plant cells to defend themselves against fungal attack by degrading haustoria and by producing callose and papilla. This strategy can also be used for the control of this disease.

Biological control and Bio-pesticides

Felber Irina and Hamner (1948) obtained most satisfactory results by a spray consisting of an aqueous dispersion of methylcellulose (0.25%) and actidion (5ppm), an antibiotic isolated from bears in which strains of *Streptomyces griseus* are cultivated. A dust formulation (1000-2000 ppm) of Na ushate, derived from usnic acid, an antibiotic from lichens proved a better treatment against this disease compared to sprays (Ark *et al.*, 1960). Davis *et al* (1960) also observed inhibition of the development of *E. polygoni* on bean when antibiotic P-9 from *Streptomyces* sp. was sprayed 2 days after inoculation. Gupta and Kumar (2014) evaluated different botanicals for the management of powdery mildew and reported that ethanol extract of *Acorus calamus* (1.0%) caused maximum inhibition of conidial germination (98.13%) followed by *Lantana camara, Urticadioica* and *Melia azedarach* while under field conditions no botanical was as effective as the standard check i.e. hexaconazole but among botanicals three sprays of *U. dioica* (2.0%) showed least disease severity and increased green pod yield. Yadav *et al.* (2017) evaluated the efficacy of bioagents, botanicals and selected fungicide for the management of powdery mildew and found that the sprays of *Trichoderma viride* (2.5%) were more effective in reducing the disease severity to 15.98% than other bioagents and botanicals though it was slightly less effective to standard fungicide carbendazim (14.27%) followed by *Pseudomonas flouresens* while garlic clove extract was least effective.

REFERENCES

Almeida, A.M.R., Binneck, E., Piuga, F.F., Marin, S.R.R., do Ribeiro Valle, P.R.Z. and Silveira, C.Z. 2008. Characterization of powdery mildews strain from soybean, bean, sunflower and weeds in Brazil using rDNA-ITS sequences. *Trop. Plant. Pathol.* 33: 22-26.

Andrus, C.F. and Hoffman, J.C. 1952. Fullgreen, a blight resistant bush snap bean. *Seed World* 66 (9): 16,18.

Anonymous, 1953. Research work and workers. *Seed World* 72(11): 80-81.

Ark, P.A., Bottini, A.I. and Thompson, J.P. 1960. Sodium ushate as an antibiotic for plant diseases. *Pl. Dis. Reptr*. 44(3): 200-203.

Arriel, E.F., Santos, J.B., Ramalho, M.A.F. and Dos Santos, J.B. 1991. The effect of powdery mildew on yield of common bean cultivars. *Pesquisa Agropecuaria Brasileira* 26(6): 849-852.

Bassi, D., Briñez, B., Rosa, J. S., Oblessuc, P. R., Almeida, C. P., Nucci, S. M., Silva, L., Chiorato, A. F., Vianello, R. P., Camargo, L., Blair, M. W. and Benchimol-Reis, L. L. 2017. Linkage and mapping of quantitative trait loci associated with angular leaf spot and powdery mildew resistance in common beans. *Genetics and molecular biology* 40(1): 109-122.

Bisht, G.S., Harish, Singh and Singh, H. 1997. Fungal diseases of useful plants in the Garhwal Himalaya and their management. *Himalayan Microbiol. Diversity Part I* pp. 225-271.

Bannet and Fiona, G.A. 1984. Resistance to powdery mildew in wheat: a review of its use in agriculture and breeding programmes. *Plant Pathol*. 33: 279- 300.

Borders, H.I. 1966. Snap bean powdery mildew control by captan-tribasic copper sulphate mixture. *Pl. Dis. Reptr*. 50 (11): 874-879.

Campbell, J.A. and Hoffman, J.C. 1952. The Regional Vegetable Breeding Laboratory introduces Contender, a new fresh market snap bean. *Seed World* 66(2): 14, 16, and 66.

Childs, J.F.L. 1940. Diurnal cycle of spore maturationin certain powdery mildews. *Phytopathology* 30: 65-73.

Ciccarone, A. 1940. Italian East Africa: plant diseases reported in 1939. *Intl. Bull. Pl. Prot.* 14(6): 117-119.

Cook, H.T. 1931. Nature of powdery mildew injury to snap beans in Virginia in 1930. *Phytopathology* 21: 116 (Abstr.).

Cook, M.T. 1931. Powdery mildew disease of snap bean. *Virginia Truck Exp. Sta. Bull.* 74: 931-940.

Cook, M.T. and Horne, W.T. 1908. Insects and diseases of vegetables. *CubaEstac. Expt. Agron. Bol.* 12: 28pp.

Davis, D., Chaiet, L., Rotbrock, J.W., Deak, J., Halmos, S. and Garber, J.D. 1960. Chemotherapy of cereal rusts with a new antibiotic. *Phytopathology* 50(11): 841-843.

De la Rocha, G. 1956. The cultivation of artichoke, string bean and garlic. *Circ. Estae. Exp. Agric. La Molina* 69:24.

Dundas, B. 1936. Inheritance of resistance to powdery mildew in beans. *Hilgardia* 10(8): 243-53.

Felber Irina, M. and Hamner, C.L. 1948. Control of mildew on bean plants by means of an antibiotic. *Bot. Gaz*. 110(2): 324-325.

Ferreira, R.V., Ramalho, MAP and Corte, H.R. 1999. Genetic control of common bean (*Phaseolus vulgaris*) resistance to powdery mildew (*Erysiphe polygoni*). *Genet. Mol. Biol.* 22: 233-236.

Galloway, B.T. 1889. Notes: Powdery mildew of bean. *J. Mycol*. 5: 214-216.

Grover, R.K. 1953. Folicolous fungi of Saugar. *Bull. Bot. Soc. Univ. Saugar* ½: 9-13.

Gupta, S.K. and Bharat, N.K. 2010. Occurrence of powdery mildew in French bean under protected cultivation in Himachal Pradesh. *J. Mycol. Pl. Pathol.* 40(3): 462-463.

Gupta, S.K. and Kumar, A. 2014. Management of powdery mildew of pea with botanicals. *J.Res. Punjab agric. Univ.* 5(1): 45-48.

Gupta, S.K. and Shyam, K.R. 1998. Control of Powdery mildew and rust of pea by fungicide. *Indian Phytopath.* 51 (2): 184-186.

Hsieh, L.C.; Chang, Y.C. and Chien, C.C. 1988. Studies on the ecology of the powdery mildew of beans. *J. Agri. Res., China* 37(1): 76-85.

Hoffman, J.C. and Kanapaux, M.S. 1952. Wade, a new all-purpose snap bean. *Seed World* 70 (2): 38-39, 46.

Janson, A. 1922. Control of powdery and downy mildew. *Deutsche Obstbau Zeit* 78(23): 224-225.

Lamarck, J.B. and Candolle, A.P. DE. 1805. *Flore Francaise Ed.* 3, V. 2. Paris.

Litzenberger, S.C. and Stevenson, J.A. 1957. A preliminary list of Nicaraguan plant diseases. *Pl. Dis. Reptr.*, Suppl. 243: 19.

Lukens, R.J. and Horsfall, J.G. 1967. Control of powdery mildew with chemical anti-sporulants. *Phytopathology* 57(4): 342.

Meena, N.K., Zacharia, S. and Patil, N.B. 2016. Management of green gram powdery mildew (*Erysiphe polygoni* DC.)by foliar sprays of fungicides and *Trichodderma* spp. *Eco. Env. & Cons.* 22(June Suppl.): S251-S-255.

Moore, W.D. 1936. Powdery mildew (*Erysiphe polygoni*) of garden snap bean. *Phytopathology* 26(12): 1135-44.

Ogilive, L. 1925. Report of the plant pathologist for the year 1925. *Rept. Dept. Agric., Bermuda* pp. 36-63.

Pauvert, 1991. Studies on bean powdery mildew in Guadeloupe. *Agronomie* 11(2): 109-113.

Raguchander, T. and Rajappan, K. 1995. Relationship between powdery mildew and meteorological factors in urd bean. *Pl. Dis. Res.* 10(1): 59.

Rezende, V.F., PattoRamalho, M.A. and Corte, H.R. 1999. Genetic control of common bean (*Phaseolus vulgaris*) resistance to powdery mildew (*Erysiphe polygoni*). *Genetics and Molecular Biol.* 22(2): 233-236.

Rheenen, H.A.V.; Hasselbach, D.E. and Muigai, S.G.S. 1981. The effect of growing beans together with maize on the incidence of bean diseases and pests. *Netherlands J. Pl.Pathol.* 87(5): 193-199.

Schwartz, H.F.; Katherman, M.J. and Thung, M.D.T. 1981. Yield response and resistance of dry bean to powdery mildew in Colombia. *Plant Dis.* 65(9): 737-738.

Singh, K., Kumar, S. and Kaur, P. 2016. Detection of powdery mildew disease of beans in India. *Oriental J. Computer Sci. Tech.* 9(3): 226-234.

Stevens, F.L. 1913. *The fungi which cause plant disease*. The Macmillan Company, New York, 754pp.

Trabanco, N., Vega, E. Perez, Campa, A, Rubiales, D. and Ferriera, J.J. 2012. Genetic resistance to powdery mildew in common bean. *Euphytica*186: 875-882.

Yadav, R.K., Kakraliya, S.S., Bajiya, M.R. and Abrol, S. 2017. Ecofriendly management of powdery mildew of green gram (*Vigna radiata* L.). *Int. J. Curr. Microbiol. App. Sci.* 6(7): 435-439.

Yarwood, C.E. 1936. The tolerance of *Erysiphe polygoni* and certain other powdery mildews to low humidity. *Phytopathology* 31(9): 845-59.

Yarwood, C.E. 1939. Control of powdery mildew with a water spray. *Phytopathology* 29(3): 288-290.

Zaumeyer, W.J. 1947. Control of bean disease. *Yearb agric. U.S. Dept. Agri*. pp. 333-337.

Zaumeyer, W.J. 1957. Control of powdery mildew of beans with two antibiotics. *Medical Encyclopaedia, New York* pp. 1015-1018.

Zaumeyer, W.J. and Meiners, J.P. 1975. Disease resistance in beans. *Ann. Rev. Phytopath.* 13: 313-334.

Zaumeyer, W.J. and Thomas, H.R. 1957. Monographic study of bean diseases and methods of their control. *US DA. Tech. Bull. No.* 868: 255.

5

RUST

Rust is an important foliar disease of French bean throughout the world. The disease occurs in severe form under favourable weather conditions and results in early leaf defoliation and a considerable reduction in the potential yield of the crop.

GEOGRAPHICAL DISTRIBUTION

French bean rust, caused by *Uromyces appendiculatus* (Pers.) Unger, is one of the prime and devastating diseases of French bean in different regions of the world. The disease was first reported in 1795 from Germany (Persoon, 1795). Since then the occurrence of the disease is common and has been described from various countries of the world like France (Malencon, 1936), North America (Fromme and Wingard, 1921; Groth and Scrum, 1977;), Mauritius (Anonymous, 1924), USA (Gilbert and Popenoe, 1924; Echandi, 1976), Central Audes (Chardon and Toro, 1927), Bulgaria (Kovacevski, 1930; Christoff, 1939), Guatemala (Johnston, 1937), East Africa (Walliace, 1939), Greece (Alexopaulos, 1940), Hawaii (Anonymous, 1943), Chile (Anonymous, 1943), Northern Iran (Esfandiari, 1947), Tangamyika (Wallace and Wallace, 1950), Nyasald (Lawrence, 1951), Caledonia (Bugnicourt *et al.,*1951), New Zealand (Brien and Dingley, 1953), Japan (Hiratsuka and Sato, 1953), Eastern Asia (Hiratsuka and Shimabukuro, 1954), New Zealand (Brien and Jacks, 1954), Europe (Rodrigues, 1955; Stavely and Batra, 1991), Holland (Weibosch and Buishand, 1955), Tucuman (Muntanola, 1955), Britain (Goodban, 1956), Kenya (Anonymous, 1957), Herzegovina (Blagojevic, 1958), Macedonia (Koleva, 1959), Germany (Masurat and Stephan, 1961), Africa (Kaiser, 1976; Allen, 1995), Cuba (Gonzalez, 1976), Australasia (Ballantyne 1978), many Latin American countries (Graham, 1978; Van Schoonhoven), Ethiopia (Assefa, 1987), Netherland (Habtu *et al.,* 1996), China (Wang *et al*., 1998), Sri Lanka (Ariyarathne and Pradeepnuwan, 2001). The disease could be found wherever wild or cultivated *Phaseolus* beans found (Laundon and Waterston, 1965).

Bean rust was first reported in India from Pusa in 1918 (Butler, 1918). Later on, the occurrence of the disease has been documented from Andhra Pradesh, Gujarat and Bombay (Butler and Bisby, 1960), Karnataka (Sharma, 1989), Tamil Nadu, Uttarakhand (Sharma, 1998), Meghalaya (Bhat *et al*., 1999) and Sikkim (Bhat, 2002). In Himachal Pradesh, the disease was recorded on pencil type of bean cv. ‘Falguni’ during 2006 in Solan area first time (Gupta *et al*., 2008).

LOSSES

Rust is a polycyclic disease which can result in significant economic losses under favourable environmental conditions. Premature leaf fall and decrease in number of pods per plant along with decreased weight are the consequences of the disease. Due to the occurrence of bean rust, varying degree of yield losses were reported in different parts of the world. 29-76 per cent losses in 5 cultivars of bean were reported due to *Uromyces appendiculatus* (Issa *et al.*, 1980) while 36.7 per cent yield losses were observed by Singh and Musyimi (1981). Yield losses ranging from 28 to 50, 8 to 33 and 13to 29 per cent in different bean cultivars like ICA- Pijao, Velasco Largo and Bonte 11, respectively due to bean rust were documented (Gonzalez and Garcia, 1996a). Up to 100 per cent yield losses have also reported in different parts of the World (Grafton *et al.*, 1985; Mohan *et al.*, 1993; Wagacha *et al.*, 2007). It was established that rust negatively affect actual yield of beans with average yield loss of 65.5, 34.4 and 14 per cent and an average yield loss of 4.25 per cent for each one per cent increase of mean severity of rust (Mersha and Hau, 2011).

In India, the losses in green pod yield have been reported from 4.7 to 69. 0 per cent depending on the cultivars, age of the crop at the time of infection and per cent disease intensity (Sharma, 1989) under Bangalore (Karnataka) conditions while in Almora, Uttrakhand, up to 60.8 per cent losses due to rust has been reported (Sharma, 1998). In Himachal Pradesh, the disease has been observed particularly on pencil type of beans and often appearing in epidemic proportions in major bean growing areas which results in extensive losses to crop (Singh and Gupta, 2019). Up to 53 per cent losses owing to rust disease on pencil type of beans were recorded (Devi *et al.*, 2019).

SYMPTOMS

The symptoms of the disease appear on all above ground parts of the plant. However, the symptoms occur occasionally on pods. The characteristic symptoms of the disease are described below:

On Leaves

Leaves are the chief plant part which is infected by the disease. Symptoms appear on both leaves surfaces but are more frequently observed on the abaxial leaf surface. Initially, rust pustules usually appear as small, slightly raised and almost white spots. These pustules increase in diameter, up to 2 mm, by forming reddish brown sori of urediniospores. These reddish brown sori are often bordered by a chlorotic halo (Plate 1a). The sori coalesce together and occupy larger areas with the passage of time. Gradually, on susceptible cultivars, ring of secondary sori may appear around the original locus of infection. Eventually leaves curl upwards, dry up, turn brown and fall pre-maturely. Telial stage appears occasionally. Telial sori are dark brown to black in colour and are linear.

On Petioles and Stems

Under high disease severity, symptoms appear on the petioles and stems also. The appearance of symptom initiates as the formation of minute elongated and raised spots (Plate 1b) which eventually enlarge in size and join together. Afterwards, these spots are exposed and brown rusty mass of spores are released.

On Pods

The symptoms appear as reddish brown, elongated raised pustules on the bean pods (Plate 1c) under extremely high disease severity.

Plate 1a. Symptoms of rust on French bean leaves

Plate 1b. Symptoms of rust on stem and petiole of French bean

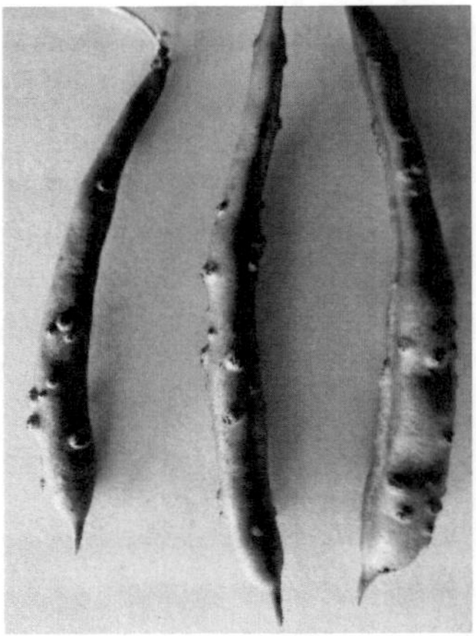

Plate 1c. Symptoms of rust on pods of French bean

THE PATHOGEN

The pathogen causing bean rust is *Uromyces phaseoli* var. *typica* Arth. and *U. appendiculatus* (Pers.) Unger. It is an autoecious rust with all its spore stages being formed on bean plant with rare occurrence of pycnial and aecial stages in nature. The urediniospores are brownish in colour, with globoid or ellipsoid shape, one celled echinulate with dimension 23-33 x 16- 23 µm (Plate 2). Urediniospore wall is golden brown in colour, 1-1.5 µm thick and have two equatorial or super equatorial pores. Teliospores are single-celled, globoid or broadly ellipsoid and with pedicel. The dimension of teliospores are 24-32 X 20-26 µm and may be smooth or possessing a few verrucose marks on the wall. The teliospores have a chestnut brown wall that is 3-3.5 µm thick and has a hyaline papilla covering the pore. Telia are derived from uredinia and are dark brown or black in colour. Pycnia emerge as yellowish dots on the leaf's upper surface. The aecia are cupulate, orange in colur and 0.25- 0.30 mm in diameter. Aeciospores are ellipsoid in shape, one-celled, and 20-26 X 10- 20 µm in size.

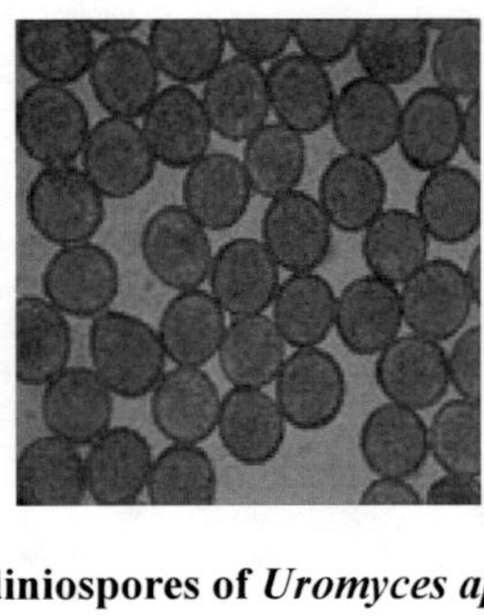

Plate 2. Urediniospores of *Uromyces appendiculatus*

VARIABILITY

In as early as in 1939, a degree of variability was observed among different isolates of *U. appendiculatus* (Dundas and Scott, 1939). On a set of differential hosts, twenty physiologic races of the fungus have been distinguished under greenhouse conditions (Harter and Zaumeyer, 1941). Four races *viz.,* race 1, 4, 7 and 12 of the fungus have also been accounted by De Menzes (1952). Chirsten and Echandi (1967) reported that race 3 and 15 in Alajuela, 3 and 25 in Turrialba, 8 and 22 in San Isdro del General, 10 and 32 in Cartago and 8 in Los Angeles de Tilaren (regions of Costa Rica) were more common from greenhouse and field studies. In Brazil, B1 and B2 were the most common races identified on the set of six differential hosts (Dalas and De Costa, 1968). Races 3, 10, 18, 19, 32, 33, 34, 36 and 37 have been reported from Cauca Valley where race 33 was encountered most frequently (Zuniga de Rodriguez and Victoria, 1975). Stavely *et al.* (1989) reported 58- 70 as the new races of the fungus in north America of which race 67 was the first to produce moderate size urediniospores on a previously resistant Cv. Compuesto Negro Chimaltenango. Occurrence of 13 different races among 15 monopustular isolates of *U. appendiculatus* obtained from 4 bean growing areas of the state of Minas Gerais, Brazil had been reported (Faleiro *et al.*, 1999). From 80 monopustular isolates, 53 races could be distinguished with a set of only eight differential hosts (Kentucky Wonder 814, Early Gallatin, 51051, NEP-2, Ecuador 299, Olathe, Mexico 309 and Redlands Pioneer) instead of 20 differential varieties (Mora *et al.*, 1992). 13 different races were reported among 15 monopustular isolates of *U. appendiculatus* obtained from four bean growing regions of Minas Gerain, Brazil (Faleiro *et al.*, 1999).

The variation among the isolates of bean rust pathogen was assessed on the basis of virulence and random amplified polymorphic DNA (RAPD) marker (Mohammed, 1994). *U. appendiculatus* diversity has traditionally been assessed using host differential cultivars with distinct single or multiple rust resistance genes (Araya *et al.,* 2004; Souza *et al.,* 2007). The 47 single pustule isolates of *U. appendiculatus* were divided into nine races using the new international classification system and binary nomenclature. Based on the virulence of the races on the host differential cultivars, hierarchical cluster analysis divided the races into two primary clusters. Ur-5, Ur-11, and Ur-CNC were shown to be the most resistant to pyramiding in Kenya (Arunga *et al.,* 2012).

HOST RANGE

The host of rust pathogen includes *Phaseolus vulgaris, P. lunatus, Dolichos lablab, D. biflorus, Vigna chinensis* etc.

DISEASE CYCLE

It is an autoceious rust, which means that its life cycle is restricted to a single host. The aecia are uncommon in nature, but have been seen in Oregon bean fields (Eastman, 1943). The sori containing urediniospores are formed in large numbers on both upper and lower surfaces of the leaf. They disseminated the disease far and wide, causing severe outbreaks. As soon as the urediniospores are developed, they will germinate. From single germinated urediniospores, one or two germ tubes emerged (Xin *et al.*, 2000). The penetration of the germ tube occurs mainly through the stomata but occasionally penetrated directly through the epidermal cell. Following pathogen penetration, a well-defined appressorium develops. When the fungus penetrates the stomata, it forms a sub-stomatal vesicle which usually results in a round inflated body. The inflated body then produces one or two primary hyphe. The fungus develops intercellularly once inside the bean leaf, and leaf cell penetration occurs once a haustorium mother cell forms at the hyphal tip in contact with a host cell. A haustorium mother cell forms behind the primary hyphe, which is where the branches arise. The secondary hyphae spread intercellularly. Urediniospores generate another generation of the same kind of spores in around 10-15 days under favourable environmental conditions, and numerous generations in a single season (Zaumeyer and Thomas, 1957).

Teliospores are formed in the same sorus when the climate becomes unfavourable. Teliospores, which require a rest period and can live for months, play a critical role in the organism's transition from one season to the next. Pycnia and aecia are formed the next season, indicating infection by basidiospores from teleutospores (Zaumeyer and Thomas, 1957, Staveley *et al.*, 1989), however no such stage has been seen in some regions. Aecia has been reported from various parts of the world, including Europe (Milbrath 1944; Wilson and Henderson 1966; Heinze 1974) and New Zealand (Brien and Jacks, 1954), whereas they were absent in warmer climates such as California, Florida, and northeastern South Africa (Harter *et al.*, 1935; Zaumeyer and Thomas 1957; Liebenberg, 2003), indicating that the chances of sexual stage occurrence are higher in temperate climates. So in warmer climate, the pathogen may overwinter as urediniospores in trash, trellis poles, or sheltered buildings, thereby contributing to disease beginning early in these places (Davison and Vaughan, 1963, Gross and Venette, 2001).

EPIDEMIOLOGY

The pathogen of bean rust is an obligatory parasite and autoecious in nature and hence the pathogen can complete its life cycle on bean plants. It isn't a seed-borne pathogen. In cooler areas, the pathogen survives in the form of teliospores on plant debris, while in other areas, the pathogen survives on continued crop growth of bean and other hosts. Cloudy, humid days that allow dew to remain on the leaves for a long time in the morning helps in better spore germination and infection. The ideal temperatures for urediniospore germination and infection are 14.5 and 17° C, respectively (Harter *et al.*, 1935). Davison and Vaughan (1963) proposed that rust overwinters in or as urediniospores in trash and on trellis poles because 16 % urediniospores of *U. phaseoli* race 33 germinated even after storage at –18° C for >600 days and produced profuse infection on bean leaves. A significant level of rust infection was observed when plants were exposed to 96 % or more RH (Harter *et al.*, 1935). Highest rate of bean rust infection was recorded under the exposure of plants to 100 % RH for 11 hours or more (Rey and Lozono, 1961). Maximum infection i.e. number of sporulating pustules per leaf were observed after 22h of high RH (Mendes and Bergamin-Filho, 1989). According to Singh and Gupta (2019), disease severity was highest at 20°C while lowest at 30°C, and didn't appear at 35°C (Fig. 1). There was also a consequent rise in disease severity and number of pustules as humidity increased from 75 to 100% (Fig. 2), with the highest disease severity (81.92 %) and number of pustules/ cm2 (70.50) at 100% RH, followed by 78.90 and 75.58 % disease severity at 95 and 90 percent RH, respectively, and the lowest (55.63 %) disease severity at 75 per cent RH. Sharma (1977) also studied the influence of number irrigations on the disease development and found that irrigation given at 3-days intervals favoured disease development compared to 18-day intervals.

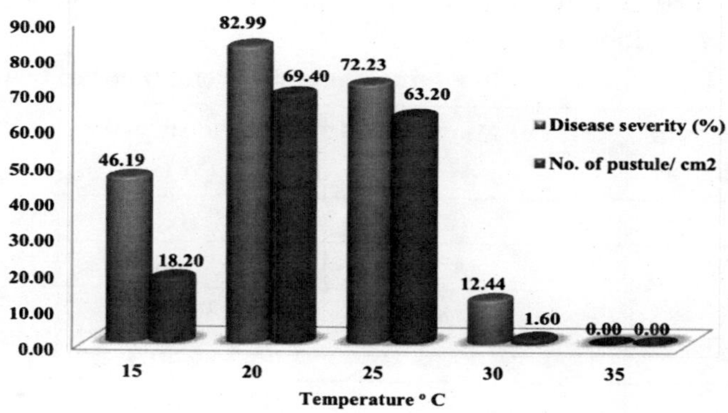

Fig 1. Effect of different temperature regimes on bean rust disease development under glasshouse conditions

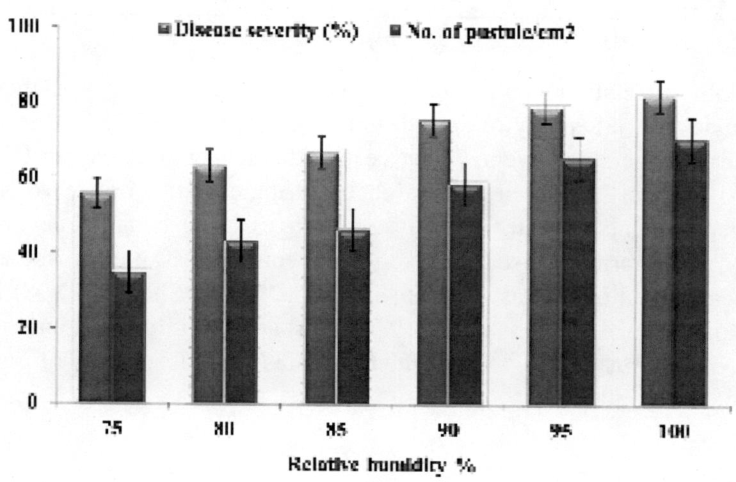

Fig. 2. Effect of relative humidity levels on bean rust disease development under glasshouse conditions

The duration of leaf wetness, mostly provided by rain and dew deposition, also has important effect on development of rust disease. Under glasshouse conditions, Singh and Gupta (2019) conducted experiments to see the effect of leaf wetness (Table 1) and durations of intermittent leaf wetness (Table 2) on disease development and reported that disease did not develope when plants were exposed to 3 hr leaf wetness. However, continuous leaf wetness of 6 hr was sufficient to commence the infection of rust. They further showed the reduction in incubation period from 243 to 168 hr and increase in disease severity from 30.0 to 83.6 percent and pustule number from 6.0 to 72.3 per cm^2 of leaf with increase in leaf wetness durations from 6 to 24 hr. However, increasing the moisture of the leaves had little influence on incubation period. Number of pustules and disease severity increased with the interruption of leaf wetness with dry periods upto two cycles (72.50 pustules/ cm^2 and 82.49% disease severity) and further extension of wet and dry periods resulted in a non-significant decrease in number of pustules and disease severity (Table-1).

Table 1. Effect of leaf wetness duration on development of bean rust disease

Duration of wetness (h)	Incubation period (h)	Disease severity (%)	No. of pustule / cm^2
3	0	0.00 (0.00)	0.00
6	243	30.00 (33.19)	6.00
9	218	53.93 (47.24)	24.67
12	178	75.15 (60.08)	63.67
24	168	83.63 (66.13)	72.33
36	168	82.42 (65.23)	69.67
48	168	81.81 (64.75)	68.33
$CD_{0.05}$	1.64	(3.41)	9.11

Figures in parentheses are arc sine transformed values

Multiple correlation coefficients between disease severity and meteorological factors under natural epiphytotic conditions suggested that 92.36 % rust severity was due to mean temperature, average relative humidity and cumulative rainfall (Singh and Gupta, 2019). Rest 7.64 % variation might be attributed to other meteorological factors which were not included in the analysis. Changes in N, P, and K levels have little impact on the severity of disease. Farm implements, insects, humus, and animals help in the local spread of the rust fungus (Sumatrini, 1998). Sharma (1998) investigated meteorological data for eight years (1986-1993) and found that extreme precipitation is detrimental to rust development, which explains the high severity of rust during the periods of low rainfall.

Table 2. Effect of intermittent leaf wetness on rust development

Total wetness (h)	Wet/dry sequence (h)								Disease severity (%)	No. of pustules /cm²
	Wet	Dry	Wet	Dry	Wet	Dry	Wet	Dry		
12	112	112	–	–	–	–	–	–	68.28 (55.71)	57.33
24	112	112	112	112	–	–	–	–	82.49 (65.31)	72.50
36	112	112	112	112	112	112	–	–	78.90 (62.67)	72.17
48	112	112	112	112	112	112	112	112	76.75 (61.19)	69.83
CD $_{0.05}$									(2.44)	8.28

Figures in parentheses are arc sine transformed value

MANAGEMENT

Cultural practices, chemicals, biological methods and host resistance are used for the management of bean rust.

Cultural practices

To prevent disease spread, limited tillage or no tillage, as well as intercropping bean with maize were recommended as the severity and rate of infection were higher in monoculture (Moreno and Mora, 1984). Some cultural practices like long crop rotations, collection and destruction of plant debris, wider plant spacing and weed-control to allow aeration and drying which would check the rapid build-up of inoculum and spread of the disease were recommended for management of bean rust under field conditions (Sharma and Sohi, 1989). Mixed cropping reduced the incidence of rust by 51 and 25 % with sole cropping and row intercropping, respectively (Fininsa, 1996). Plant density and disease occurrence are linked directly (Flores Revilla *et al.,* 1994). Mishra *et al.* (1998) tested seven sowing dates, between 12 October and 23 November, of Arka Komal variety in Orissa. 9th November was considered to be the best sowing date due to the lowest rust incidence.

Host resistance

Extensive cultivation of the new cultivars may make them susceptible due to the prevalence of different races of rust pathogen. More than 35 physiologic races of the rust fungus *viz., U. appendiculatus* are present in Australia, Brazil, Hawaii and the USA. Golden Gate Wax and Brown Kentucky Wonder 298 have been used in breeding for resistance against four races (Dundas, 1942). During rust epidemic of 1946 in Mississippi, only two lines, Virginia 119 and No.780 remained free from infection (Foster, 1947). Frazier and Hendrix (1949) developed a new rust resistant pole French bean named Hawaiian Wonder from the cross Bountiful X Lualualei. Seminole is a bush bean variety having multiple disease resistance including rust (Anonymous, 1953). Cultivars/lines like Redlands Greenleaf and Redlands Pioneer (Anonymous, 1966), Jackson Wonder and Strikton (Makraw *et al.*, 1973), Brazil K-12973 (Sokol *et al.*, 1977), Catu and Carioca (Issa, 1985a) Serva Negra, Ag497 and Agrorrico (Azevedo and Kushalppa, 1986), Gresham (Anonymous, 1989), PV 136, PV 88, PV 89 and PV 3 (Prakasam and Thumburaj, 1991), white seeded Enganador, Chevere and CC 25-9B (Faure *et al.*, 1992), Negro Inifap (Villar-Sanchez and Lopez-Salinas, 1993), Negro Cotaxtle 91 (Becerra Leor *et al.*,1994), IBRN-6, IBRN-11, Hawaiian Wonder, Bush Bean S-9, Bush Blue Lake and Contender (Ghimire *et al.*,1995), lines 83-10 and 83-11 with stringless pods (Guu,1995), Ouro Negro (Faleiro *et al.*,1996), IIHR 220, stringless bean (Mohan *et al.*, 1997), Mimoso Rasteiro (Paula *et al.*, 1998), SVM-1, Hans (Sharma, 1998) and IC 47643, PI 201489 and PI 247761 (Bhat *et al.*, 1999) have been found to be resistant in different parts of the world. Cv. Ouro Negro showed immunity to 3 out of 4 races tested and PI 181395 was the best source of resistance to rust, as it was immune to all the four races of *U. appendiculatus* (Faleiro *et al.*, 1998). A single, dominant gene controls the rust resistance inheritance (Aghora *et al.,* 2007). Line 'Arka Anoop' developed by breeding programme is resistance to rust and has high pod yield and good pod quality. Cultivars *viz.,* Alpa Trey, Sing Tamey and LC (Local Collection) were found resistant to ban rust in Himachal Pradesh (Sharma *et al.*, 2014).

Chemical control

For effective management of this disease, fungicides with different modes of action should be incorporated in a spray program. Different workers recommended a variety of chemicals for the management of the disease. Straib (1943) suggested the disinfection of stakes with formalin (0.1%) before reuse. Preventive application of zineb (5%) with a mist blower or as a spray (0.35%) proved efficient (Cheo, 1955). Due to the difficulty of attaining under leaf coverage with non-systemic fungicides, systemic fungicides are more effective. Various systemic fungicides like oxycarboxin, carboxin, triforine, bitertanol, triadimefon, tridemorph, hexaconazole and HWG 1608 were reported effective against the rust and possibly eradicated the already established infection of *U. phaseoli* in bean plants after three sprays (Jaffer, 1971; Rolim*et al.,* 1981; Singh and Musymi, 1979a; 1984; Prakasam and Thamburaj, 1991; Gonzalez and Garcia, 1996b; Sumartini, 1998).

Twice application of hexaconazole, maneb and tebuconazole before flowering reduced rust infection drastically and increased seed yield by > 200 kg/ha (Becerra Leor *et al.*, 1994). Three to four sprays of hexaconazole (0.1%) effectively managed the disease in Sikkim and Himachal Pradesh (Bhat, 2002, Devi *et al.*, 2019). Thiophanate methyl or chlorothalonil alone or in combination with other fungicides decreased severity on leaves and increased pod yields (Balardin, 1994). Mancozeb was the most effective fungicide in managing the disease (Sharma, 1998).

Sharma *et al.* (2018) assessed various systemic and non-systemic fungicides against the bean rust under *in vitro* and in field conditions and obtained lowest urediniospore germination of 0.69% in zineb + hexaconazole compared to 45.26% germination in control. Similar trend was obtained in germ tube length. Further, all fungicides could significantly reduced rust severity in comparison to untreated check (49.33%) under field conditions. Azoxystrobin (0.1%) was significantly most efficient in reducing the disease severity to 4.64% and also increased green pod yield by > 97%. Additionally, azoxystrobin also exhibited some phytotonic effect which delayed the crop senescence and attributed to the superior crop yield.

Different fungicides were evaluated against bean rust and azoxystrobin, difenoconazole + azoxystrobin, pyraclostrobin + metiram, kresoxim - methyl, fluopyram + tebuconazole, trifloxystrobin + tebuconazole, propiconazole + difenoconazole resulted in complete inhibition in germination of urediniospore (Devi *et al.,* 2020). Difenoconazole + azoxystrobin gave up to 96 h post-infection activity while non-systemic fungicides i.e. mancozeb and chalorothalonil were found least effective. Further, four foliar sprays of azoxystrobin (0.1%) at 10 days interval was effective and reduced the disease index to 3.81% and increased green pod yield by > 100%.

Certain antibiotics like distamycin (100 ppm), griseofulvin (30 ppm) and chloramphenicol have also been reported effective in managing rust of beans. Addition of glycerine (1 %) and IAA (1 ppm) increased the efficacy of griseofulvin, which was the only material displaying systemic activity (Baldacci and Betto, 1958).

Biological control

Bio control fungus *Verticillium lecanii* has been recommended for the management of rust due to its ability to penetrate the hyphae and invasion of urediniospores of *U. appendiculatus* (Allen, 1982). *V. lecanii* reduced the disease incidence and severity to 56 and 17% as compared to 98 and 85%, respectively, in agrochemical package while seed yield was slightly lower in biocontrol than agrochemical package possible due to the use of fertilizer in latter package (Carrion *et al*., 1999). APPL-1, FF-1, FF-5 and FF-6 isolates of *Bacillus subtilis* have the potential to be used for bean rust management (Baker *et al*., 1983, 1985; Mizubuti*et al*., 1995). *B. subtilis* isolates released a thermo stable substance(s) which reduced rust pustule by >95 per cent (Centurion *et al*., 1994). Isolate B 206 was best against the bean rust on whole plant assay.

Plant origin pesticides

The hazardous chemicals used to combat the ravages of plant diseases needs to be replaced by less detrimental compounds preferably from natural resources. Some plant extracts could be a possible substitute to manage plant pathogens due to the presence of antifungal compounds in them (Fabry *et al*., 1996, Okemo *et al*., 2003). Plants with pest killing properties have been recorded as early as the Indian Regveda during the second millennium BC (Hoddy, 1991). Preventive sprays of neem extract provided excellent control of bean rust, however, when it is used as erradicant it performed poorly (Locke, 1990). Natural oil based fungicides such as neem offered a good option as they control some pathogens at very low doses and induce little or no resistance in target fungi besides having excellent spreading and leaf surface adhesion characteristics with little impact on the environment (Wicks *et al.*, 1999; Martin *et al.*, 2005). Inclusion of stinging nettle (*Urtica massaica*), papaya (*Carica papaya*) and neem oil in the bean rust programme has been suggested (Monda *et al.*, 2009).

Efficacy of neem based products against bean rust has been reported (Singh and Narayana, 2002, Bhat, 2002). Out of thirteen aqueous plant extracts evaluated against urediniospore germination of bean rust pathogen, Panchparni extract, bulb extract of *Allium sativum*, leaf extract of *Eupatorium odoratum* and *Agratum conyzoids* were found effective as they significantly reduced the urediniospore germination compared to control (Kamble and Patil, 2017). Neem formulations like neemajal, achook and nimbecidine caused complete inhibition in germination of urediniospore at 5% whereas aqueous extracts of four botanicals like *Allium cepa, Zingiber officinale, Allium sativum* and *Melia azedarach* caused complete inhibition at 20% (Devi *et al.*, 2019). Further, neemajal (2%) was found more effective in reducing rust severity to 26. 25% compared to 70.35% in control followed by nimbecidine, achhok and *A. sativum* under field conditions.

Systemic resistance inducer

Plants that are infected by diverse pathogenic invaders develop a physiological immunity known as systemic acquired resistance (SAR). SAR can be caused by a wide range of organisms, pests, or chemicals delivered to the plant from the outside. Various chemicals and environmental stressors can cause SAR in plants, which can then be used as a trigger to subsequent pathogen attack. This method is less expensive and less harmful to the environment. Rather than building immunity against the trespasser, there is no direct effect on the pathogen. Treatment of bean leaves with salicylic acid (SA) or 2, 6-dichloro-isonicotinic acid (DCINA) induced resistance against the rust fungus and resulted in decreased density of rust-pustules (Rauscher *et al.* 1999). Application of saccharin in broad bean induced resistance to the rust fungus *Uromyces viciae- fabae* and provided systemic protection to rust infection (Boyle and Walters, 2005). Bion was also used to induce resistance in different bean cultivars against *U. appendiculatus*. A single spray of benzothiadiazole (BTH) at 0.3 mM concentration 7 days before inoculation was sufficient to complete control of the bean rust (Iriti and Faoro, 2003).

To explore the mechanism of BTH induced resistance against bean rust, light and transmission electron microscopy showed the absence of particular ultrastructural alterations in most of the penetrated cells, even at late stages of infection, indicating that BTH treatment did not induce host cells to respond with a hypersensitive reaction (HR) but phenylalanine ammonia lyase (PAL) mRNA is persistently transcripted in BTH treated plants since the 6th hr after treatment (Maffi *et al.*, 2011). BTH directly activates the plants own defenses against rust. Sprays of bean plants with Saccharin and Ulvan in glasshouse conditions and challenge inoculation with *U. appendiculatus* six days after spray provided an intermediary level of protection as they reduced the number, diameter and sporulation of rust pustules in susceptible bean plants of cv. Guara (Delgado *et al.*, 2013). Devi *et al.* (2020) evaluated the effectiveness of eight SAR chemicals against bean rust under pot conditions. Change in the biochemical constituent's viz., total phenol, peroxidase activity, polyphenol oxidase and phenyl ammonia lyase was reflected. Maximum increase in the activity biochemicals was recorded in in potassium dihydrogen phosphate followed by BTH after challenge inoculation. Least disease severity (8.50 %) was recorded in plants sprayed with potassium dihydrogen phosphate compared to control. Analogically, potassium dihydrogen phosphate sprays were also found effective in reducing the disease severity to 30.71 % followed by BTH with disease severity of 41.23 % under field conditions.

REFERENCES

Aghora, T.S., Mohan, N., Somkumar, R.G. and Girija, G. 2007. Breeding French bean (*Phaseolus vulgaris*) for resistance to rust (*Uromyces phaseoli*RebenWint.). *J.Hortic. Sci.*2: 104-107.

Alexopoulos, C.J. 1940. Some fungi from Greece. *Mycologia*32: 336-358.

Allen, D.J. 1982. *Verticillium lecanii* on the bean rust fungus, *Uromyces appendiculatus. Trans. British Mycol.* Soc. 79(2): 362-364.

Allen, D.J. 1995. An annotated list of diseases, pathogens and associated fungi of the common bean (*Phaseolus vulgaris*) in eastern and southern Africa. *Phytopathological Papers* 34, CAB Intl, Wallingford, UK.

Anonymous. 1924. Botanical division: Investigation of plant diseases. *Ann. Rep. Dept. Agric. Mauritius for the year 1923*:10-11.

Anonymous. 1943a. Soil and plant investigations: Fruits and nuts, vegetable crops and diseases. *Rep. Hawaii Agric. Exp.Stn.*42: 85- 111.

Anonymous. 1943b. The principal diseases of origin which were the object of consultation during half year January to June 1943. *Boletin de Sanidad VegetalChile*3: 38-39.

Anonymous. 1953. Research work and workers. *Seed World* 72(11): 80-81.

Anonymous. 1957. *Annual Report, Dept. Agri., Kenya*2: 237.

Anonymous. 1966. Division of Science Services, *Biol. Branch, Plant Pathol. Sec. Rep. Dep. Agric. NSW,* 1964-65: 75-80.

Anonymous. 1989. French bean (*Phaseolus vulgaris). Pl. Varities J.* 2(2): 15-16.

Araya, C.M., Alleyene, A.T., Steadman, J.R., Eskridge, K.M. and Coyne, D.P. 2004. Phenotypic and genotypic characterization of *Uromyces appendiculatus* from *Phaseolus vulgaris* in the Americas. *Plant Dis.* 88: 830-836.

Ariyarathne, H.M. and Pradeepnuwan, E.A. 2001. A simple method to screen bean seedlings against bean rust. *Annals of the Sri Lanka Dept. Agric.*3: 1-7.

Arunga, E.E., Ochuodho, J.O., Kinyua, M.G. and Owuoche, J.O. 2012. Characterization of *Uromyces appendiculatus* isolates collected from snap bean growing areas in Kenya. *African J. Agric. Res*. 7(42): 5685-5691.

Assefa, H. 1987. Haricot bean diseases and their importance in Ethiopia. *Ethiopian J. Agric. Sci.*9: 55-66.

Azevedo, L.A.S. de and Kushalappa, A.C. 1986. Monocyclic parameters of bean rust caused by *Uromyces phaseoli* var. *typica* on *Phaseolus vulgaris* varieties. *Fitopatol.Brasileira* 11: 609-616.

Baker, C.J., Stavley, J.R., and Mock, N. 1985. Biocontrol of bean rust by *Bacillus subtilis* under field conditions. *Plant Dis.* 69(9): 770-772.

Baker, C.J., Stavley, J.R., Thomas, C.A. Sasser,M. and Macfall, J.S. 1983. Inhibitory effect of *Bacillus subtilis* on *Uromyces phaseoli var. typica* and on development of rust pustules on bean leaves. *Phytopathology* 73(8) 1148-1152.

Balardin, R.S. 1994. Effect of different fungicides on rust and powdery mildew control in common bean, "Carioca" and "GoianoPrecoce" cultivars. *CienciaRural* 24: 261-264.

Baldacci, E. and Betto, E. 1958. Tests of the fungicidal effect of some antibiotics applied by coverage and systemically. *Agricotura ital.* 58(NS 13): 289-301.

Ballantyne,B.J. 1978. The genetic bases of resistance to rust, caused by *Uromyces appendiculatus* in bean (*Phaseolus vulgaris*). Ph.D. Dissertation. University of Sydney, Sydney, New South Wales, Australia, 262p.

Becerra Leor, E.N., Lopezsalinas, E. and Acosta, G. J. A. 1994. Genetic resistance and chemical control of bean rust in the humid tropics of Mexico. *RevistaMaxicana de Fitopatologia*12(1): 35-42.

Bhat, M.N. 2002. Chemical control of rust and angular leaf spot of French bean in Sikkim. *Pl. Dis. Res.*17: 90-92.

Bhat, M.N., Singh, A.K. and Medhi, R.P. 1999. Reaction of French bean (*Phaseolus vulgaris*) cultivars/accessions to rust pathogen *Uromyces appendiculatus*. *Indian J. Agric. Sci.*69: 849-850.

Blagojevic, M. 1958. Plant parasites and pests in the region of Herzegovina during 1954-1956. *Zasht. Bilja*. 45: 57-65.

Boyle, C. and Walters, D. 2005. Induction of systemic protection against rust infection in Broad bean by saccharin: effects on plant growth and development. *New Phytologist* **167**: 607-612.

Brien, R.M. and Dingley, M. 1953. First supplement to 'A revised list of plant diseases recorded in New Zealand'. *New Zealand J. Sci. Technol.*34: 50-55.

Brien, R.M. and Jacks, H. 1954. French-bean rust (*Uromyces appendiculatus*) in New Zealand. *New Zealand J. Sci. Technol.* 36: 280-284.

Bugnicourt, F., Cochic, F. and Dandant, R. 1951. *A catalogue of the animal and plant parasites of the cultivated plants of New Caledonia*. 120p.

Butler EJ. 1918. *Fungi and Diseases in Plants*. Thacker Sprink and Co., Calcutta VI, 547p.

Butler, E.J. and Bisby, G. R. 1960. *The Fungi of India*. (Revised by R. S. Vasudeva 1960). ICAR Publication, New Delhi India, 552pp.

Carrion, G., Romero, A. and Rico-Gray, V. 1999. Use of *Verticillium lecanii* as a biocontrol agent against bean rust. *Fitopatologia* 34(4): 214-219.

Centurion, M.A.P., Kimati, H. and Pereira, G.T.1994. Machanism of action of antagonisits selected for the biological control of bean rust (*Uromyces phaseoli* (Reben.) Wint.). *Cientifica (Jaboticobal)* 22(2): 163-175.

Chardon, C.E. and Toro, R.A. 1927. Plant diseases notes from the Central Audes. *Phytopathology* 17: 147-153.

Cheo, P.C. 1955. The control of bean rust. *Tijdschr. Plziekt* 6I, 1:26.

Christen, R.and Echandi, E. 1967. The more common races of rust *Uromycesappendiculatus* in Costa Rica and evaluation of the rust resistance of some French beancultivars. *Turrialba* 17(1): 7-10.

Christoff, A. 1939. Revision and notes on the parasitic flora of Bulgaria. *J. Agric. Exp. Stn. Bulgaria* 9: 77-85.

Davison, A.D. Vaughan, E.K. 1963. Longevity of uredospores of race 33 of *Uromyces phaseoli* var. *typica* in storage. *Phytopathology* 53(6): 736-737.

De Menezes, O.B.1952. Identification of four physiologic races of bean rust. *Dusenia* 3(4): 309-312.

Delgado, D.Z, Mateus, B.F. and Marciel, J.S. 2013. Effectiveness of saccharin and ulvan as resistance inducers against rust and angular leaf spot in bean plant (*Phaseolus vulgaris*). *Plant Prot*. 47: 67-73.

Devi, Banita, Gupta, S.K. and Singh, G. 2019. Yield loss assessment in French bean due to rust (*Uromyces appendiculatus*) in Himachal Pradesh. *Pl. Dis. Res*. 34 (1): 41-44.

Devi, Banita, Gupta, S.K., Singh, G. and Prasad, P. 2020. Efficacy of new generation fungicides against French bean rust caused by *Uromyces appendiculatus*. *Phytoparasitica*. doi.org/10.1007/s 12600-020-00820-9.

Devi, Banita, Singh, G. and Gupta, S.K. 2019. Evaluation of biopesticides for effective and ecofriendly management of French bean rust. *Pl. Dis. Res*. 34 (2): 113-118.

Devi, Banita, Singh, G., Dash, A.K. and Gupta, S.K. 2020.Chemically induced systemic resistance in the inhibition of French bean rust. *Curr. Plant Biol.* Doi. org/10.1016/j.cpb.2020.100151.

Devi, Banita. 2019. Variability studies in bean rust (*Uromyces appendiculatus* (Pers.) Unger) and its ecofriendly management. Ph.D Thesis submitted to Department of Plant Pathology, Dr. Y.S.Parmar University of Horticulture & Forestry, Nauni, Solan, 130pp.

Dundas, B. 1942. Breeding beans for resistance to powdery mildew and rust. *Phytopathology* 39(9): 828.

Dundas, B. and Scott, G.W. 1939. Physiologic strains of bean rust. *Phytopathology* 29(9): 820-821.

Eastman, J.W. 1943. Report of Provincial Plant Pathologist. *British Columbia Dept. Agric. Ann. Rept.* 1943: 59-63.

Echandi, E. 1976. Principal fungus disease of bean (*Phaseolus vulgaris*) in the American tropics in different ecological zones. *Fitopatol. Brasileira*1: 171-177.

Esfandiari, E. 1947. The diseases of cultivated plants and fruit trees in the subtropical regions of Northern Iran. *App.Entomol.Phytopathol.*5: 21.

Fabry, W, Okemo, P. and Ansorg, R. 1996. Activity of East African medicinal plants against *Helicobacter Pylori. Chemotherapy* 42:315-7.

Faleiro, F.G., Paula Jr. T. J. De, Barros, E. G. De, Frietos, M. A. S. and Moreira, M. A. 1996. Common bean resistance to *Uromyces appendiculatus* from zona da Mata region of the state of Minas Gerais, Brazil. *Fitopatol. Brasileira* 21(1) : 123-125.

Faleiro, F.G., Vinhadelli, W.S., Ragagnin, V.A., Paula, T.J.Jr., Moreira, M.A. and Barros, E.G. 1998. Resistance of the common bean to four races of *Uromyces appendiculatus*. *Revista Ceres*. 263: 11-18.

Faleiro, F.G., Vinhadelli, W.S., Ragagnin, V.A., Zambolin, L., Paula, T.J.Jr., Moreira, M.A. and Barros, E.G. 1999. Identification of physiological races of *Uromyces appendiculatus* in the state of Minas Gerais, Brazil. *Fitopatol. Brasileira* 24(2): 166-169.

Faure,B., Ponce, M. and Garcia, Y.E. 1989. Evaluation of common bean (*Phaseolus vulgaris*) varieties with black and white seeds. *Ciencia y Tecnica en la Agricultura* 8(1): 31-41.

Fininsa, C. 1996. Effect of intercropping bean with maize on bean common bacterial blight and rust diseases. *Int. J. Pest Mgt*. 42(1): 51-54.

Flores Revilla C., Nieto Angel, D. and Torre Almaraz, R. de la.1994. Effect of varieties, foliar nutrients and seeding rates on disease severity and yield in French bean (*Phaseolus vulgaris* L.) in Huejotzingo. *Puebla Revista Mexicana de Fitopatol.*12: 69-74.

Foster, H.H. 1947. Reaction to rust under Misssippi conditions, of certain lines and vars. of pole snap beans. *Pl. Dis. Reptr*. 31(10): 378-383.

Frazier, W.A. and Hendrix, J.W. 1949. Hawaiian Wonder, new rust resistant pole green bean. *Circ. Hawaii Agric. Exp. Stn*. 28: 7.

Fromme FD and Wingard SA. 1921. Varietal susceptibility of beans to rust. *J. Agric. Res.* 21: 385-404.

Ghimire, S.R., Pradhanang, P.M. and Lohar, D.P. 1995. Use of fungicides and identification of resistant varieties for the management of bean rust and anthracnose diseases in common beans (1992/93). *Working Paper LumleReg. Agric. Res. Centre* 95/76: 42.

Gilbert, W.W. and Popenoe, C. H. 1924. *Diseases and insects of garden vegetables. United States Dept. Agric. Farmers Bull.* 37: 46.

Gonzalez M and Garcia E. 1996b. Evaluation of losses due to the rust on bean (*Phaseolus vulgaris* L.) in four sowing times in Cuba. *AgronomiaMesoamericana* 7: 95-98.

Gonzalez M. 1976. *Researchon the behavior of bean varieties to the causal agent of rust* (*Uromyces phaseoli* var. *typica* Arth.). Ten years of scientific collaboration between Cuba and the German Democratic Republic Diez AnosDe colaboracioncientifica Cuba RDA: 26-32.

Gonzalez, M. and Garcia, E. 1996a. Evaluation of fungicides for the control of bean rust (*Uromyces appendiculatus*). *AgronomiaMesoamericana* 7: 86-89.

Goodban, J.W.D. 1956. *Ann. Rep. Agric. Dept., British Honduras, for the year 1955*: 55.

Grafton, K.F, Weiser, G.C, Little, F.L.J. and Stavely, J.R. 1985. Influence of resistance of two races of leaf rust in dry edible bean. *Crop Science* 25: 537-539.

Gross, P.L. and Venette, J.R. 2001. Overwinter survival of bean rust urediniospores in North Dakota. *Plant Dis.* 85: 226-227.

Growth, J.V. and Mogen, B.D. 1978. Completing the life cycle of *Uromyces phaseoli* var. *typica* on bean plants. *Phytopathology* 68: 1674-1677.

Gupta, S.K., Jarial, K. and Kumar, M. 2008. Occurrence of French bean rust in Himachal Pradesh. *J. Mycol. Pl. Pathol.*38: 352-353.

Guu, J. W. 1995. Development of rust resistant and stringless bean lines. *Bull. Taichung Distt. Agric Dept. Stn. No.*46: 11-18.

Habtu, A., Sache, I. and Zadoks, J.C. 1996. A survey of cropping practices and foliar diseases of common beans in Ethiopia. *Crop Prot.*15: 179-186.

Harter, I.L., Andrus, C.F. and Zaumeyer,W.J. 1935. Studies on bean rust caused by *Uromyces phaseoli* var. *typica. J.Agric. Res.* 1(9): 737-759.

Harter, L.L. and Zaumeyer, W.J. 1941. Differentiation of physiologic races of *Uromyces phaseoli* var. *typical* on bean. *J. Agric. Res*. 62(12) :717-731.

Heinze, K.1974. Leitfaden der Schadlingbekampfung; band I: Verlag, Stuttgart, FederalRepublic of Germany. Cited in Stavely and Pastor-Corrales, 1989.

Hiratsuka, N. and Sato, S. 1953. Contributions to the rust flora of Eastern Asia. VII. Nagaoa. *Mycol. J. Nagao Inst.*3: 57-100.

Hiratsuka, N. and Shimabukuro, S. 1954. Uredinales of the Ryukyu Islands. Contributions to the rust flora of Eastern Asia. *Sci. Bull. Fac. Agric. Univ. Ryukyus* 1: 1-56.

Hoddy, E. 1991. Nature's bitter boon: The neem tree-a substitute for pesticides. *Dev. and Comm.* 5: 29-30.

Iriti, M. and Faoro, F. 2003. Benzothiadiazole (BTH) induces cell -death independent resistance in Phaseolus vulgaris against *Uromyces appendiculatus. J. Phytopathol.* 15: 171-180.

Issa, E. 1985a. Resistance to four diseases in cultivars of French bean, *Phaseolus vulgaris* L., bred in Sao Paulo. *Biologico*51: 113-117.

Issa, E., Cruz, B.P.B., Wattanabe, K. and Arruda, H.V. 1980. Damage done by bean (*Phaseolus vulgaris*) diseases. *Biologica* 46: 135-140.

Jaffer, A.A. 1971. Effect of fungicides on bean rust (*Uromyces appendiculatus*) (Pers.) *Lev. Misc. Rep. Pestic. Res. Inst.* 765:8pp.

Johnston, J.R. 1937. The rust fungi in Guatemala. *Rev. Agric., Guetamala*14: 473-478.

Kaiser, W.J. 1976. Important diseases and pests of bean (*Phaseolus vulgaris*), lima bean (*Phaseolus lunatus*) and pigeon pea (*Cajanus cajan*) in Africa. *African J. Plant Prot.* 1: 97-102.

Kamble, S.K. and Patil, B.J. 2017. Evaluation of some phytofungicides against *Uromyces appendiculatus* F. Strauss. *Biosci. Discovery* 8: 507-520.

Koleva, S. 1959. A contribution to the knowledge of parasitic flora in the People Republic of Macedonia. *Zasht. Bilja.* 51: 107-117.

Kovacevski, I.C. 1930. *French Bean Diseases in Bulgaria.* Govt. Printing Office, Sofia. 46p.

Laundon, G.F. and Waterston, J.M. 1965. *Uromyces appendiculatus. CMI, Descriptions of Pathogenic Fungi and Bacteria No. 57.*

Lawrence, E. 1951. Report of the acting Director of Agriculture. *Rep. Dep. Agric., Nysald.* 23p.

Liebenberg, M.M. 2003. *Breeding for resistance to rust of dry bean (Phaseolus vulgaris) in South Africa.* PhD Thesis. University of the Free State, Bloemfontein, South Africa. 152p.

Locke, J.C. 1990. Activity of extracted neem seed oil against fungal plant pathogens. In: *Proc.Res.Plant Workshop on Botanical pest control project*, 6-10 August 1984. Los Banos, Philippines. 132-136pp.

Macraw, W.M., Nasser,S.H., Sidky, S.T. AND Faris, F.S. 1973. Producing a variety of beans, resistant to the rust disease. 1. Screening local and imported varieties against the disease. *Agric.Res.Rev.* 51(3): 145-152.

Maffi, D., Iriti, M., Pigni, M., Vannini, C. and Faoro, F. 2011. *Uromyces appendiculatus* infection in BTH-treated bean plants: ultrastructural details of lost fight. *Mycopathologia* 171: 209-221.

Malencon, G. 1936. Moroccon mycological notes. *Rev.Mycologie*1: 257- 275.

Martin, B, Hernandez, S., Silvarrey, C., Jacas, J.A. and Cabaleiro, C. 2005. Vegetable, fish and mineral oils control grapevine powdery mildew. *Phytopathol.Mediter.* 44:169-179.

Masurat, G. and Stephan, S. 1961. The occurrence of the most important diseases and pests of crop and garden plants in 1960 in the German Democratic Republic. *Nachrbe. Dtsch. Pfl. Schdienst, Berl., N.F.* 15: 125-158.

Mendes, B.M.J. and Bergamin-Filho, A. 1989. Influence of temperature, wetness duration and leaf type on the quantification of monocyclic parameters of bean rust. *J.Phytopathol.*126: 183-189.

Mersha, Z. and Hau, B. 2011. Reciprocal effects of host and disease dynamics in the bean rust pathosystem. *J. Plant Dis. Prot.* 118: 54-62.

Milbrath, J.A. 1944. Studies on the control of bean rust. *Phytopathology* 34:936.

Mishra, H.N., Mohapatra, A.K, Sahu, S. and Mohapatra, A.K.B. 1998. Performance of French bean under different dates of sowing in Eastern Ghat high land zone of Orissa. *Annl. Agric. Res.*19: 83-88.

Mizubuti, C.S.G., Maffia, L.A., Muchovej, J.J., Romeiro, R.S. and Batista, U.G. 1995. selection of isolates of *Bacillus subtilis* with potential for the control of dry bean rust. *Fitopatol.Brasileira* 20(4): 540-544.

Mohammed, M. 1994. Bean rust pathogen variability measured by virulence and the random amplified polymorphic DNA (RAPD) technique. Ph.D. thesis. The University of Nebraska-Lincoln 139pp.

Mohan, N. Aghora, T.S. and Girija, G. 1993. Sources of resistance to rust (*Uromyces phaseoli*) in French bean. In: Golden Jubilee Symposium-Horticltural Research-changing Scenario, May 24-28, 1993. Bangalore, Karnataka, India.

Mohan, N., Aghora, T.S., Somkwar, R.G. and GirijaGaneshan. 1997. Sources of resistance to rust in French bean (*Phaseolus vulgaris). Indian J. Agri. Sci.* 67(2): 85-86.

Monda, E.O., Asanga, D.O., Okemo, P.O., Ndegwa, A. and Mariita, M.R. 2009. Efficacy of selected plant extracts against bean rust disease (*Uromyces appendiculatus*) on French beans (*Phaseolus vulgaris* L.). *J. Trop. Microbiol.Biotechnol.* 5: 1607-4106.

Mora, N.O.A., Vieira, C. and Zambolim, L. 1992. Differential bean varieties for determination of physiological races of *Uromyces phaseoli* var. *typica. Revista Ceres* 224: 391-404.

Moreno, R.A. and Mora, L.E. 1984. Cropping pattern and soil management influence on plant disease 2. Bean rust epidemiology. *Turrialba* 34(1): 41-45.

Muntanola, M. 1955. Cryptogamic parasites of peas (*Pisum spp.*) and beans (*Phaseolus spp.*) in the province of Tucuman. *Rev. Agron. Noroeste Argent*1: 283-299.

Okemo P, Harsh PB and Vivanco M. 2003. *In vitro* activity of *Maesalanceolata* extracts against plant pathogens. *Fitoterapia* 74:312-316.

Paula, T.J. Jr., Pinto, C.M. F., Silva, M. B. da, Nietsche, S., Carvalho, G. A. de and Faleir, F.G. 1998. Resistance of snap bean cultivars to anthracnose, angular leaf spot and rust. *Revista Ceres* 45(258): 171-181.

Persoon, C.H. 1795. Observations Mycologicae. In: *Usteri, P., Annalen der Botanic.* 3: 1-43.

Prakasam, V. and Thamburaj, S. 1991. Efficacy of fungicides in the control of rust disease of French bean caused by *Uromyces phaseoli* (Reben.) Wimb. *Indian J. Mycol.Pl.Pathol.* 21(2): 158-159.

Rausher, M., Adam, A.L., Wirtz, S., Guggenheim, R., Mendgen, K. and Deising, H.B. 1999. PR-1 protein inhibits the differentiation of rust infection hyphae in leaves of acquired resistant broad bean. *Plant J.* 19: 625-633.

Rodrigues, J.C.J. 1955. Racasfisiologicas de *Uromyces appendiculatus* (Pers.) Link. *AgronomiaLusitana*17:263-274.

Rolim, P.R.R., Brignani Neto, F., Roston, A.J. and Oliveira, D.A. 1981. Chemical control of bean (*Phaseolus vulgaris* L.) diseases. Rust (*Uromyces phaseoli* (Pers.) Wint. var. *typica* Arth. *Biologico*47: 201-205.

Sharma, A.K. 1998. Epidemiology and management of rust disease of French bean. *Veg. Sci.* 25: 85-88.

Sharma, N., Gupta, S.K., Bharat, N.K. and Sharma, M. 2014. Evaluation of French bean germplasm against bean rust (*Uromyces appendiculatus*) *Pl. Dis. Res.* 29: 115-117.

Sharma, N., Sharma, Savita, Gupta, S.K. and Sharma, Monica. 2018. Evaluation of fungicides against bean rust (*Uromyces appendiculatus*). *Pl. Dis. Res.* 33(2): 174-179.

Sharma, S.R. 1977. Annual Report, IIHR, Bengalore, 1976.

Sharma, S.R. 1989. Yield losses caused by rust in French beans. *Pl. Dis. Res.*4: 92-94.

Sharma, S.R. and Sohi, H.S. 1989. Diseases of French bean in India and their control. *Indian Rev. Life Sci.*9: 153-168.

Singh, A.K. and Narayana, B.M. 2002. Control of French bean rust through fungicides and a neem-based formulation. *Indian Phytopath.* 55: 241-243.

Singh, Gurvinder and Gupta, S.K. 2019. Role of temperature, relative humidity and rainfall in the development of French bean rust (*Uromyces appendiculatus*). *Indian Phytopath.* DOI 10. 1007/s 42360-019-00133-w.

Singh, J.P. and Musyimi, A.B.K. 1984. Chemical control of bean rust in Kenya. *East African Agric. Forestry J.l* 45: 207-209.

Singh, J.P. and Musymi A.K.B. 1979. Control of rust and powdery mildews by a new systemic fungicide, Bayleton. *Pesticides* 13(4): 51-53.

Sokol, P.F., Gracia, R.M.,Zimina,T.A. and Pivovarov, V.F.1977. some results of studies on the improvement of horticultural plants under Cuban conditions. *InforemCientifico- Tecnico Instituto de Investigaciones Fundamentales en Agricultura Tropical No*. 39: 15-19.

Souza, T., Alzate-Marin, A.L., Dessaune, S.N., Nunes, S.E, Quiroz, V.T., Moreira, M.A. and Barros, E.G. 2007. Inheritance study and validation of SCAR molecular marker for rust resistance in common bean. *Crop Breed. Appl. Biotech.* 7: 11-15.

Stavely JR, Steadman JR and McMillan Jr. RT. 1989. New pathogenic variability in *Uromyces appendiculatus* in North America. *Plant Dis.* 73: 428-432.

Stavely, J.R and Batra, L.R. 1991. Recently identified new pathogenic variability in *Uromyces appendiculatus*. *Phytopathology* 81:704.

Stavely, J.R. and Pastor-Corrales, M.A. 1989. Rust. In: *Bean Production Problems in the Tropics*. 2nd Ed. HF Schwartz and MA Pastor-Corrales, (Eds.). Centro Internacional de Agricultura Tropical (CIAT), Cali, Colombia.

Striab, W. 1943. Investigation on the biology and control of the bean rust (*Uromyces phaseoli* var. *typica* (Pers.) Wint.). *Gartenbavwirs*17: 397-445.

Sumartini. 1998. Rust disease in French bean and its control. *JurnalPenelitian and PengembanganPertanian* 17: 149-153.

Villar- Sanchez, B.and Lopez-Salinas, E.1993. Negro Inifap, a new var. of French bean for Chiapas state and similar tropical regions. *RevistaFitotecnia Mexicana* 16(2): 208-209.

Wagacha, J.M, Muthomi, J.W., Mutitu, E.W. and Mwaura, F.B. 2007. Control of bean rust using antibiotics produced by *Bacillus* and *Streptomyces* species - translocation and persistence in Snap Beans. *J. Appl. Sci. Environ. Mgt.* 11:165-168.

Wallace GB and Wallace M. 1950. Tanganyika fungus list: Recent records no. XIII. *Mycological Circular of Department of Agriculture Tanganyika* 26: 8.

Walliace GB. 1939. French bean diseases and bean fly in East Africa. *East African Agric. J.*3: 170-175.

Wang, S., Feng, D., Li, B. and Xu, Z. 1998. The identification of rust resistance in Chinese snap bean germplasm resources. *Ann.Rep. Bean Improv. Cooper.* 41: 179-180.

Weibosch, W.A. and Buishand, T. 1955. National trial of some dwarf bean varieties. *Meded. Proefst. Groenteell.* 2: 36.

Wicks, T.J., Hitch, C., Campbell, K. and Hall, B. 1999. Control of grapevine powdery mildew with mineral oil: an assessment of oil concentration and spray volume. *Australian J. Grape and Wine Res.*5: 61-65.

Wilson, M. and Henderson, D.M. 1966. *British rust fungi*. Cambridge Univ. Press, XVIII.Cited in Ballantyne 1978.

Xin, F.G., Gouren, Z., Baodong, L., Jing, H.H. and Jine, Z. 2000. Microscopic and ultramicroscopic investigation into the infection process of snap bean rust (*Uromycesappendiculatus*). *Acta Phytopathol. Sinica* 30(1): 53-59.

Zaumeyer, W.J. and Thomas, H.R. 1957. A monographic study of bean diseases and methods for their control. *United States Dept. Agric. Tech. Bull.*868: 255.

Zuniga de Rodriguez, J.E. and Victoria K, J.I. 1975. Determination of physiologic races of bean rust (*Uromyces phaseoli var. typica*) Arth in the Cauca Valley. *Acta Agronomica* 25(1/2/3/4): 75-85.

6

ANGULAR LEAF SPOT

It is one of the most destructive disease of French bean particularly in mid hill conditions where combination of moderate temperatures and high humidity favours the development of the disease.

GEOGRAPHICAL DISTRIBUTION

All French bean growing areas of the world reported the disease. The disease was reported for the first time from Italy in the year 1878 (Saccardo, 1878). Since then, the disease has been reported from Argentina (Spegazzini, 1882), Austria (Bubak and Kabat, 1905), USSR (Jaczewski, 1912), Netherland (Schoevers, 1920), Germany (Keissler, 1923), Kenya (McDonald, 1926), Australia (Noble, 1929), U.S.A. (Gardner and Mains, 1930), Yugoslavia (Skorik, 1930), Eritrea (Petri, 1930), Siberia (Abramoff, 1931), Tanganyika (Wallace, 1932), Guatemala (Palm, 1932), Trinidad (Thorold, 1932), Brazil (Muller, 1934), Spain (da Camara, 1936; Landeras *et al.,* 2017), Bulgaria (Kovachevsky, 1936), Belgium (Hendrickx, 1939),South Rhodesia (Hopkins, 1940), Palestine (Ryass, 1943), Peru (Garcia and Stevenson, 1944), Mozambique (de Carvalho, 1948), Turkey (Bremer *et al.*, 1948), New Caledonia (Bugnicourt *et al.*, 1951), Queensland (Simmonds, 1952), England (Deighton, 1953), Columbia (Orejuela, 1953), Mexico (Willams and Alfanso, 1955), Fiji (Morwood, 1955), Nicaragua (Litzenberger and Stevenson, 1957), Ireland (Ryan, 1965), Venezuella (Diaz Pollanco and Renaud, 1966), Tanzania (Hocking, 1967), New Zealand (Hill, 1982) and Senegal (Defranco and Collingwood, 1983).

In India, the disease for first time was reported from Nilgiri hills in South India (Srinivasan, 1953). Subsequently, the disease was recorded in Solan, Himachal Pradesh (Sohi*et al.*, 1963), Chaubattia, Uttranchal (Bose and Sindhan, 1972), Bangalore, Karnataka (Ponnappa*et al.*, 1976) and Tadong, Sikkim (Srivastava and Gupta, 1994). Since than the disease has been appearing regularly in varying intensities in all bean-growing areas.

LOSSES

Angular leaf spot is one of the most destructive diseases of bean crop in tropical as well as subtropical countries (Mahuku *et al.*, 2003). There are two types of losses due to the disease. Firstly, there is decrease in photosynthetic ability due to the damage of foliage which results in subsequent green pod loss. Secondly, the green pods become unfit for market due to infection. Varying degree of losses has been reported in different parts of the world. 50 per cent yield losses under epiphytotic condition from USA had been reported in 1950s (Cardona-Alvarez and Walker, 1956). In Tanzania, Hocking (1967) reported the disease to be highly destructive leading to the death of the plants in exceptionally wet season whereas in Mexico, it has been considered as a main fungal disease with upto 90 per cent incidence (Campos and Fucikovsky, 1981). In Brazil, Sarotorato (1988) also observed severe yield losses due to this disease.

In India, the total loss of about 40-70 per cent including damaged and unmarketable pods has been estimated due to this disease (Singh and Saini, 1980). In Bangalore, Karnataka, 80 percent yield loss and early death of the plants were reported (Ponappa *et al.*, 1976). In Sikkam, up to 20-25 per cent yearly loss in seed yield was reported due to this disease (Srivastava and Gupta, 1994). In various commercially grown French bean cultivars in Himachal Pradesh, the losses in green pod and seed yields were maximum in cv. Contender (64-69%) while in Kentucky Wonder, SVM-1 and Lakshmi, the loss amounted to over 50 per cent (Gupta and Mathew, 2005).

SYMPTOMS

All the green parts of the plants i.e., leaves, petioles, stems, pods and seed are susceptible to the pathogen and produce varied symptoms after pathogen infection. The detailed symptoms are described below:

On leaves

The symptoms appear as circular spots on cotyledonary leaves, while 3-5 angled spots, in between veins and veinlets, appear on true leaves (Plate 1a, 1b). The spots vary from 0.5 to 8.5 mm in size. The spots are dark grayish on upper surface and lighter gray on the lower surface of the leaves. Subsequently, the colour of spots changes to reddish brown and finally to dark brown colour with the progress of time. On the lower leaf surface, light reddish discolouration appears on the veins and veinlets, which delimit such spots. Gray moldy coating covers the spots on the lower surface of the leaves. Presence of coremia bearing large number of spores is visible on the spots on close observation. Many of such spots coalesce covering a large portion of the leaf giving it dirty brown appearance, whereas the remaining part turns yellowish. Such severely infected leaves show upward curling and prematurely defoliate.

On petioles and stems

Elongated spots, dark brown in colour appear both on petioles and stems. These are concentrated towards the point of leaf attachment on the petiole while on the stem, at the point of origin of branches. Under severe conditions, complete defoliation occurs resulting in pre-mature death of the plants (plate 1d). Coremia formation is rare on these parts under natural conditions.

On pods and seeds

Pods are infected during all the developmental stages of the plant. Lesions are smooth, usually circular and rarely slightly elongated on the pods. These spots are initially superficial having reddish brown centre with well-defined ashy black borders (Plate 1c). Deeper tissues of the spots are involved with the passage of time. In the central reddish-brown portion, coremia and spores are formed. No seeds or shrivelled seeds are produce in severely infected pods. In late infections i.e. infection at maturity stage, the seeds underneath the pod lesions show yellowish brown discolouration which is more remarkable on hilum area but can occur anywhere on the surface of seed.

THE PATHOGEN

Saccardo (1878) was the first to describe the pathogen as *Isariopsis griseola* on *P. vulgaris* in Northern Italy and Southern America, which was completely accepted at that time. Later, Ellis (1881) described the causal fungus on beans as *Graphium laxum* Ell. Subsequently, Saccardo (1886) described it to be synonymous to *Isariopsis laxa* (Ell.) Sacc. Ellis and Everhart (1893) observed the fungus on dead bean leaves collected at New Jersey, U.S.A. and described it as *Cercospora columnare* of the Italian exsiccati with other exsiccati of American origin at the National Fungus Collections, United States Department of Agriculture, revealed identical characteristics in respect of spore shape and size. Accordingly, the fungus was named *Isariopsis griseola* Sacc. on priority basis. Ferraris (1909) reported that the genus *Isariopsis* was identical to the genus *Phaeoisariopsis* and proposed the name *Phaeoisariopsis griseola* (Sacc.) Ferraris for the fungal pathogen of angular leaf spot disease of beans. However, this generic change was not completely accepted at that time. According to Hocking (1967) and Ponnappa *et al* (1976) from Tanzania and India, respectively, *Phaeoisariopsis griseola* (Sacc.) Ferr. should be accepted as the correct name. Finally, in the last review of the genus (Crous *et al.*, 2006 and 2013), it was named *Pseudocercospora griseola* (Sacc.) Crous & Braun, although many references still use the old nomenclature. The fungus belongs to belongs to the phylum *Ascomycota*, class *Dothideomycetes*, order *Capnodiales*, family *Mycosphaerellaceae* (Crous *et al.*, 2006).

Plate 1a. Symptoms on upper and lower surface of leaves

Plate 1b. Symptoms on cotyledonary leaves

Plate 1c. Infected pods Plate 1d. Completely infected plant

Plate 1. Symptoms of angular leaf spot of French bean

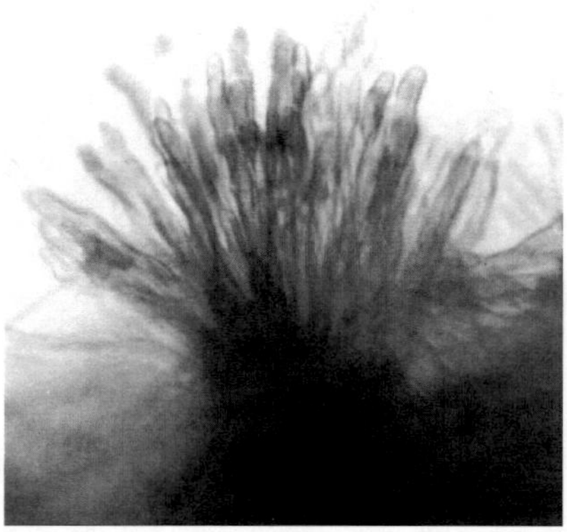

Plate 2. Coremia of *Pseudocercospora griseola* (syn: *Phaeoisariopsis griseola*)

On the host tissue, the fungus produces a stromatic structure in the substomatal cavity. Conidiophores are laterally joined together in the form of synnemata or coremia, which are darker at the base and lighter towards the tips, and either curved flexuous or septate (Plate 2). The coremia measure 128-680 μm in length and 28-69 μm in thickness. Conidia are light grey to hyaline, cylindrical to spindle shaped and straight to slightly curved. Conidia are borne singly on the tips of columnar hyphae, which are mostly 1-3 (rarely 4-6) septate, with dimensions ranging from 50-60 x 7-8μm (Miles, 1917; Benlloch, 1944; Srinivasan; 1953; Zaumeyer and Thomas, 1957). Karanja *et al.* (1994) also reported cylindrical to spindle shaped brown coloured conidia of average dimensions ranging from 17-95 x 7-24 μm and 2-3 septate. No sexual stage of the fungus has yet been observed.

PHYSIOLOGY OF PATHOGEN

Very little information is available on the physiology of *P. griseola*. The fungus can grow and sporulate profusely on potato dextrose agar (PDA) amended with one per cent bean leaf extract, honey peptone agar, baby food medium and V 8 juice agar (Cardona-Alvarez and Walker, 1956; Santos Filho *et al.*, 1976; Campos and Fucikovsky, 1981). The most favorable temperature for the growth is 24°C and pH 5-6 (Cardona-Alvarez and Walker, 1956). PDA containing 1 per cent bean seed extract at 20-25°C temperature and RH 80-85 per cent supported best growth of *P. griseola* (Gupta and Shyam, 1998).

VARIABILITY

The disease is of widespread occurrence throughout the world and leads to significant reductions in yield. Continuous monitoring of the pathogen variability is important for breeding programmes for the development of resistant varieties/ cultivars (Nietsche *et al.*, 2002). The prevalence of variation among different isolates of *P. griseola* belonging to different geographical areas of the world is well documented in literature. Cardona-Alvarez (1956) reported that single spore isolates of *P. griseola* might exhibit variations for colony structure, coloration and quantity of sporulation while conidia produced by Malaga (Spain) isolates were mostly tri-septate measuring 39.3-70.0 x 5.6-7.0 μm, where as extreme limits ranged from 28.0-31.8 x 5.6-8.4 μm (uni-septate) to 67.2-80.3 x 5.6-8.4 μm (quinqueseptate) (Benlloch, 1944). Conidial measurements i.e. 44.2-95.2 x 6.8-11.9 μm (on host) and 37.4-81.4 x 6.8-10.2 μm (in culture) differ from those reported by various workers elsewhere as 50.0-60.0 x 5.0-8.0 μm, 39.3-70.0 x 5.6-7.0 μm, 50.75.0 x 7.5, 34-83.0 x 5.0-9.0 μm and 55.0-57.0 x 5.2 μm (Saccardo, 1886; Miles, 1917; Benlloch, 1944; Srinivasan, 1953; Hocking, 1967).

Villegas (1959) working with 30 isolates of *P. griseola* collected from different bean production sites in Colombia reported the existence of 13 pathogenic races. On the basis of morphological characters viz., length and width of synnemata, conidiophores, conidia and conidial septation, existence of a new strain of *P. griseola* from Himachal Pradesh, India had been described (Sharma and Sohi, 1978). Similarly, Lacerda *et al.* (1994) identified five races (viz., G1, G2, G3, G4 and G5) from Pennambuco state, Brazil of which G1 was the most common and G3 most virulent. Earlier, Hocking

(1967) recovered a Tanzanian isolate of *P. griseola*, which produced circular lesions and speculated the role of single mutation within natural isolates.

Using isozyme analysis Correa-Victoria (1987) divided 55 isolates of *P. griseola* from Latin America, the United States and Africa into two groups of which all African isolates displayed the same electrophoretic pattern and were associated with large seeded bean cultivars originating in the Andean region of South America, while a second electrophoretic pattern was associated with small seeded bean cultivars from Central and North America. Similarly, Guzman *et al.* (1995) using random amplified polymorphic DNA markers differentiated 62 isolates of *P. griseola* from Malawi, USA and Brazil into two major groups (Andean and Mesoamerican). Corrales and Jara (1995) further suggested that these races have evolved with large seeded Andean cultivars while Mesoamerican races have evolved with small or medium seeded varieties from Central America. However, Boshoff *et al.* (1996) failed to detect any genetic variation among local South African isolates tested using isozyme analysis and postulated that genetic variation, if any, could possibly be evident in large populations of local and foreign isolates. Sartorato (2002) identified seven different pathotypes and classified them as Middle American pathotypes. Pathotype 63-63 overcame resistance genes present in all differential cultivars and also the resistance gene(s) present in cv. AND 277. Similarly, twenty-six different pathotypes were identified among 72 isolates which demonstrated the great variability among the pathogen in Brazil (Nietsche *et al.*, 2002). The virulence test and molecular marker analysis, using the random primer OPAA 11 confirmed the only presence of Mesoamerican phenotype pool in Brazil. Among 112 *P. griseola* isolates collected from various locations in Central America revealed broad polymorphic diversity as indicated by RAPD markers while on 12 differential bean genotypes, fifty pathotypes were identified (Mahuku, 2002). Comparison of virulence phentypes and RAPD profiles to known Andean *P. griseola* isolates confirmed that all isolates belonged to Mesoamerican group. Pairwise comparison between individual RAPD loci showed that the majority were in gametic phase linkage disequilibrium, revealing that *P. griseola* maintains a genetic structure that is consistent with asexual reproduction (Mahuku *et al.*, 2002). Gupta *et al.* (2006) studied morphological, cultural and pathogenic variation among twelve isolates of the pathogen and reported slight variation in conidial size and diametric vegetative growth, however, the isolates varied significantly in their aggressiveness on cultivars Contender and Jawala and Isolate I_{10} was found to be most aggressive.

The pathogenic variability among 48 isolates of *Pseudocercospora griseola*, collected from the State of Minas Gerais, Brazil was evaluated (De Silva *et al.*, 2008). A set of 12 international differential cultivars were inoculated with 48 isolates of *P. griseola* under greenhouse conditions. Ten pathotypes *viz.*, 55-15, 63-7, 63-15, 63-23, 63-25, 63-27, 63-31, 63-47, 63-55 and 63-63 of *P. griseola* were identified, exhibiting huge pathogenic variability. Of the 48 isolates, four pathotypes namely 55-15, 63-15, 63-25 and 63-27 had not earlier been reported in the State of Minas Gerais. All 48 isolates, except 55-1547 pathotype, showed a compatible reaction with all Andean group cultivars. All 48 isolates were highly pathogenic in both Andean and Mesoamerican cultivars, and hence are grouped as Mesoamerican pathotypes. Among all the pathotypes, pathotype 63-63 of the fungus was the most prevalent, and has the potential to breakdown the resistance genes of all differential cultivars.

HOST RANGE

The pathogen infects the members of family *Leguminosae* and hence has a very narrow host range. In 1931, Abramoff reported *I. griseola* to cause angular brown spot on soyabean in Russia. Out of six species of legumes namely *Phaseolus lunatus* L., *Dolichoslablab* L., *D.* sp., *Indegofera sumatrama* Gaertn. *Cajanus* sp. and *Vigna sinensis* L., only *P. lunatus* L. was attacked by this pathogen (Cardona-Alvarez and Walker, 1956). In India, among thirteen leguminous plants tested, only *Phaseolus multiflorus, P. vulgaris* and *P. lunatus* were found susceptible (Verma, 1979).

DISEASE CYCLE

The fungus perpetuates in the infected plant debris in fields where the crop is grown regularly year after year in the same field. Some reports suggests the seed borne nature of the pathogen. Seed and soil borne nature of the fungal pathogen has been reported in India (Sohi and Sharma, 1974; Sindhan and Bose, 1980). Primary source of infection is the infected seeds. The infected seed after germination give rise to a plant with circular brown spots on the cotyledonary leaves. These spots on the under-surface bears numerous dots like structures known as 'Coremia' which bears condiophores and conidia of the fungus (Fig. 1). The conidia may be disseminated from the debris by rain splashes or along with windblown soil particles and from sporulating lesions by wind currents. The pathogen infects host tissue by entering through stomata and advancing intercellularly in the mesophyll and palisade parenchyma. After infection i.e. within 9 days, the fungus develops intracellularly throughout necrotic lesions and stromata develop in the substomatal cavity and sporulation then may occur during periods (24 to 48 h) of continuous moisture. The conidia cause secondary infection and cycle continues. In severe infections, the leaves become yellow and premature leaf defoliation occurs. Infection from pods also goes to the developing seed.

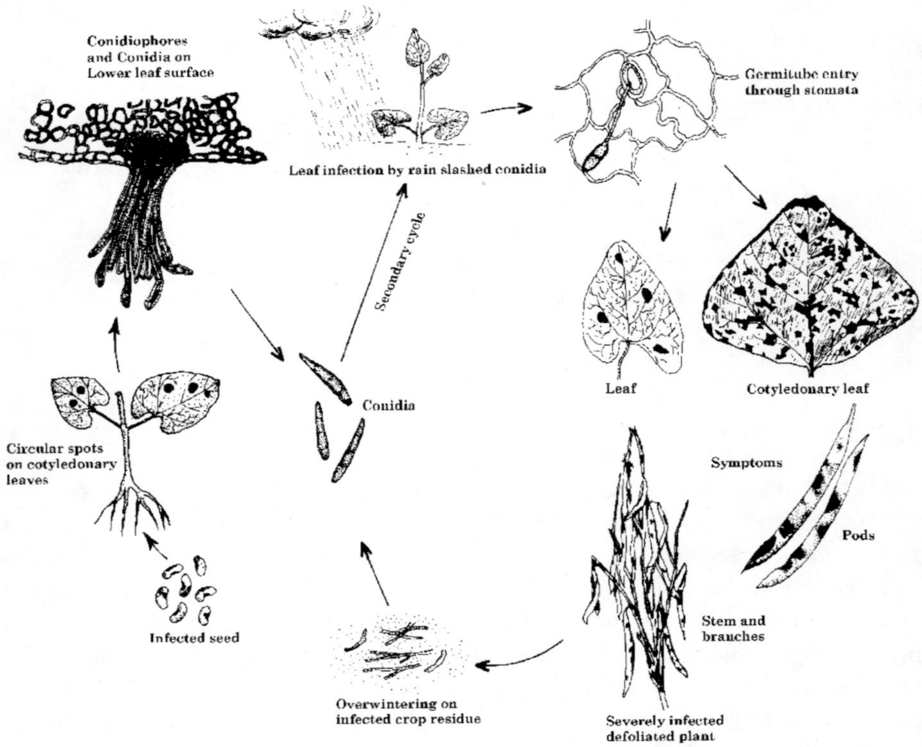

Fig. 1. Disease cycle of angular leaf spot of French bean

EPIDEMIOLOGY

The pathogen *Pseudocercospora griseola* (syn: *Phaesariopsis griseola*), overwinters on plant debris as well as in infected seeds (Sohi and Sharma, 1974; Sindhan and Bose, 1980). The conidia serve as a source of primary inoculums and remain viable for 6 and 8 months on plant debris in laboratory and field conditions, respectively (Sindhan and Bose, 1980). Infected seeds could also serve as other source of primary inoculum. In most of the cultivars, hilum of seed is the infestation site but in some cultivars both hilum and seed coat seem to be infected (Saettler and Corea, 1988). In young seedlings, disease initially results from the planting of infected seed causing spots on cotyledonary leaves. The conidia splashed by rain, conidia from plant debris or windblown conidia with soil particles serves as the source of secondary spread of the disease in most of bean growing regions. Infected leaves become major inoculum source for secondary spread of the disease. However, the rate and degree of secondary spread of the disease is dependent on the interactions of all factors of disease development like climate, pathogen and host. Moderate temperatures and moist conditions favour the arrival of viable inoculum and successful infections on bean plants in the field.

Effect of temperature

Optimum temperature for fungal growth and sporulation is 21-24°C, for spore germination is 18-24°C and for disease development is 24°C (Sindhan and Bose, 1980; Sarotorato, 1988). However, Bassaniezi *et al.* (1998) suggested temperature ranging between 24.2 to 28.3° C as most favorable for disease development. Gupta *et al.* (2000c) also reported that optimum temperature for disease development ranged between 20 to 25° C with 35.2 and 45.4 lesions/ leaf respectively.

Relative humidity

High relative humidity ranging between 95-100 per cent is essential for artificial infection (Llanos, M.C. 1958). However, relative humidity of more than 90 per cent is optimum for sporulation, spore germination and development of symptoms (Sindhan and Bose, 1980, Sartorato, 1988). Gupta *et al.* (2005) reported that rise in relative humidity from 85.7 to 89. 9 per cent at moderate temperatures (25± 1° C) exhibited a steep increase in development of angular leaf spot disease (Fig. 2).

Effect of Leaf wetness

Abnormally wet season, monoculture and use of frequent irrigation resulted in severe disease development (Barros *et al.*, 1958a). The effect of leaf wetness durations at optimum temperature (Table-1) was studied and it was found that the existence of continuous leaf wetness for 3 h was enough to initiate the disease but increase in leaf wetness duration from 3 to 9 h resulted in a gradual decline of incubation period from 12 to 9 days, while there was progressive increase in lesion number per leaf with increase in wetness duration upto 24 h (Mathew *et al.,* 1998).

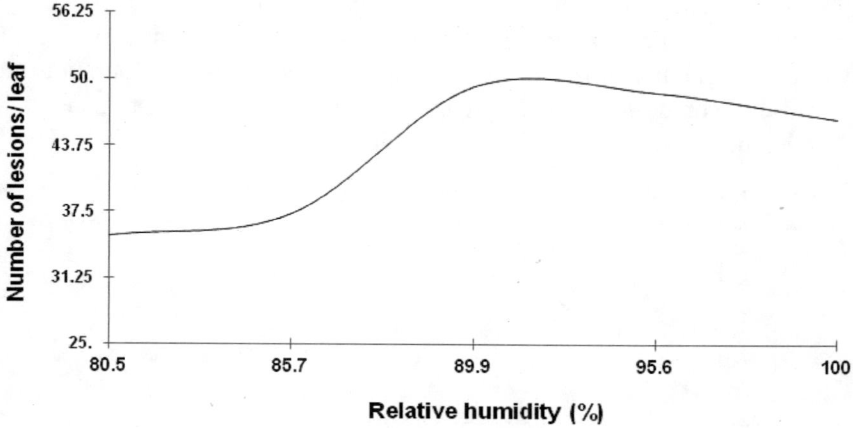

Fig. 2. Effect of relative humidity on development of angular leaf spot of French bean

Table 1. Effect of different leaf wetness durations on angular leaf spot of French bean

Leaf wetness (hours)	Average number of spots per leaf	Incubation period (days)
3	9.22	12
6	24.25	12
9	33.25	9
12	38.33	9
18	40.50	9
24	49.25	9
48	46.25	9
CD (P≤0.05)	1.5	

Evaluated at temperature 24±1°C

Duration of continuous leaf wetness upto 24 h is generally uncommon in nature. Since the majority of rainfall and/ or dew events are of short duration and evaporation rates are high, the effect of intermittent leaf wetness on disease development was studied and it was found that the number of lesions increased with the interruption of leaf wetness with dry periods upto 3 cycles (Mathew *et al.,* 1998). This suggested that the pathogen is not only adapted to survive under dry periods during the infection process but the process itself is enhanced by alternate dry and wet cycles.

Diurnal periodicity of conidial release in P. griseola

Diurnal periodicity of conidial release of *P. griseola* revealed a distinct rhythm at different times of the day (Fig. 3) with maximum conidial density being observed between midnight and early morning hours (Gupta *et al.* 2005). Optimum formation and release of conidia were observed at temperature 23-24° C, more than 90 % high relative humidity, occurrence of 20.00 to 8.00 h of dew and dark period.

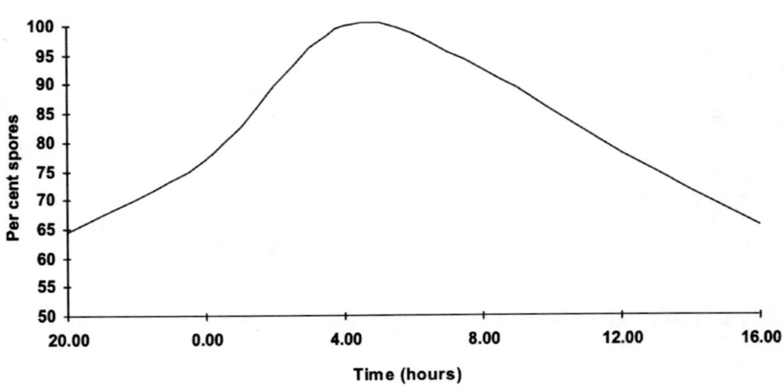

Fig. 3. Diurnal periodicity curve of conidial release in *P. griseola*

Role of meteorological factors on disease development

Effect of environmental factors *viz.,* effect of rainfall, soil moisture and relative humidity, on development of angular leaf spot disease was studied and highly significant and positive simple correlation coefficients between disease severity and rainfall (0.7596), soil moisture (0.7070) and relative humidity (0.5253) were found. Further, partial correlation coefficients between disease severity and rainfall (0.5270) and soil moisture (0.4475) showed significant positive correlations (Gupta *et al.*, 2005). The multiple regression equation $Y_1 = -5.3781 + 0.0079\ X_1 + 0.0177\ X_2 + 0.0494\ X_3 + 0.0308\ X_4 + 0.3237\ X_5$ which indicated that the a unit change in rainfall could influence the disease severity upto an extent of 0.0494 units, followed by soil moisture (0.3237units), relative humidity (0.0308units) and soil temperature (0.0177 units) and mean air temperature (0.0079 units). However, more pronounced effect of relative humidity compared to the mean air temperature and rainfall on development of angular leaf spot disease was reported by Adikshita and Kansal (2018). But Chhetry *et al.* (1994) reported non-significant linear correlation and regression coefficients of disease incidence and severity with microclimatic factors. However, Barros *et al.* (1958) observed that repeated cultivation of the crop, frequent irrigation or abnormally wet season resulted in severe outbreak of the disease.

MANAGEMENT

Cultural practices

Various cultural practicesviz., use ofdisease free seed and removalofdiseasedleaves and seedlings (Ploper, 1980; Trutmann and Kayitare, 1991); 3-4 year crop rotation (Anonymous, 1940; Ploper, 1980; Correa-Victoria and Saetler, 1987); destruction of diseased plant material (Anonymous, 1940; Benlloch, 1944; Wallace, 1951); deepploughing (Anonymous, 1940; Hocking, 1967); widerspacing (Navarro *et al.*, 1981; Gupta *et al.*, 2000b) and altering of sowing dates (Bhardwaj *et al.*, 1994, Gupta *et al.*, 2000) and staking of pole type varieties (Gupta *et al.*, 2000b) have been recommended from time to time for the control of this disease.

Severity of angular leaf spot of bean in different cropping systems (bean grown alone or in combination with maize, sweet potato or Cassava) was studied and it was reported that severity of the disease was maximum when maize was included and lowest with sweet potato and Cassava (Moreno, 1977). However, the effectiveness of intercropping with maize to reduce the severity of the disease in a multilocational 2 years trials in Tanzania had been reported (Kikoka *et al.*, 1989). Boudreau (1993) and Gupta *et al.* (2000) reported the similar observations from Kenya and India, respectively. Soil amendment with poultry manure, composted *Urtica* sp. and composted *Lantana camara* treatments were found effective to reduce the severity of angular leaf spot disease of bean (Joshi *et al.*, 2009).

Host resistance

The causal fungus, *P. griseola* is highly variable and a diverse source of resistance genes is required to manage this disease (Mahuku *et al.*, 2003). Varying degree of resistance is present in French bean cultivars against the fungal pathogen (Buruchara *et al.*, 1988). Cultivars/ lines like Albama No.1, Café, California Small White, Case Knife, Epicure, Mexico blank, McKastan, Navy bean, Negro Costa Rica, Scotia Rojo Chico, Mexico 11, Mexico 12 and Cauca 27 (Olava, 1958); Carpoata, Negro Tacana and Manteiga Maravilha were documented as resistant whereas VIII-B, Amarillo 153, Jampa, P-524, Negro Purisima and Amapolo were reported to be tolerant ((Brock, 1951; Santos Filho *et al.*, 1976; 1978; Campos and Fucikovsky, 1981; Lopez *et al.*, 1997; Paula *et al.*, 1998). Among black and non-black coloured bean cultivarsscreened in five locations in Minas Gerais, Araujo *et al.* (1989) reported that non-black cv. Ouro and black colouredcvs. Rio Tibagi and Milionario 1732 possess resistance to *P. griseola.* Of the 1441 accessiones in the core collection, only 2.2 per cent were resistant to both Andean and Mesoamerican races, 28 per cent were resistant to Andean and 9 per cent to Mesoamerican races (Mahuku *et al.*, 2003).

In India, cvs./lines viz., EC 38921, EC 44621, PLB 148, Kentucky Wonder, Canadian Round, Canadian Long Red, Rajmah Sel. R 6, R. 7, Biaser 3, EC-44784, IC 47651-6, EP 146 (BAT-482), HPR-54, HPR-63, HPR-92, HPR-93, HPR-111, HPR-232, HPR-260, HPR-299, HPR-332, HPR-326, Hans, Nagar Local, SVM-1, JK-8, JK-12, PDR-5, PDR-14, Him-11, Him-12, IIHR 42-2, IIHR 901, EC 97830, EC 44758 and NIC 14402, SL-4-PL-1, HIM-1, Bountiful, JKRO-3, L-27, L-28, L-8 and Local Big Beans were found resistant against angular leaf spot disease (Singh and Sharma, 1975; Bhardwaj *et al.*, 1992; Anonymous, 1992; Srivastava *et al.*, 1995; Mohan *et al.*, 1995; Gupta *et al.*, 2000a, Bashir *et al.,* 2006).

The disease progress (r) was about a third in CIAT lines BAT 67(20) P and BAT 9601 (20P) of that in G 2858 and Albama-1 (Buruchara *et al.*, 1988). Later the effectiveness of supplementing local bean mixture with 25 per cent or more resistant lines (BAT 76 and A 285) which resulted in significant reduction of angular leaf spot severity in Africa (Pyndji and Trutmann, 1992; Trutman and Pyndji, 1994). Single dominant gene conferred resistance in line BAT 332 against angular leaf spot (Caixeta *et al.*, 2003). They further identified two RAPD markers, OPAA07950 and OPO12950, linked in coupling phase at 5.10 and 5.83 cM of this gene, respectively.

A set of 37 released varieties of common bean in Democratic Republic of Congo were screened using virulent Andean and Mesoamerican isolates and four resistant bean varieties *viz.,* ARA 4, COD MLV 059, MLV 224/94B, LSA 144 and Mexico 54 were identified (Kijana *et al.*, 2017).

Chemical control

The angular leaf spot disease could be effectively managed by using various fungicides in India and other countries. However, any delay in routine application of fungicides either due to occurrence of continuous rains or other reasons, could result in heavy build up of the disease. Initiation of sprays at proper timing is of primary significance to attain desirable level of disease management.

Seed treatment with Germisan, Ceresan dry, Agrosan, Vitavax benomyl and carbendazim, seed dipping in mercuric chloride solution (0.1%) for 30 minutes (Milatovic, 1958; Sohi and Sharma, 1969a; Sindhan, 1984; Trutman*et al.*, 1992; Rajappan and Yesuraja, 1999; Bhat, 2002) and foliar sprays of fungicides like Thiovit, wettablesulphur, Dithane M-45, Zincocap, Bordeaux mixture (Sohi and Sharma, 1969b), Zineb (Barros *et al.*, 1958b; Corbett and Siddiqi, 1966), Derosal, Saprol (Giroto, 1976), triforine (Fortugno, 1977), carbendazim (Sharma and Bisht, 1997), triademorph, Fernacol, Ferbam (Sindhan, 1984; Rajappan and Yesuraja, 1999) benomyl (Giroto, 1976, Fortugno, 1977; Gonzalez *et al.*, 1977; Singh and Sharma, 1976; Issa *et al.*, 1982, 1983; Hidalgo and Araya, 1993; Singh *et al.*, 1995; Bhat, 2002), chlorothalonil (Gonzalez *et al.*, 1977; Castro *et al.*, 1991; Vierra *et al.*, 1998), mancozeb (Issa *et al.*, 1982; Castro *et al.*, 1991; Sharma and Bisht, 1997; Vierra*et al.*, 1998; Rajappan and Yesuraja, 1999), triphenyltin acetate, triphenyltin hydroxide (Castro *et al.*, 1991) and propiconazole and hexaconazole (Bhat, 2002) have been found to be effective for reducing the disease and increasing the yield. However, Oliveira *et al.* (1992) reported stannic triphenyl (fentin) acetate, bitertanol and tebuconazole as most effective to reduce the disease compared to Bavistin while, Srivastava and Gupta (1994) reported that captan, thiophanate methyl, carbendazim, ziram and captafol were superior to mancozeb, copper oxychloride and triademorph and applications of captan and zineb were the most economical. Combi fungicide sprays benomyl + captafol (Issa *et al.*, 1982), chlorothalonil + copper oxychloride (Issa *et al.*, 1983) and thiophanate methyl + thiram (Rodrigues *et al.*, 1987) were also effective for the management of the disease.

Persistence of protective action of fermate, Parzate, Fuclasin, Manzate, Orthocide 50, Oxicol Ultra, Shell copper fungicide and thiolate upto 9 days (Cardenas, 1958). Applications of benomyl, thiophanate methyl + thiram, and triforine were effective when applied upto 15 days prior to inoculation with the fungus and had curative effect when applied after inoculation under glasshouse conditions (Rodrigues *et al.*, 1987). Seed dressing with carbendazim (0.25%) followed by three foliar sprays (0.1%) at 15 day interval were found effective in the management of this disease (Bhardwaj and Thakur, 1992). Different fungicides have also been screened against this disease under laboratory, poly house and field conditions both as seed treatment alone and in combination with foliar sprays. All the fungicides both systemic and non-systemic, used as foliar spray alone provided good control of the disease when applied before infection, carbendazim (0.05%) being the best. A combination of carbendazim (seed treatment + foliar spray) with bitertanol (seed treatment + foliar spray) was reported to be most efficacious for the reduction of disease severity and and increasing the green pod yields compared with seed treatment or foliar spray alone (Mathew *et al.*, 1998). Fungicides like azoxystrobin, chlorothalonil, thiophanate methyl+ chlorothalonil, propiconazole, tebuconazole, triphenyltin hydroxide provided more than 70 per cent control of the disease (Picinini and Fernandas, 2000). Epoxiconazole singly or in combination with pyraclostrobin or tebuconazole showed high efficacy against this disease (Rava,

2002). Various strobilurin fungicides like azoxystrobin and trifloxystrobin and EBI's like tebuconazole and propiconazole were found effective in the management of this disease (Oliveira and de Oliveira, 2003, Sartorato and Rava, 2003). Four sprays of carbendazim(0.1%), hexaconazole (0.05%) and propiconazole (0.05%) at 10 days interval were found effective in the management of this disease (Shukla and Sharma, 2009; 2016). Sectin WG, Cabrio Top WG, Amistar Top SC, Nativo WG and Amistar SC were found most effective for significant mycelial growth inhibiton of the fungal pathogen i.e. *P. griseola* and four periodic sprays of Sectin (0.2%) at 15 days interval under field conditions were found most efficient to reduce the severity of the disease and to enhance the yield of green pod yields of beans (Adikshita and Kansal, 2016).

Ergosterol Biosynthesis Inhibitor (EBI) fungicides have longer post-infection and eradicant potential and are systemic. EBI fungicides are used in different spray schedules for effective management crop disease. EBI fungicides are environment friendly and also cost effective. Mode of action of EBI's like hexaconazole, penconazole and bitertanol along with mancozeb (protectant) and carbendazim (systemic) against the angular leaf spot disease was studied in polyhouse conditions by providing artificial weather conditions for infection (Mathew *et al.*, 1998) and following observations on the activity of EBI fungicides against the angular leaf spot disease were made:

(a) **Protective activity:** In this strategy, a chemical toxic barrier is created between the host and pathogen before infection. Carbendazim provided complete control of the disease upto 120 h whereas mancozeb upto 96 h. EBI's exhibited protective activity upto 72 h pre-inoculation period. Thus EBI's are poor in protective activity. Efficacy of hexaconazole (0.1%) in protective spray programme against this disease has also been reported from Sikkim (Bhat, 2002).

(b) **Post-infection activity:** In post-infection spray programme, fungicides are applied after the occurrence of suitable weather for infection as recorded by weather monitoring devices. Bean plants were inoculated with the fungal pathogen and were kept in moist chamber for 24, 48, 72 and 96 h after inoculation. These plants were then sprayed with EBI fungicides. All the fungicides exhibited post-infection activity upto 72 h. However, it was only 24 h in case of carbendazim and mancozeb. This longer postinfection activity can be utilized in planning spray programmes where routine application of fungicides is delayed.

(c) **Pre-symptom activity:** Sometimes it is not possible to carry on sprays even after 3-4 days of recording of suitable weather for infection, possibly due to prolonged rains. In such cases certain fungicides are applied 2-3 days prior to the symptom expression as determined (Table 1). Mancozeb had little effect and produced sporulating spots while EBI's and carbendazim produced reddish brown non-sporulating spots when applied 7th and 8th day of inoculation.

(d) **Post-symptom activity:** Post-symptom antisporulant activity of fungicides is sometimes necessary for reducing the disease pressure in already infected fields and secondary build up of the inoculum. Fungicides like carbendazim, hexaconazole, flusilazole and bitertanol had high level of antisporulant activity and their residue persisted for more than 10 days, which can be utilized in planning different spray programme.

Biocontrol agents

Adhikshita and Kansal (2016) evaluated different bioagent against *P. griseola* and reported that *Trichoderma harzianum* proved most inhibitory to the pathogen under *in vitro* conditions while testing its antagonistic efficacy through dual culture or through its toxic substances.

Biopesticides

Very limited work on this aspect has been carried out on this aspect. Adhikshita and Kansal (2016) evaluated different botanicals under *in vitro* conditions against the mycelial growth *P. griseola* and reported that garlic bulb extract @ 10 per cent caused significant mycelia growth inhibition (84.62%) of the pathogen followed by neem seed kernel extract. The antifungal activity of *A sativum* may be due to sulphur containing compounds S-methylL-cysteine sulphoxide (Whitemore and Naidu 2000).

Integrated disease management

Careful selection of varieties and adjustment of sowing dates along with judicious fungicidal treatments proved very efficient for the reduction of the disease severity and to increase the crop yield (Bhardwaj *et al.*, 1994). Application of mulches either pine needles or eucalyptus leaves, alone and in combination with sprays of carbendazim (0.1%) on mulch and foliage reduced the disease severity and increased pod yields (Mathew *et al.*, 1998). Sprays of carbendazim on mulches and foliage were highly effective in reducing the disease severity and increasing green pod yields yet sprays of carbendazim on mulches alone were also found effective and reduced the severity to 20.85 and 22.15 per cent in pine needle and Eucalyptus leaves mulch, respectively, and increased green pod yields by 90.5 and 106 per cent with additional advantage of residue free pods. Application of molybdenum (Mo) in combination with fungicide spray applied once (at early growth stage) or twice in the bean floweing period (25-45 days after sowing) provided substantial control of this disease (Jesus Junior *et al.*, 2004).

REFERENCES

Abramoff, I.N. 1931. Fungal diseases of Soybeans in the Far East 3-84 pp. Diseases and Pests of Soybean in the Far East. *Pamphlet Far Eastern Pl. Prot. Stat., Vladivostock* 120 p.

Adikshita and Kansal ,S. 2016. Evaluation of fungicides against angular leaf spot of French bean caused by *Phaeoisariopsis griseola. International Journal of Farm Sciences* 6(4): 1-5.

Adikshita and Kansal, S. 2016. *In vitro* evaluation of botanicals and bioagents against *Phaeoisariopsis griseola* (Sacc) Ferraris causing angular leaf spot of French bean. *International Journal of Farm Sciences* 6(3): 254-260.

Adikshita and Kansal, S. 2018. Studies on the epidemiological parameters of Angular leaf spot (ALS) of French bean caused by *Phaeoisariopsis griseola. International Journal of Chemical Studies* 6(4): 2520-2523.

Anonymous, 1940. Plant Diseases. Notes contributed by the Biological Branch. *Agric. Gaz. N.S.W.* 51(2):79-82.

Anonymous, 1992. *Package of practices for Vegetable Crops in Himachal Pradesh.* Directorate of Extension Education, Dr. Y.S. Parmar Univ. of Hort. & Forestry, Solan,pp.144.

Araujo, G.A.de A., Viera, C. and Chagas, J.M. 1989. Performance of bean (*Phaseolus vulgaris* L.) cultivars in the Zona da Mata area of Minas Gerais V. *Revista Ceres* 36:382-390.

Barros, O., Cardona, C., Cardenosa, R. and Skiles, R.L. 1958a. Angular leaf spot of beans in Colombia. *Pl. Dis. Reptr.* 42: 420-424.

Barros, O., Cardona, C., Cardenosa, R. and Skiles, R.L. 1958b. The control of angular leaf spot of beans in Colombia. *F.A.O. Plant Prot. Bull.* 6: 97-101.

Bashir, S., Mushtaq Ahmad, Vinay Sagar and Mohd-Najeeb. 2006. Identification of sources of resistance in French bean to *Phaeoisariopsis griseola* causing angular leaf spot disease. *Indian Journal of Plant Protection*: 34(2): 221-224.

Bassaniezi, R.B., Amorim, L., Filho, A.B., Hau, B. 1998. Effects of bean line pattern mosaic virus on the monocyclic components of rust and angular leaf spot of *Phaseolus* bean at different temperatures. *Plant Pathol.* 47: 289-298.

Benlloch, M. 1944. A new bean disease, *Phaeoisariopsis griseola* (Sacc.) Ferr. *Bol. Pat., Veg. Ent. Agric. Madr.* 13: 27-32.

Bhardwaj, C.L. and Thakur, D.R. 1992. Fungicidal management of angular leaf spot of French bean. *Indian J. Mycol. Pl. Pathol.* 22 (3):232-235.

Bhardwaj, C.L., Kalia, N.R. and Singh, L.N. 1992. Reaction of French bean germplasm against angular leaf spot. *Indian J. Mycol. Pl. Pathol.* 22 (3):270-271.

Bhardwaj, C.L., Nayital, S.C., Verma, S. and Kalia, N.R. 1994. Effect of sowing date, variety and management of angular leaf spot (*Phaeoisariopsis griseola*) on yield of French bean (*Phaseolus vulgaris*). *Indian J. Agric. Sci.* 64(5):336-338.

Bhat, M.N. 2002. Chemical control of rust and angular leaf spot of French bean in Sikkim. *Pl. Dis. Res.* 17(1): 90-92.

Bordreau, M.A. 1993. Effect of intercropping beans with maize on the severity of angular leaf spot of beans in Kenya. *Plant Pathol.* 42(1):16-25.

Bose, S.K. and Sindhan, G.S. 1972. Leaf spot of French bean caused by *Isariopsis griseola* Sacc. and its control. *Prog. Hort.* 4: 69-75.

Boshoff, W.H.P., Swart, W.J., Prtorius, Z.A., Liebenberg, M.M. and Crous, P.W. 1996. Isozyme variability among isolates of *Phaeoisariopsis griseola* in Southern Africa. *Plant Pathol.* 45 : 344-349.

Bremer, H., Ismen, H., Karel, G., Ozkan, H. and Ozkan, M. 1948. Contributions to the knowledge of the parasitic fungi of Turkey. Part III. *Rev. Fac. Sci. Univ. Istanbul, Ser. B*. 13(1):1-53.

Brock, R.D. 1951. Resistance to angular leaf spot among varieties of beans. *J. Australian Inst. Agri. Sci.* 17: 25-30.

Bubak, F. and Kabat, J.E. 1905. Vierter Beitrag Zur Pitzflora Von Triol. *Osterr. Bot. Ztschr*. 55:73-79.

Bugnicourt, F., Cohic, F. and Danant, R. 1951. *A catalogue of the animal and plant parasites of the cultivated plants of New Caledonia*, 120 p.

Buruchara, R.A., Gathuru, E.M. and Mukunya, D.M. 1988. Disease progress of angular leaf spot caused by *Isariopsis griseola* Sacc. and its implications on resistance of some bean (*Phaseolus vulgaris*L.) cultivars. *Acta Horticulturae* 218: 321-328.

Campos, A.J., Fucikovsky, Z.L. 1981. Study of some aspects of angular spot caused by *Isariopsis griseola* Sacc. on bean. *Fitopatologia* 16(1):16-19.

Cardenas, C.H. 1958. Persistence of the protective action of some fungicides on bean (*Phaseolus vulgaris* L.). *Acta Agron. Palmira* 8(1-2):77-100.

Cardona-Alvarez, C. 1956. Angular leaf spot of bean. Ph.D. Dissertation, University of Wisconsin, 49 p.

Cardona-Alvarez, C. and Walker, J.C. 1956. Angular leaf spot of bean. *Phytopathology* 46: 610-615.

Castro, J.L. de, Ito, M.F., Dudienas, C., Bulisani, E.A., D'Artegnan-de-Almeida, L. and De-Castro, J.L. 1991. Fungicide activity on two bean cultivars in Capao Bonito, Sau Paulo State, Brazil. *Bragantia* 50(2):309-321.

Chhetry, G.K.N., Mishra, R.R. and Sharma, G.D. 1994. Effect of microclimatic factors on epidemiology of leaf spot disease caused by *Colletotrichum lindemuthianum* and *Phaeoisariopsis griseola* of French bean (*Phaseolus vulgaris*). *Indian J. Agric. Sci*. 64(9):667-670.

Corbett, D.C.M. and Siddiqi, M.A. 1966. Fungicidal control of plant diseases in Malawi. *PANS B* 12: 37-38.

Corrales, M.A. and Jara, C.E. 1995. Evolution of *Phaeoisariopsis griseola* with common bean in Latin America. *Fitopatologia Colombiana* 19(1):15-24.

Correa-Victoria, F.J. 1987. Pathogenic variation, production of toxic metabolites and isozyme analysis in *Phaeoisariopsis griseola* (Sacc) Ferr. Michingan, USA, Michigan State University, Ph.D. Thesis.

Correa-Victoria, F.J. and Saettler, A.W. 1987. Angular leaf spot of red kidney beans in Michigan. *Plant Dis*. 71(10):915-918.

Crous, P.W., Liebenberg, M.M., Braun, U. and Groenewald, J.Z. 2006. Re-evaluating the taxonomic status of *Phaeoisariopsis griseola*, the causal agent of angular leaf spot of bean. *Stud Mycol* 55: 163-173.

Crous, P.W., Braun, U., Hunter, G.C., Wingfield, M.J., Verkley, G.J.M., Shin, H.D., Nakashima, C. and Groenewald, J.Z. 2013. Phylogenetic lineages in *Pseudocercospora*. *Stud Mycol* 75: 37-114.

da Camara, E. de S. 1956. Contributions to the mycoflora of Portugal XI. *Bot. Agric. Lisbon* 1:88.

de Carvalho, T. 1948. Preliminary report of the diseases observed in plants and insects with phytopathological annotations. *Colonia de Mocambique, Reparticao de Agricultura Seccao de Micologia*, 84 p.

de Silva, K. J. D., de Souza, E. D., Sartorato, A. and Freire, C. S. 2008. Pathogenic variability of isolates of *Pseudocercospora griseola*, the cause of common bean angular leaf spot, and its implications for resistance breeding. *J. Phytopathology* 156: 602-606.

Defranco, M. and Collingwood, E.F. 1983. *Isariopsis griseola* on dwarf bean. *FAO Plant Prot. Bull.* 31(4): 168.

Deighton, F.C. 1953. Plant Pathology Section. *Rep. Dept. Agric. S. Leone,* 1951, 21-23.

Diaz Pollanco, C. and Renaud, C.J. 1966. The most important fungal diseases of French bean in the central region of *Venezuela. Agron. Trop.* 16(2):129-139.

Ellis, J.B. 1881. New species of North American Fungi. *Torrey Botanical Club Bulletin* 8: 64-66.

Ellis, J.B. and Everhart, B.M. 1893. New species of North American fungi from various localities. *Acad. Nat. Sci. Philadelphia Proc.* 128-172.

Ferraris, T. 1909. Obsservazionimicologichesu species del gruppohyphales (Hyphomycetae). *Ann. Mycol.* 7: 273-286.

Fortugno, C. 1977. Angular leaf spot (*Isariopsis griseola* Sacc.) of bean. *IDIA* 317/320: 26-33.

Garcia, R.C. and Stevenson, J. A. 1944. The peruvian fungus flora. A preliminary list of fungi attacking plants in Peru. *Estac. Exp. Agric. La Molina*, 1942, 112p.

Gardner, M.W. and Mains, E.B. 1930. Indiana Plant Diseases. *Proc. Indiana Acad. Sci.* 39: 85-99.

Giroto, R. 1976. Evaluation of fungicides in the control of anthracnose and angular leaf spot of bean. *IDIA* 313/314: 38 p.

Gonzalez, L.C., Gutierrez, R., Cascante, F. and Portilla E, 1977. Control of leaf diseases of bean (*Phaseolus vulgaris*) with limited use of fungicides. *Agronomia Costarricense* 1(2): 107-118.

Gupta, S.K, Mathew, K.A and Shyam, K.R. 2000c. Effect of abiotic environmental factors on the development of French bean angular leaf spot. *J. Mycol. Pl. Pathol.* 30: 271 (Abstr.).

Gupta, S.K. and Mathew, K.A. 2005. Assessment of yield losses in French bean cultivars due to angular leaf spot (*Phaeoisariopsis griseola*). *J. Mycol. Pl. Pathol.* 35: 270-272.

Gupta, S.K. and Shyam, K.R. 1996b. Annual Progress Report of ICAR Adhoc Research Scheme "Studies on role of meteorological factors on initiation and development of angular leaf spot (*Phaeoisariopsis griseola* (Sacc.) Ferr.) of bean (*Phaseolus vulgaris* L.) and its management." Department of Mycology and Plant Pathology, YSPUHF, Nauni, Solan. 14 p.

Gupta, S.K., Mathew, K.A. and Shyam, K.R. 2000b. Management of angular leaf spot of French bean through cultural practices. In: Proceedings of Indian Phytopathological Society-Golden Jubilee International Conference on "*Integrated Plant Disease Management for sustainable Agriculture*" Vol. II, Indian Phytopathological Society, IARI, New Delhi . pp. 701-704.

Gupta, S.K., Mathew, K.A., Shyam, K.R. and Sharma, A.K. 2000a. Evaluation of French bean germplasm against angular leaf spot. *Indian Phytopath*. 53(4): 488-489.

Gupta, S.K., Mathew, K.A. and Shyam, K.R. 2005. Influence of inoculum density and environmental factors on angular leaf spot of French bean. *Pl. Dis. Res*. 20(1): 25-28.

Guzman, P., Gilbertson, R.L., Nodari, R.l, Johnson, W.C., Temple, S.R.; Mendala, D., Mkandawire, A.B.C. and Gupta, P. 1995. Characterization of variability in the fungus *Phaeoisariopsis griseola* suggests coevolution with the common bean (*Phaseolus vulgaris*). *Phytopathology* 85(5):600-607.

Hendrickx, F.L. 1939. Phytopathological observation at the station of Mulungu in 1938. *Publ. Inst. Nat. Etud. Agron. Congo. Belge,* pp. 117-128.

Hidalgo, R. and Araya, C.M. 1993. Optimum growth stage of common bean for chemical control of anthracnose (*Colletotrichum lindemuthianum*) and angular leaf spot *(Isariopsis griseola*) in San Carlos, Costa Rica. *Agronomic Costarricense* 17(1):75-80.

Hill, C.F. 1982. New plant disease records in New Zealand, 1974-1982. *New Zealand J. Bot.* 20(4): 355-359.

Hocking, D. 1967. A new virulent form of *Phaeoisariopsis griseola* causing circular leaf spot of French bean. *Pl. Dis. Reptr.* 51: 276-278.

Hopkins, J.C.F. 1940. Annual Report of the branch of Plant Pathology for the year ending 31st December, 8 p.

Issa, E., Sinigaglia, C. and Oliveira, D.A. 1982. Chemical control of angular leaf spot of bean (*Phaseolus vulgaris* L.) due to *Isariopsis griseola* Sacc. *Biologico* 48(12): 299-303.

Issa, E., Sinigaglia, C. and Oliveira, D.A. 1983. Chemical control of anthracnose (*Colletotrichum lindemuthianum* (Sacc. &Magn.) Scrib.) and two other diseases of bean (*Phaseolus vulgaris* L.). *Biologico* 49(4): 89-95.

Jaczewski, A.A. 1912. Russia Biuro Po MikalogiiiFetopatologiiEzhegodnikAbiedienii. *Ann. Rpt. Dis. and Enemies of Useful Plants* 6:157-159.

Jesus-Junior, W.C.-de, Vale, F.X.R., Coelho, R.R., Hau, B., Zambolim, I. and Berger, R.D. 2004. Management of angular leaf spot in common bean (*Phaseolus vulgaris* L.) with molybdenum and fungicide. *Agron. J.* 96: 665-670.

Joshi Deeksha., Hooda, K.S., Bhatt, J.C., Mina, B.L. and Gupta, H.S. 2009. Suppressive effects of composts on soil-borne and foliar diseases of French bean in the field in the western Indian Himalayas. *Crop Protection* 28 (7): 608-615.

Karanja, T.W., Mwang'ombe, A.W. and Mibey, K. 1994. The effect of media and light regimes on cultural and morphological characteristics and sporulation of *Phaeoisariopsis griseola* Deighton. *East African Agri. For. J.* 59(3):241-251.

Keissler, K. 1923. Revision of some of Fautrey's fungi. *Ann. Mycol.* 21(1-2):70-83.

Kijana, R., Abang, M., Edema, R., Mukankusi, C. and Buruchara, R. 2017. Prevalence of angular leaf spot disease and sources of resistance in common bean in eastern Democratic Republic of Congo. *African Crop Science Journal* 25(1): 109 - 122.

Kikoka, L.P., Katunzi, A.L. and Teri, J.M. 1989. Evaluation of effect of cropping systems on bean diseases and yield. In *Proc. Int. Pest Mgt. in Trop. and Subtrop. Crop. Syst.* 89(3):853-858.

Kovachevsky, I.C. 1936. Parasitic fungi new for Bulgaria (Fourth contribution). *Trav. Soc. Bulg. Sci. Nat.* 17:13-24.

Landeras, E., Trapiello, E., Braña, M. and González, A. J. 2017. Occurrence of angular leaf spot caused by *Pseudocercospora griseola* in *Phaseolus vulgaris* in Asturias, Spain. *Spanish Journal of Agricultural Research* 15 (3): e10SC03.

Lecerda, J.T., Coetho, R.S.B., Mariano, R.L.R. and Menezes, M. 1994. Pathogenic variability of *Isariopsisgriseola* on bean in the state of Pernambuco. *Summa Phytopath.* 20:93-96.

Litzenberger, S.C. and Stevenson, J.A. 1957. A preliminary list of Nicaraguan plant diseases. *Pl. Dis. Reptr. Suppl.* 243:19 p.

Lopez, S.E., Acosta, G.J.A., Becerra, L.E.N., Frayre, V.G., Orozco, S.H. and Beebe, S.E. 1997. Registration of Negro Tacana common bean. *Crop. Sci.* 37:1022.

Mahuku, G.S., Jara, C., Cajiao, C. and Beebe, S. 2003. Sources of resistance to angular leaf spot (*Phaeoisariopsis griseola*) in common bean core collection, wild *Phaseolus vulgaris* and secondary gene pool. *Euphytica* 130: 303-313.

Mahuku, G.S., Jara, C., Cuasquer, J.B. and Castellanos, G. 2002. Genetic variability within *Phaeoisariopsis griseola* from Central America and its implications for resistance breeding of common bean. Plant Pathology

Mathew, K. A., Gupta, S.K. and Shyam, K.R. 1998. New strategies in Fungicidal management of angular leaf spot (*Phaeoisariopsis griseola*) of French bean. *J. Mycol Pl. Pathol.* 28(2): 123-133.

McDonald, J. 1926. Report of the Mycologist, Kenya Colony. *Dept. Agr. Ann. Rpt.*, 1925, 141-148.

Milatovic, I. 1958. Control experiments against *I. griseola* on beans. *ZashtBilja* 45: 39-42.

Miles, E. 1917. Some diseases of economic plants in Puerto Rico. *Phytopathology* 7: 345-348.

Mohan, N., Aghora, T.S. and Somkuwar, R.G. 1995. Reaction of French bean germplasm to angular leaf spot (*Isariopsis griseola* Sacc.). *Legume Res*. 18:215-216.

Moreno, R.A. 1977. Efecto de diferentes sistemas de cultivo sobre la severidad de la mancha angular del frijol (*Phaseolus vulgaris* L.) caused for *Isariopsis griseola* Sacc. *Acta Agron. Palmira* 7:164-169.

Morwood, R.B. 1955. A preliminary list of plant diseases in Fiji. *Rept. Dept. Agric.* Fiji, 9 p.

Muller, A.S. 1934. Brazil : Preliminary list of diseases of plants in the state of Minas Gerias. *Inst. Agr. Internatl. Bull. Pl. Prot*. 8:193-198.

Navarro, A.R., Puerta-E, O.D. and Isaza, L, 1981. Combined effect of population density and chemical control of leaf diseases and yield of bean. *FitopatologiaColombiana* 10(1/2): 7-9.

Nietsche, S., Borem, A., Alzate Marin, A.L., Costa, M., Rocha, R.C., Caixeta, E.T., Barros, E.G. de and Moreira, M.A. 2002. Genetic variavbility of pathogenicity of *Phaeoisariopsis griseola* in Brazil. *Summa Phytopathol*. 28: 331-335.

Noble, R.J. 1929. Plant diseases observed in New South Wales. *Int. Bull. Plant Prot.* 3:6-7.

Olave, L.C.A. 1958. Resistance of several bean varieties and lines to *I. griseola*. *Acta Agron. Palmira* 8: 197-219.

Oliveira, S.H.F. de and de Oliveira, S.H.F. 2003. New fungicides and spraying schedules in the control of anthracnose and angular leaf spot of common bean. *Summa Phytopathol.* 29: 45-48.

Oliveira, S.H.F., Barros, B.C. and Castro, J.L. 1992. Effect of fungicides on the control of bean foliar diseases and seed quality. *Summa Phytopathol*. 18(2):178-184.

Orejuela, C.G. 1953. Plant pathology in Colombia *F.A.O. Pl. Prot. Bull*. 1(7):97-100.

Palm, B. 1932. Plant Diseases from Guatemala. *Zeitschr. fur Pflanzenkrankh*. u. *Pflanzenschutz* 42(1):11-17.

Paula, T.J. Jr, Pinto, C.M.F., Silva, M.B. da, Nietsche, S., Carvalho, G.A. de and Falerio, F.G. 1998. Resistance of snap bean cultivars to anthracnose, angular leaf spot and rust. *Revista Ceres* 45(258):171-181.

Petri, L. 1930. Review of phytopathological records in 1930. *Boll. R. Staz. di Pat. Veg. N.S.* 11(1): 1-50.

Picinini, E.C. and Fernandes, J.M. 2000. Chemical control of angular leaf spot and anthracnose of beans. *FitopatologiaBrasileira* 25: 92-94.

Ploper, L.D. 1980. Angular spot of bean (*Phaseolus* spp.) in Tucuman province. *Revista Industrial Y Agricola de Tucuman* 57(1): 119-124.

Ponnappa, K.M. Hiremath, P.C. and Sulladmath, V.V. 1976. Occurrence of *Phaeoisariopsisgriseola* (Sacc.) Ferr. on French bean. *Curr. Sci.* 45: 836-837.

Pyndji, M.M. and Trutmann, P. 1992. Managing angular leaf spot of common bean in Africa by supplementing farmer mixtures with resistant varieties. *Plant Dis*. 76 (11):1144-1147.

Rajappan, K. and Yesuraja, I. 1999. Management of angular leaf spot of French bean through fungicides. *Indian Phytopathology* 52(2): 177-178.

Rava, C.A. 2002. Efficiency of fungicides to control anthracnose and angular leaf spot in common beans. *Summa Phytopathol.* 28: 65-69.

Rodrigues, C.H., Zambolim, L. and Martins, H.C.P. 1987. Effectiveness of fungicides in the control of angular leaf spot (*Isariopsis griseola*) of bean (*Phaseolus vulgaris*). *Fitopatologia Brasileira* 12(1):40-45.

Ryan, E.W. 1965. New or uncommon plant diseases and pests. *Plant Pathol.* 14: 141-142.

Ryass, T. 1943. Material for the study of the deuteromycetes of Palestine. *Palest. J. Bot. J.* 3(1):22-51.

Saccardo, P.A. 1878. Fungi veneti novi criticivel Mycologiae venetae addendi *Isariopsis griseola* Sacc. *Michelia* 1: 273 p.

Saccardo, P.A. 1886. *Isariopsis laxa* (Ellis) Sacc. *Sylloge Fungorum* 4: 435 p.

Saettler, A.W. and Corea, F.J. 1988. Transmission of *Phaeoisariopsis griseola* by bean seed. *J. Seed Tech.* 12(2): 133-142.

Santos Filho, H.P., Ferraz S, and Sediyama, C.S. 1976. Isolation and sporulation *in vitro* of *Isariopsis griseola* Sacc. *Experimentiae* **22(7):** 175-193.

Santos Filho, H.P., Ferraz S. and Sediyama, C.S. 1978. Effect of the time of inoculation of *Isariopsis griseola* Sacc. on three bean cultivars. *Fitopatologia Brasileira* 3(2): 175-180.

Sartorato, A. 1988. Angular leaf spot. *Piracicaba Brazil, Asasociacao Brasileira para Pesquisa de Potassa e do Fosfato*, pp. 491-501.

Sartorato, A. 2002. Identification of *Phaeoisariopsis griseola* pathotypes from five States in Brazil. *Fitopatologia-Brasileira* 27: 78-81.

Sartorato, A. and Rava, C.A. 2003. Chemical control of angular leaf spot of common bean. *Summa Phytopathol.* 29: 202-204.

Schoevers, T.A.C. 1920. Nieuwe Ziekten, Waarop Gelet Moet Worden. *Tijdschr. over Plantenziekten* 26:208-211.

Sharma, A. K. and Bisht, K. K. S. 1997. Assessment of losses in seed yield of French bean due to angular leaf spot disease and its management in the Uttrakhand region of U.P. *Vegetable-Science*; 24(2): 150-152

Sharma, R.D. and Sohi, H.S. 1978. Evidence of a new strain of *Phaeoisariopsis griseola* Sacc. from India. *Curr. Sci.* 47:348-349.

Shukla, Arti and Sharma, H R. 2009.Fungicidal management of angular leaf spot and rust of bean (*Phaseolus vulgaris*). *Journal of Plant Disease Science* **4**(2): 222-223.

Shukla, Arti and Sharma, H R. 2016.Evaluation of fungicides in bush bean cultivars against major diseases. *Journal of Plant Disease Science* **11**(2): 207-211.

Simmonds, R. 1952. Science branch. *Rept. Dept. Qd. 1950-51*, pp.46-49.

Sindhan, G.S. 1984. Evaluation of fungicides against occurrence of angular leaf spot of French bean caused by *Phaeoisariopsis griseola. Indian J. Pl. Pathol.* 2(1): 39-42.

Sindhan, G.S. and Bose, S.K. 1980. Prepetuation of *Phaeoisariopsis griseola* causing angular leaf spot of French beans. *Indian Phytopath.* 32: 252-254.

Singh, A.K. and Saini, S.S. 1980. Inheritance of resistance to angular leaf spot (*Isariopsis griseola Sacc.*) in French bean (*Phaseolus vulgaris* L.). *Euphytica* 29:175-176.

Singh, B.M. and Sharma, Y.R. 1975. Screening of bean lines for resistance to angular leaf spot caused by *Isariopsis griseola. Indian Phytopath.* 28: 435-436.

Singh, B.M. and Sharma, Y.R. 1976. Screening of fungicides to control angular and floury leaf spots of bean. *Indian J. Mycol. Pl. Pathol.* 6: 148-151.

Singh, R.Y., Tripathi, S.P. and Yadav, G.R. 1995. Evaluation of different fungicides against various leaf spot diseases of beans. *Recent Hort.* 2:111-112.

Skoric, V. 1930. Diseases of economic plants in the University farm 'Maksimir'. *Yugoslavia Zagreb Mem. Cons. Faculty of Agriculture et For. University of Roy.*, 1929,pp. 738-746.

Sohi, H.S. and Sharma, R.D. 1969a. Studies on chemical control of angular leaf spot of beans. *Indian Phytopath.* 22: 157-158.

Sohi, H.S. and Sharma, R.D. 1969b. Studies on chemical control of angular leaf spot of beans caused by *Isariopsis griseola*Saccardo. *Bull. Indian Phytopath. Soc.* 5:40-45.

Sohi, H.S. and Sharma, R.D. 1974. Mode of survival of *Isariopsis griseola* Sacc. the causal agent of angular leaf spot of bean. *Indian J. Hortic.* 31: 110-113.

Sohi, H.S., Sharma, S.L. and Verma, B.R. 1963. *Annual Report of the section of Plant Pathology, Himachal Pradesh for the year 1962-63.* 30 p.

Spegazzani, C. 1882. Fungi Argentini. Soc. Cient. *Argentina An.* 13:11-35.

Srinivasan, K.V. 1953. *Isariopsis griseola* Sacc. on *Phaseolus vulgaris* L. *Curr. Sci.* 22: 20.

Srivastava, L.S. and Gupta, D.K. 1994. Management of angular leaf spot of 'rajma' through fungicides in Sikkim. *Indian Phytopath.* 47(2):211-213.

Srivastava, L.S., Gupta, D.K., Dhiman, K.R. and Singh, G. 1995. Source of resistance in French bean (*Phaseolus vulgaris*) to angular leaf spot (*Phaeoisariopsis griseola*) in Sikkim. *Indian J. Agril. Sci.* 65(4):305-307.

Thorold, C.A. 1932. A further preliminary list of Trinidad fungi. *Imp. Col. Trop. Agr. Mycol. ser. Mem* 5 (9):30.

Trutmann, P. and Kayitare, E. 1991. Disease control and small multiplication plots improve seed quality and small farm dry bean yields in Central Africa. *J. Appl. Seed Prod.* 9: 36- 40.

Trutmann, P. and Pyndji, M.M. 1994. Partial replacement of local common bean mixtures by high yielding angular leaf spot resistant varieties to conserve local genetic diversity while increasing yield. *Ann. Appl. Biol.* 125(1):45-52.

Trutmann, P., Paul, K.B. and Cishabayo, D. 1992. Seed treatments increase yield of farmer varietal field bean mixtures in the Central African highlands through multiple disease and bean fly control. *Crop Prot.* 11(5):458-464.

Van Rheenen, H.A., Hasselbach, O.E. and Muigai, S.G.S. 1981. The effect of growing beans together with maize on the incidence of bean diseases and pests. *Neth. J. Pl. Path.* 87:193-199.

Verma, B.R. 1979. Studies on angular leaf spot of beans (*Phaseolus vulgaris*L.) caused by *Isariopsis griseola* Sacc. Ph.D Thesis, HPKVV, College of Agriculture, Solan. 221 p.

Vierra, R.F., Paula, T.J. Jr. and Berger, P.G. 1998. Effect of fungicides and cartap insecticides on the control of bean foliar diseases during the winter-spring season. *Summa Phytopath.* 24(1):17-22.

Villegas, J.M. 1959. Variabilidad del *Isariopsis griseola* Sacc. agente caused de la mancha angular del frijol (*Phaseolus vulgaris* L.). Thesis Ing. Agr., Univesidad de Cldar, Mcnizales, Colombia 61 p.

Wallace, G.B. 1932. III. Preliminary list of fungi or diseases of economic plants in Tanganyika territory. *Kew Roy. Bot. Gard. Bul. Misc. Inform,* pp. 28-40.

Wallace, G.B. 1951. Diseases of French beans or dwarf beans. *Pamph. Dept. Agric., Tanganyika* 51: 11.

Weevers, L.O. and Zaumeyer, W.J. 1956. Angular leaf spot of bean found in Maryland. *Pl. Dis. Reptr.* 40: 1092.

Whitmore BB and Naidu AS 2000. Thiosulphate. In: *Natural food antimicrobial system* (AS Naidu ed), Boca Roton, FL, CRC Press, pp 265-380.

Williams, D.Y. and Alfonso, C.M. 1955. Bean diseases of importance in Mexico in 1955. *USDA Plant Reptr*. 40: 222-223

Zaumeyer, W.J. and Thomas, H.R. 1957. Monographic study of bean diseases and methods for their control. *USDA Tech. Bull.* 868:255 p.

7

PHYTOPHTHORA BLIGHT

This is an important disease of both French and lima beans and in favourable weather causes significant losses.

GEOGRAPHICAL DISTRIBUTION

After its first report in 1923, the disease has spread in many bean-growing areas infecting mainly the lima bean. The disease has been reported from New Jersey, Colorado, California, South Central and South Eastern Pennsylvania, Long Island and Delaware. *Phytophthora phaseoli* on bean has been reported from Belgian Congo (Hendrix, 1939). The same sp. was very severe throughout the New Jersey state. A severe outbreak was also observed in Italy (Casarini *et al.*, 1961). Venkatakrishnaiya (1950) reported the occurrence of *Phytophthora* spp. on French bean for the first time in India. In Himachal Pradesh, *P. nicotianae var. nicotianae* is prevalent in Kangra, Kullu, Mandi, Solan and Chamba districts, where the incidence varies from 26.09 to100 per cent on plant basis and 18.82 to 87 per cent on pod basis (Malik, 1992; Gupta *at al.*, 2020).

LOSSES

The disease causes losses both by reducing grain yield and rendering the green pods unfit for table purpose. The crop losses may range from 50 to 90 per cent undersuch favourable conditions. The first report of the disease was made by Adams (1924) from Delaware where late crop received up to 50 per cent pod infection. In Colorado, losses due to this disease on lima beans were 60 to 75 per cent (Harter and Zaumeyer, 1942). Five to 80 per cent infection has been reported by Miller and O'Brien (1949) under field conditions. New Jersey was the chief sufferer. A severe attack of Phytophthora was noticed on bean cultivars Contender, HPR-31, Palampur Sel. and Jawala (40-74%) during two consecutive years 1988 and 1989 in Palampur Valley of HP (Malik, 1992).

SYMPTOMS

On lima beans, the downy mildew fungus (*Phytophthora phaseoli*) occurs on the pods either in patches or covering them completely. The fungus grows through the pod wall into the grains resulting in drying and ultimate blackening of the pods. On the leaves, white mycelial growth is sparse, but veins are often twisted, purplish or otherwise distorted. Symptoms are also visible on young flowers and shoots.

On French bean, symptoms of the diseases usually remain confined to pods only and all other aerial parts remain free from infection. However, rotting of collar portion of French bean plant has also been reported in protected cultivation (Gupta *et al.*, 2020). The rotting of the pods start from the distal end with conspicuous white cottony growth of the fungus on the affected portion. Usually one third to two third portion of the pods is rotten and the remaining portion remains healthy. The young green diseased pods show twisting and curling. In advanced stages of infection, the pods finally become dry and remained hanging to the plant.

On collar part, dull white to light ash grey coloured thick mycelial growth adhered all over the collar portion just below the stem and above the roots is observed (Plate-1). At later stage decaying of lower stem is also observed showing girdling by the mycelial growth. The plants show wilting, drooping and stunted symptoms and ultimately are killed.

THE PATHOGEN

The disease is caused by several species of *Phytophthora*. Initially the disease was reported to be caused by *P. terrestris*Sherb. (Sherbakhoff, 1917). Later it was reported to be caused by *P. parasitica*Dast. (Venkatakrishnaiya, 1950; Zaumeyer and Thomas, 1957), *P. irritabilis* (Mantari and Deshpande, 1966), *P. Phaseoli* Thaxter (Crispin *et al.*, 1976) and *P. nicotianae* pv. *nicotianae* (Malik, 1992). *P. nicotianae* var. *parasitica, P. dechsleri* and *P. capsici* and *Phytophthora phaseoli* Thaxter have been reported on French bean and Lima bean (Ribeiro, 1978, Erwin *et al.*, 1983; Davidson *et al.*, 2002). The important species of *Phytophthora* causing infection of French bean and lima beans could be distinguished on morphological characters (Table 1). In this chapter, only *Phytophthora* blight caused by *P. nicotianae* pv. *nicotianae* is described.

Phytophthora nicotianae var. *nicotianae* consisted of cottony white, hyaline, coenocytic and branched mycelium. The average hyphal width was 4.40 μm (2.30-6.90 μm). The mycelial swelling in culture was quite common. The branching of sporangiophores is sympodial (Plate 2). The fungus produced spherical to ovoid sporangia. The spherical sporangia measured 40.30 μm (28.50-50.12 μm) and ovoid sporangia 52.43 × 36.40 μm (32.45-67.85 × 26.55-44.25 μm).

Plate 1. Symptoms of collar rot (*Phytophthora nicotianae* var. *nicotianae*) on French bean plants

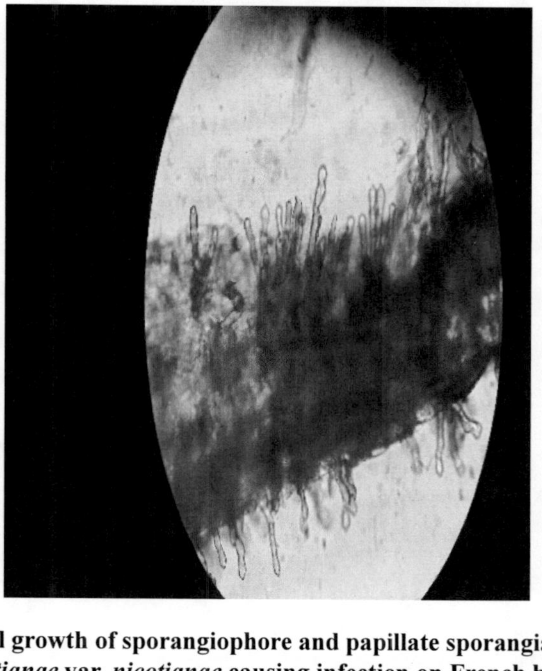

Plate 2. Sympodial growth of sporangiophore and papillate sporangia of *Phytophthora nicotianae* var. *nicotianae* causing infection on French bean

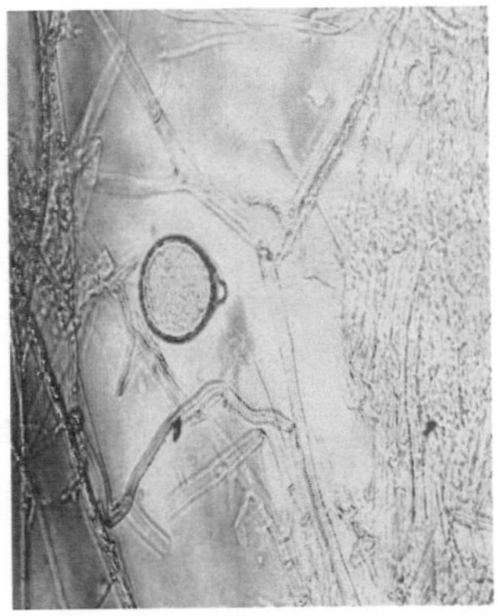

Plate 3. Sporangium of *Phytophthora nicotianae* var. *nicotianae*

The hlength: breadth (L : B) ratio of ovoid sporangia was 1.44 : 1. Sporangia possessed very prominent hemispherical papillae measuring 4.12 µm (3.46-5.38 µm) at apex (Plate 2). They did not shed easily in water. The sporangia germinated both by means of zoospores and directly by means of germ tube formation. Oospores were abundantly formed in culture, were golden brown in colour, thick walled and measured 30.50 µm (21.60-40 u) in diameter. Anthredia position was amphigynous. Terminal chlamydospores were produced in culture and measured 32 µm (16.50-43.50 µm).

Table 1. Comparative morphological characters of *P. nicotianae* var. *parasitica*, *P. dechsleri* and *P. capsici*

Morphological characteristics	*P. nicotianae* var. *parasitica*	*P. dechsleri*	*P. capsici*
Hyphae	Light brown to blackish brown, Arachnoid branched with several hyphal swellings not borne in clusters	Light grey to grey, branched with irregular to rounded, clustered or in chains hyphal swellings	Hyaline, branched
Sporangiophores	Sympodial branched with several nodular swellings	Sporangiophores are usually narrow but sometimes widen just below the sporangium and form a sympodium.	Short simple hyphal branches

Sporangia	Sporangia, hyaline to light grey ellipsoidal to pyriform with papilla (20-48x11-60 μm (average 15.5x51.18 μm) with long pedicel (average 392 μm), Non-caducous	Sporangia vary in shape nonpedicillate, obpyriform-ovoid to elongate sporangia, nonpapillate. Non-caducous. Size in general 36 to 70x 26 to 40μm (average 39x24 μm). Sporangia have internal and external proliferations,	Sporangia rarely produced, hyaline, ovoid to pyriform or sometimes round to lemon shaped, 29.7-59.8x33.0-72.5μm (average 38.4x45.5 μm), nonpedicillate with a predominant, hemispherical papilla at the apex

(Source: Gupta *et al.,* 2020)

VARIABILITY

Literature on this aspect is too meagre. Wester and Jorgensen (1959) isolated a new race B from pods of var. U.S. 355 lima beans in New Jersey. This race was virulent to both Thaxter and U.S. 355 vars. of lima beans. Isolates from various greenhouse plants varied in pathogenicity with respect to tobacco and Kalanchoe leaves, but all attacked bean (*Phaseolus vulgaris*) fruits (Siradhana *et al.*, 1968). Thompson and Blount (1976) identified a new race D, which was isolated from infected leaves of lima bean (*P. lunatus*) cvs. In experimental field plots at Beltsville, Md. This race is distinguished from races A, B and C by its reaction with PI 195342.

HOST RANGE

P. nicotianae var. nicotianae has been reported to infect pod/fruit of bell pepper, tomato, brinjal, French bean (cv. Jawala, VL Bonni, Contender, Master Piece, Pusa Parvati), broad bean and red pepper (*Capsicum frutescens*) by Malik (1992).

DISEASE CYCLE

The pathogen being soil borne and seed borne in nature persists in infected pods, seed or other plant debris. The pathogen over winters in the form of oospores. Sporangia have also been detected in seed samples of Contender, Pedi Local and HPR 31 vars. of French bean in Palampur and Bajaura areas of H.P. Under favourable conditions, oospores germinate to form mycelium or the sporangium. The sporangium may germinate directly or indirectly through zoospores, depending upon the temperature. Free water helps in the spread of the infection. At an optimum temperature of 20°C, symptoms appear after 4 days of infection. Sporangia on the host tissue constitute the source of secondary infection. Under unfavourable conditions, oogonium and anthredia are produced from the mycelium. After plasamogamy, karyogamy and meiosis, oospores are formed, which are hard and resistant structures, serving as the source of primary infection.

EPIDEMIOLOGY

The pathogen survives in the form of oospores in soil as well as on infected plant debris. It was demonstrated by Critopolous (1955) that the pathogen could survive for five months in moist soil in the absence of the host. In tests with lima bean seedlings, inoculation for four days at 100% RH and 65-70°F, and then at 50% RH for 24 h showed that the fungus remained viable for over 100 days with no loss of pathogenicity (Wester *et al.*, 1958). For rapid development of symptoms, the optimum temperature is 25°C at which only 2.5 days are required (Cox, 1954). Free water is essential for sporangial germination. The rate of zoospore germination and germ tube elongation increased with temperature up to 25°C and then declined (Hyre and Cox, 1953). In *P. nicotianae var. nicotianae*, maximum radial growth was observed at 30°C with excellent sporangial formation on lima bean agar. Optimum pH was 6.0 Colonization decreased as the temperature and or number of exposures was increased, but it didn't cease. With 1-7 exposures to 35°C, the lesions were significantly shorter than were the check lesions. In other tests, colonization was significantly less after 6 h daily exposure to 33°C than after 4 h daily exposures. The mean length of stem lesions was 20 and 34 mm, respectively; when 6 exposures were made (Malik, 1992). Temperatures of 24-28°C were unfavourable for oospore development because of rapid development of bacteria, which destroyed the downy mildew fungus within 10 days or less. The findings suggest that bacteria may prevent oospore development above 21°C through early destruction of the downy mildew fungus. At lower temperatures, where bacteria are absent or inactive, oospores develop below the epidermis of lima bean seedlings (Drechsler and Wester, 1962).

FORECASTING

Forecasting methods developed for any disease are helpful not only in planning various spray programmes but also reduce the indiscriminate use of fungicide for management of a disease. The Horticultural Crops Research Branch, Delaware developed a method for forecasting downy mildew outbreaks on lima bean from temperature and rainfall data. A day is considered favourable for infection by *P. phaseoli* when a day moving mean temperature, ending on that day is $<26°C$, with a minimum of 7°C or above, when there has been a 10 day total rainfall, also ending on that day of 1.2 in. or more. Initial disease appearance is expected after about 8 consecutive favourable days, if there is a prospect of favourable weather continuing (Hyre, 1957).

For this disease, another method of forecasting given by Scarpa and Raniere (1964) involves prevalence of a persistent mid-tropospheric air mass resulting in accumulation of 40 or more hrs of dew point. After 1962, successful warnings were issued when dew point was 20.5°C or more for 12 consecutive hrs.

MANAGEMENT

Cultural practices

Use of recommended cultural practices is of immense value these days owing to pollution and health hazards resulting from deposits of chemical sprays. Cultural practices like use of seed produced in disease free areas, plan atleast 2-3-year crop rotation in affected areas, collect and burn infected plant debris and maintain proper drainage of the field.

Host resistance

Disease resistant varieties if available in a crop serve as the most reliable tool in the management of a disease. In this direction, not much work has been done in French bean, however, there are some valuable references relating to lima bean. Four lines of lima bean showing a high degree of resistance (PEJ Nos. 164155 and 163580, 8747 and 4931) were identified (Thomas *et al.,* 1952). In the field, sporulation on resistant lines was scanty. In another study, Valenta and Sester (1962) reported vars. Thaxter and Piloy as resistant to race A of P. phaseoli. The resistance is associated with a phenolic acid in the sap, which is toxic to zoospore germination. Hybrids G_1 and G_2 of *Phaseolus lunatus* proved immune to strains A and B of *P. phaseoli* (Wester, 1968). Wester *et al.* (1972) found accessions having multiple resistance to strains A, B and C of *P. phaseoli*. On the other hand, Thomas and Blount (1976) reported breeding line CRB2BC of *P. lunatus* as the only line resistant to all the 4 races of *P. phaseoli* (strains A, B, C and D).

Paul and Sood (1990) observed that eight lines of French bean viz., HPR 27, HPR 84, HPR 94, HPR101, HPR 140, HPR 222 and HPR 270 were free of *Phytophthora* blight under Himachal Pradesh conditions.

Chemical control

Clayton (1928) obtained best control of downy mildew of lima bean with Bordeaux mixture (4:4: 50) under severe attacks of the disease. Two pre-bloom sprays at 7-10 days intervals followed by 3-6 sprays at thick setting give good results. Under wet conditions, spraying should be continued up to one week of picking. Use of copper sprays and copper dusts have also been advocated (Anonymous, 1943). Zaumeyer and Wester (1965) reported effective control of downy mildew by streptomycin (100 ppm). Similarly, Zaumeyer (1957) obtained satisfactory control of this disease with antibiotic anisomycin. Crossan *et al.* (1957) on the other hand observed excellent control by two applications of manzate (2 lb/100 gal) within one month of harvest at 21 days interval. Three applications per week of a wettable dust containing 65 per cent zineb (0.5%) + an adhesive (0.2%), starting 21 days after seeding gave satisfactory control of *P. phaseoli* on Ibundu var. of French bean (Anonymous, 1959). Spray applications of combi fungicides like metalaxyl + mancozeb and cymoxanil + mancozeb can be effectively used for the management of this disease under field conditions. Recently available Strobilurin fungicides can also be used before harvesting being eco-friendly fungicides (Jensen, 1997; Gupta and Jarial, 2010). For the management of diseases caused by *Phytophthora* spp., foliar sprays of metalaxyl + mancozeb, cymoxanil + mancozeb were found very effective. Four sprays of metalaxyl + mancozeb (0.25%) at fortnightly intervals were found highly effective to manage the Phytophthora leaf blight and fruit rot of bell pepper, caused by *P. nicotianae var. nicotianae,* followed by copper oxychloride and cymoxanil + mancozeb (Gupta and Jarial, 2010). Foliar spray of fungicide fenamidone + mancozeb, famoxadone + cymoxanil, metalaxyl-M + mancozeb and metiram + pyraclostrobin were found effective to manage the buckeye rot disease of tomato caused by *P. nicotianae var. parasitica* under field conditions (Shridhar *et al.,* 2018).

REFERENCES

Anonymous. 1943. Divisions of plant pathology and seed investigation. *Rep. N.Y. st. agric. Exp. Stn*. Pp. 52-60.

Anonymous. 1959. Annual report for the year 1958. *Pub. Inst. Nat. agron. Congo belge* 529 p.

Casarini, B., Quaglia, A. and Silvestrti, G. 1961. *La 'peronospora' del Fagidodilima.* (Peronospora of lima bean) *medit.* 1: 47-48.

Chandrani, A.N. 1989. Studies on bio deterioration of French bean (*Phaseolus vulgaris* L.) seeds in Himachal Pradesh. M.Sc. Thesis. Submitted to HPKVV, Palampur (H.P.), India.

Clayton, E.E. 1928. Spraying experiments with bush lima beans. *New York (Geneva) Agric. Exp. Stn. Bull.* 558: 22 p.

Cox, R.S. 1954. Effect of temperature on the development of downy mildew of lima bean. *Phytopathology* 44(6): 325-327.

Crossan, D.F., Lloyd, P.J., Hyre, R.A. and Henberger, J.W. 1957. Control of downy mildew of lima beans. *Pl. Dis. Reptr.* 41(3): 156-159.

Erwin, D.C., Burtnicki-Garcia, S. and Tsao, R.H. 1983. *Phytophthora its biology, taxonomy, ecology and pathology*. APS Press, The American Phytopathological Society, St. Paul, Minnesota, 55121 USA, 391pp.

Harter, L.L. and Zaumeyer, W.J. 1942.Downy mildew of lima bean in Colorado. *Phytopathology* 32(5): 438.

Drechsler, C. and Wester, R.E.1962. Effect of temperature on development of oospores of *Phytophthora phaseoli*Thaxt. on lima bean seedlings. *Phytopathology* 52: 164.

Gupta, S. K., Gupta, A.K. and Spehai, R. S. 2020.Occurrence of Collar rot in French bean grown under protected cultivation conditions. *Pl. Dis. Res*. 35 (2): 184-185.

Gupta, S.K. and Kumud, J. 2010. Comparative efficacy of two strobilurins in management of leaf blight and fruit rot (*Phytophthora nicotianae* var. *nicotianae*) of bell pepper, *Indian Phytopathology* 63(2):216-218.

Hendrix, F.L. 1939. Phytopathological observations at the Station of Mulungu in 1938. *Publ. Inst. Nat. Etud. Agron. Congo Belge* pp. 117-128.

Hine, R.B. and Aragaki, M. 1963. Pathgenicity, vitamin nutrition and cultural characteristics of *Phytophthora parasitica* from carnation, and hosts in Hawaii. *Phytopathology* 53: 1194-1197.

Hyre, R.A. 1957. Forecasting downy mildew of lima bean. *Pl. Dis. Reptr.* 41: 7-9.

Hyre, R.A. and Cox, H.S. 1953. Factors affecting viability and growth of *Phytophthora phaseoli*. *Phytopathology* 43(8): 419-425.

Jensen, H. 1997. Amistar a broad spectrum fungicide. *14th Danish Plant Protection Conference on Pests and Diseases* 8: 59-71.

Malik, P.K. 1992. Studies on Phytophthorta blight of French bean (*Phaseolus vulgaris* L.). M.Sc. Thesis, Deptt. Pl. Pathol., HPKVV, Palampur. pp. 72 + viii + III.

Mantari, J.M. and Deshpande, K.B. 1966. Two new species of *Phytophthora* from India. *Sydowia* 20: 28-32.

Miller, P.R. and O'Brien Moreil. 1949. The plant disease warning service in 1949. *Pl. Dis. Reptr. Suppl.* 188: 297-314.

Paul, Y.S. and Sood, A.K. 1990. Occurrence of Phytophthora blight of French bean (*Phaseolus vulgaris* L.) in India. *Indian J. Mycol. Pl. Pathol.* 20: 81-82.

Ribeiro, O.K. 1978. *A source book of the genus Phytophthora* pp. 14, Straws and Cramer Gmbh, D-6945 Hirschberg 11, ISBN 3-7682-1200-9, Germany, 417pp.

Scarpa, M.J. and Raniere, L.C. 1964. The use of consecutive hourly dew points in forecasting downy mildew of lima beans. *Pl. Dis. Reptr.* 48(2): 77-81.

Sharma, S.L., Chowfla, S.C. and Sohi, H.S. 1977. Chemical control of buckeye rot of tomato. *Indian J. Mycol. Pl. Pathol.* 6: 130-134.

Siradhana, B.S., Ellett, CV.W. and Schmitthener, A.F. 1968. Pathogenic and cultural variation in *Phytophthora nicotianae var. parasitica* from greenhouse plants. *Phytopathology* 58(5): 718-719.

Shridhar, B.P., Sharma, M., Sharma, S. K. and Gupta, S. K. 2018. New generation fungicides for the management of buckeye rot of tomato. *Indian Phytopathology* 71: 621-625.

Thomas, C.A. and Blount, V.L. 1976. Race 'D' of *Phytophthora phaseoli*. *Pl. Dis. Reptr.* 60(4): 308.

Thomas, H.R., Jorgensen, H. and Wester, R.E. 1952. Resistance to downy mildew in lima beans, and its inheritance. *Phytopathology* 42(1): 43-45.

Valenta, J.R. and Sisler, H.D. 1962. Evidence for a chemical basis of resistance of lima bean plants to downy mildew. *Phytopathology* 52(10): 1030-1037.

Venktakrishnaiya, N.S. 1950. *Phytophthora parasitica* on French bean, *Phaseolus vulgaris* L. *Indian J. Agric. Sci.* 20: 391-394.

Waterhouse, G.M. 1963. Key to the species of *Phytophthora* de bary. *Commonwealth Mycol. Inst. Mycol. Pap.* 92: 22.

Wester, R.E. 1968. New lima beans resistant to 'B' strain of *Phytophthora phaseoli*. *Pl.Dis. Reptr.* 52(7): 563-5654.

Wester, R.E., Brechsler, C. and Jorgensen, H. 1958. Effect of freezing on viability of the lima bean downy mildew fungus (*Phytophthora phaseoli* Thaxt.). *Pl. Dis. Reptr.* 42(4): 413-415.

Wester, R.E., Fisher, V.J. and Blount, V.L. 1972. Multiple resistance in lima beans to downy mildew (*Phytophthora phaseoli*). *Pl.Dis. Reptr.* 56(1): 65-66.

Zaumeyer, W.J. 1957. Control of downy mildew of lima beans with two antibiotics. *Reprinted Antibiotics Annual* pp. 1015-1018.

Zaumeyer, W.J. and Wester, R.E. 1965. Control of downy mildew of lima beans with streptomycin. *Pl. Dis. Reptr.* 40(9): 776-780.

8

FLOURY LEAF SPOT

Floury leaf spot also known as Farinose spot is an important disease of beans in relatively high altitudes in the tropics where moderate temperatures along with plentiful moisture in the form of rain or dew is prevalent.

GEOGRAPHICAL DISTRIBUTION

The disease has a worldwide distribution particularly in areas experiencing moderate temperatures and high rainfall. The occurrence of this disease was first reported from Ecuador in 1950 (Petrak, 1950). Subsequently it was reported from Brazil, Colombia, Nicaragua, Zambia Uganda, Tanzania, Kenya, Ethiopia, North West India, Malaya, New Guinea, Venezuela, Honduras, Panama, Guatemala and the Dominican Republic (Petrak, 1950; Cardona and Skiles, 1958; Cardona and Renaud, 1962; Deighton, 1967; Wellman, 1977). Floury Leaf Spot is one of the most important diseases of bean in Ethiopia and occurred with high prevalence of 78.7% (Degu *et al.*, 2020).

In India, this disease appears in an epidemic form in the northwestern Himalayan region and its occurrence was first reported from Solan during the year 1964 (Sohi *et al.*, 1965). Subsequently, it was reported from Uttar Pradesh (Tripathi, 1985) and Sikkam (Srivastava and Gupta, 1989).

LOSSES

Very few reports on estimates of yield losses caused by this disease are available in literature. Srivastava and Gupta (1989) reported about 65% infection of lower leaves, which were defoliated prematurely. The premature defoliation phase of this disease is most damaging which reduces yield considerably.

Plate 1. Symptoms of Floury leaf spot of French bean on lower leaf

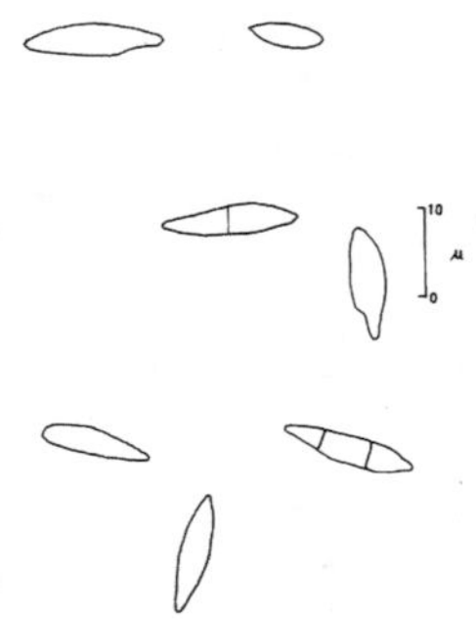

Plate 2a. Conidia of *Ramularia phaseoli*

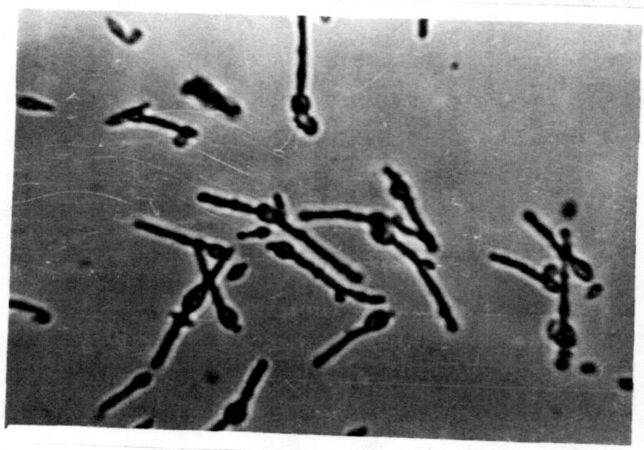

Plate 2b. Germinating conidia of *R. phaseoli*

(Courtesy: Dr. K.D. Sharma)

SYMPTOMS

The disease appears on lower older leaves as yellowish spots and progress upward. In severe infection, leaves show yellow mottling and in older infection, the spots become brown with yellow margins. These spots are generally 2-3 mm in diameter, but several spots merge together to affect most of the area. On the under surface, white fluffy growth of the fungus conidiophores can be seen which give to the leaf the appearance of having been sprinkled with coarse flour, hence the disease is named as floury leaf spot (Plate 1). Infected leaves turn brown and finally shed prematurely. Pre-mature defoliation causes heavy yield losses.

THE PATHOGEN

The fungus responsible for this disease is *Mycovellosiella phaseoli* (Drummond) Deighton (Syn.: *Ramularia phaseolina* Pertak). It is an obligate parasite. On the under surface of host, colorless conidiophores emerge in tufts from the stomata and interwine in clusters around the leaf hairs. Erect, more or less divergent, somewhat tapering towards the tip, simple or irregularly dichotomous, a few short, erect, single conidiophores were also found scattered over the leaf surface. Conidia are variable in shape and size mostly ellipsoid or ovoid, tips acute or obtuse, base papillate at the point of attachment, straight hyaline, commonly aseptate, rarely one septate and measure 5.24-21.4 x 3.5- 5.3 µm (Plate 2).

DISEASE CYCLE

Very little information is available in literature regarding the disease cycle. The pathogen cannot be cultured (Sharma and Kaushal, 1990) and no perfect stage is known. The fungus is not a seed borne and perpetuate in infected plant debris from one season to another (Sharma and Kaushal, 1990). Conidia loose germination with the passage of time and conidia after one year of dry storage germinate poorly but when supplemented with sugars (0.1%) and amino acids like DL- methionine, L- leucine, Valine and DL- threonine, good germination of one-year-old conidia was obtained (Sharma *et al*., 1998). This indicates the possibility that conidia can remain viable in crop debris. The initial occurrence of lesions on lower most leaves on the plants grown in disease prone areas indicates that soil or debris may act as source of primary inoculum. Overwintering conidia may be splashed by rain to lower leaves, which in the presence of free water germinate on host surface and cause infection. Conidia produced on primary spots are blown by wind and cause secondary infections.

EPIDEMIOLOGY

The fungus perpetuates on infected plant debris in the soil and serve as source of primary inoculum since the first lesion appear on lower most leaves of the plants (Sharma and Kaushal, 1990). The disease is more common in altitudes of 1500 to 2000 m with heavy rainfall (Chupp and Sherf, 1960). Sharma *et al*. (1998) studied the effect of temperature and different humidity levels on the germination of *M. phaseoli* and reported that conidia start germination within 6 h of incubation at 20 and 25°C but the optimum temperature for spore germination and germ tube elongation was found to be 20°C (Plate 2b; Fig. 1). None of the conidia germinated even after 48 h of incubation at 25°C at any level of relative humidity.

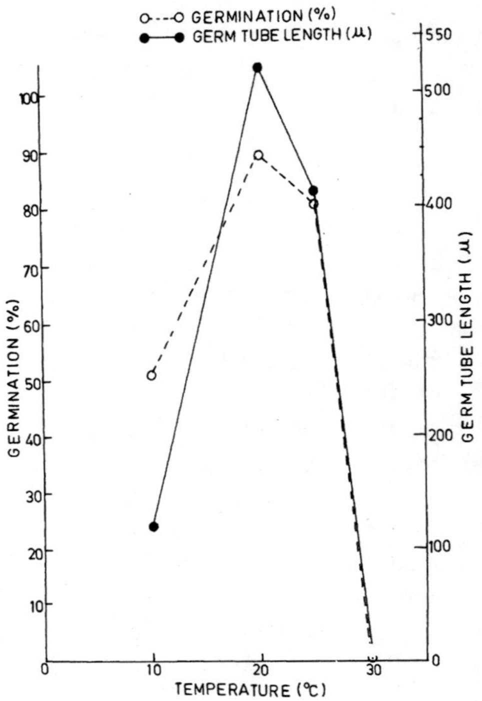

Fig. 1. Effect of temperature on the germination and germtube length of conidia of *Ramularia phaseoli* (incubated for 24 hr.)

(Courtesy: Dr. K.D. Sharma)

MANAGEMENT

Since the pathogen is obligate parasite and overwinters in infected plant debris, the disease can be managed by using cultural, chemical and resistant varieties.

Cultural practices

Cultural practices like collection and destruction of infected fallen leaves, two-year crop rotation with non-leguminous crops and wider plant spacings (Gupta and Mathew, 1999) along with weed free cultivation can be useful in reducing the disease.

Host resistance

French bean cultivars/ lines viz., PIB 326, EC 30020, Mantulan, 12-A, Sel.-P-2-3 and EC 10037 have been reported resistant (Singh *et al.,* 1974; Kapoor *et al.*, 1988) which can be successfully utilized in breeding programs for developing resistant varieties. Sugha *et al* (1998) reported that White Local developed minimum disease as compared to other genotypes and observed that all the genotypes irrespective of their being pole or bush types were equally susceptible to the disease which are contrary to the observations of Kapoor *et al* (1988) who reported pole types as resistant and bush types as susceptible to this disease. Kumar *et al.* (2005) reported two hybrids of *P. vulgaris* (contender) x *P. acutifolius* (G 40015 and G 40020), five of *P. vulgaris* x *P. coccineus* (Pc1 and Pc2) and five of *P. vulgaris* (Him1) x *P. coccineus* were resistant to floury leaf spot. Lemessa and Tesfaye (2005) evaluated 70 genotypes of French bean in Ethiopia and EMP-233, EMP-212, G-10843 and Dicta-65 were found to be resistant against floury leaf spot disease.

Chemical methods

Sprays of various fungicides like benomyl, carbendazim and cercobin are effective in controlling the disease (Singh and Sharma, 1976).

Integrated disease management

Two foliar sprays of *Pseudomonas fluorescens* strain RPB 14, 1st at seedling stage (10^7cfu ml^{-1}) and 2nd at flowering (10^6cfu ml^{-1}), followed by single spray of carbendazim (0.05%) at pod developmental stage was found effective in the management of this disease (Mondal, 2004).

REFERENCES

Cardona, C. and Renaud, J. 1962. La mancha harinosa, nueva enfermedad de la caraota en Venezuela. *Agron Trop. (Venezuela)* 17: 213-214.

Cardona-Alvarez, C. and Skiles, R.L. 1958. Floury leaf spot (mancha harinosa) of bean in Colombia. *Pl. Dis. Reptr.* 42: 778-780.

Chupp, C. and Sherf, A. 1960. *Vegetable diseases and their control.* John Wiley and Sons, Inc., New York, 693pp.

Degu, T., Yaregal, W. and Gudisa,T. 2020. Status of common bean (*Phaseolus vulgaris* L.) diseases in Metekel Zone, North West Ethiopia. *J Plant Pathol Microbiol* 11:494. doi: 10.35248/2157-7471.20.11.494.

Deighton, F.C. 1967. Floury leaf spot of French bean caused by *Ramularia phaseoli* (Drummond) Comb. nov. *Trans. British Mycol. Soc.* 50: 123-127.

Gupta, S.K. and Mathew, K.A. 1999. Fungal and bacterial diseases of beans. In: *Diseases of Horticultural crops- Vegetables, Ornamentals and Mushrooms* (Eds. L.R. Verma and R.C. Sharma). Indus Publishing Co., New Delhi pp. 301-335.

Kapoor, K. S., Gill, H.S. and Sharma, S.R. 1988. Screening of French bean against floury leaf spots. *Indian Phytopath. 41*: 650-651.

Kumar, A., Kaushal, R. P. and Sharma, R. 2005. Transfer of resistance to floury leaf spot and bacterial blight in French bean from alien germplasm. *Indian Phytopathology* 58(2): 136-142.

Lemessa F. and Tesfaye, A. 2005. Evaluation of bean (*Phaseolus vulgaris*) genotypes for multiple resistance to angular and floury leaf spot diseases. *Tropical Science* 45(2): 63-66.

Mondal, K.K. 2004. Integrated management strategies for fuscous blight and floury leaf spot of rajmash. *Indian Phytopath.* 57(2): 135-139.

Petrak, F. 1950. Beitrage zur Pilsflora von Ekuador. *Sydowia* 4: 450-587.

Sharma, K. D. and Kaushal, R. P. 1990. Studies on spore viability and perpetuation of *Mycovellosiella phaseoli* (Drummond) Deighton field. *Indian J. Plant Pathol.* 8(1 & 2): 28-32.

Sharma, K.D., Kaushal, R.P. and Singh, B.M. 1998. Effect of bioregulators on spore germination of *Mycovellosiella phaseoli* (Drummond) Deighton, a floury leaf spot pathogen of French bean. *Indian Phytopath.* 51(1): 43-46.

Singh, B. M. and Sharma, Y. R. 1976. Screening of fungicides to control angular and floury leaf spots of beans. *Indian J. Mycol. Pl. Pathol.* 6: 148-151.

Singh, B.M., Agarwal, V.D. and Sharma, Y.R. 1974. Screening of bean cultivars to floury leaf spot caused by *Ramularia phaseoli. Indian Phytopath.* 27: 260-261.

Sohi, H.S., Sharma, S.L., Verma, B.R. and Sachdeva, K.B. 1965. Floury leaf spot of bean caused by *Ramularia phaseolina* Pertak. *Indian Phytopath.* 18: 384.

Srivsastava, L.S. and Gupta, D.K. 1989. Farinose spot of French bean-a new disease. *Current Sci.* 58(14): 802.

Sugha, S. K., Sharma, O. P. and Sharma, P. N. 1998. Performance of 'vegetable type' French bean genotypes against floury leaf spot disease. *Pl. Dis. Res.* 13(1): 85-86.

Tripathi, K.C. 1985. Occurrence of floury leaf spot of French bean in Uttar Pradesh hills. *Indian Bot. Rep.* 4: 175.

Wellman, F. L. 1977. *Dictionary of Tropical American crops and their diseases*. Pp. 312-321. The Scarecrow Press Inc. Metuchen, New Jersey, 495p.

9

CHARCOAL ROT

The disease is also known as ashy stem blight, root rot, ashy grey stem, collar rot and Macrophomina rot.

GEOGRAPHICAL DISTRIBUTION

The disease has a worldwide distribution, being present in the areas where vegetable crops are cultivated intensively and results in huge losses. The disease was first time reported in 1905 from France (Maublang, 1905). Later it has been reported from several countries like Georgia, New Zealand, Rome, Palestine, Kenya, Italy, and Greece. In southern United States, it has been considered as a one of the major limitation in French bean cultivation. In India the occurrence of the disease has been reported from Himachal Pradesh (Sohi and Sharma, 1965) and Rajasthan (Agarwal and Jain, 1978).

LOSSES

The disease is aggravated under water stress and high temperatures conditions (Songa and Hillocks, 1996). Haskell and Wood (1926) have reported *Macrophomina phaseolina* to cause up to 65 per cent losses in Mississippi, U.S.A. Similarly, 60 per cent losses in terms of plant deaths were encountered in New Zealand at Tifton (Anonymous, 1942). In Yugoslavia, the disease has been reported to cause 10- 20 percent of the plants to wilt and die (Perisic, 1964).

SYMPTOMS

The symptoms of this disease can appear in any stage of plant growth but they are more conspicuous on young seedlings. Very young seedlings affected by this disease turn black and collapse (Plate 1a, b). During ashy stem blight phase, numerous pycnidia appear on the plant surface-causing stem to appear pale grey (Plate 1c, d). On the stem, reddish brown discolouration is common. Chlorosis of the aerial parts followed by sudden withering, stunting and root dwarfing and in final stage ash coloured stem between first to third node and even the underground part occurs.

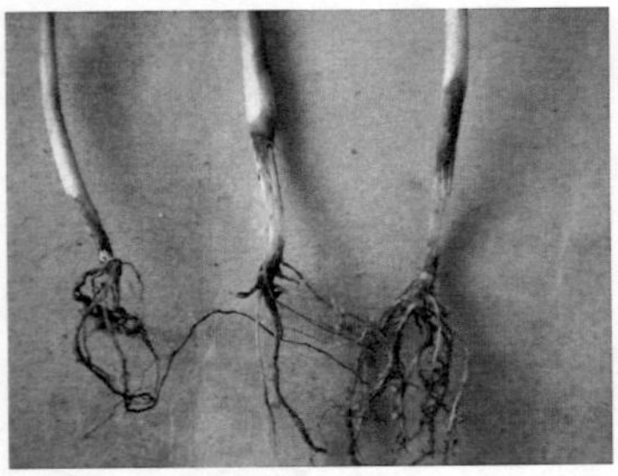

Plate 1 a, b. Symptoms of root rot caused due to infection of *Macrophomina phaseolina*

Plate 1c. Symptoms of charcoal rot on stem

Plate 1d. Pycnidia on stem surface

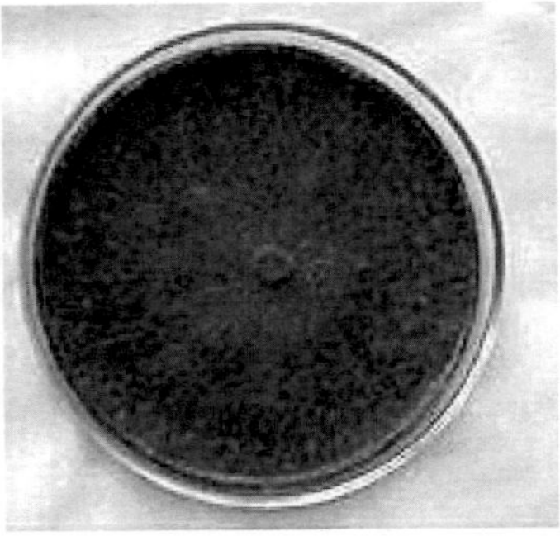

Plate 1e. Culture of *Macrophomina phaseolina*

Courtesy: Dr. Ritu Mawar

THE PATHOGEN

The fungus is readily recognizable by its dense, black sclerotia, 120 to 200 µm in diameter (Makie, 1939). The sclerotia arise by cellular proliferation at any point in the hyphae after these have darkened. At maturity, these are typically spherical and smooth, though may become gibboseas a result of swelling of peripheral cells. They appear hairy owing to dark hyphae emerging from them. Though typically spherical, they may be oblong, discoid, or nearly pyriform. The hyphae of microsclerotia are 7 – 8 µm wide. Sclerotia formation begins with branching and interwining of adjacent hyphal filaments, an increase in the size of associated cells and compaction of the sclerotial mass. The mature pycnidia are sub globose with a reticulate appearance, a short neck and a circular ostiole. The pycniospores are hyaline and one celled. The pycniospores are 14 –33.6 µm long and 5.9 -11.8 µm wide. The conidia develop on short, simple, hyaline conidiophores covering the wall of the pycnidium central cavity.

VARIABILITY

Not much literature is available on this aspect so far. Mayek Perez *et al.* (1997) studied variation *in vitro* cultural characteristics and virulence to common bean (*Phaseolus vulgaris*) of 15 *M. phaseolina* isolates from different hosts and geographical areas of Mexico. There were significant variations among the isolates in development and *in vitro* cultural characteristics, as well as in their virulence to bean. Isolates from other hosts such as maize, sesame, sorghum, and soybean infected bean, indicating non - specificity of the pathogen.

HOST RANGE

The pathogen *M. phaseolina* has a very wide host range. Luttrell (1946) reported it on cowpea, lima bean (*Phaseolus lunatus*), *Cassia inciting, Crotalaria intermedia, Lespendeza* sp., *Lactuca* sp., coffee weed (*Ditremexa occidentalis*), sweet potato, cantaloupe, tomato, maize and sorghum besides French bean. Reichert (1930) found this pathogen infecting tobacco, sesame, potato, sweet melon, pepper and tomato in addition to bean. On the other hand, Sarejanni and Cortaz (1935) have reported citrus, cotton, egg- plant and large flowered - chrysanthemum including beans as the host range for the pathogen. The pathogen also harbours on *Brassica compestris, B. geniculata*, and *Malva parviflora* (Ghaffar and Zentmyer, 1968). Maize and soybean are also amongst the host species (Mayek Perez *et al.*, 1997).

DISEASE CYCLE

The fungus overwinters as sclerotia in the soil and in infected seed (Agarwal and Jain, 1978). Survival is longer in dead host tissues than in free conditions in the soil. The fungus is a poor competitor in soil but readily colonizes plant debris. Population of the fungus increases when susceptible hosts are continuously grown in the field. When infected seed is sown in the field it also introduced the fungus in the field. The sclerotia germinate and the mycelium cause primary infections.

M. phaseolina is a common soil and seed-borne pathogen, found both on and inside the infected seed (Abawi and Pastor-Corrales, 1990). Transmission of the pathogen via infected seeds is known to be a means of spreading *M. phaseolina* into new areas. Another adverse effect of this seedborne pathogen is that infected seeds may fail to germinate, or they may produce infected seedlings that can transmit the pathogen and augment the inoculum potential in the soil. An important preventive management tool for reducing the spread of pathogens into new regions is accurate seed health testing. Molecular tools have greatly facilitated the unequivocal identification of fungal pathogens (Kaur *et al.*, 2012). A rapid, visual, and sensitive loop-mediated isothermal amplification (LAMP) assay for *M. phaseolina* detection in common bean seeds based on sequence-characterized amplified regions (Rocha *et al.*, 2017). The detection limit was 1 pico gram of *M. phaseolina* genomic DNA per reaction, and no cross reactions were observed with DNA isolated from five other seed-borne pathogens. The LAMP assay detected both common bean seeds naturally infected by *M. phaseolina*, and seeds artificially contaminated with the pathogen. The limit of detection was one infected seed per lot of 400 seeds. This detection method was possible without prior fungal DNA extraction, by soaking seeds 2-6 h.

EPIDEMIOLOGY

The pathogen is seed borne (Agarwal and Jain, 1978, Gupta *et al.*, 1991). It also survives in the form of sclerotia in soil, free or embedded in the host tissue. Survival is longer in the dead host tissues than in the soil. Sclerotia formed in plant tissues showed 1 to 27 per cent viability after 3 years (Watanabe, 1973). Dry soils favour longer survival of the sclerotia. It is also capable of living saprophytically on dead organic tissues. Dead stem tissues in soil containing antagonistic micro- organisms were invaded by *M. phaseolina* (Meyer *et al.*, 1973). At soil moisture of 60 to 100 per cent moisture holding capacity, sclerotium population declined 96 to 99 per cent compared with populations in dry soils (Dhingra and Sinclair, 1975). Micosclerotia have also been observed on or beneath the seed coat, in the inner layers of seed coat, on cotyledons and embryo (Agarwal and Jain, 1978). Survival of sclerotia is reduced significantly at high bulk density and high soil moisture, with implied reduced oxygen level (Gangopadhyay *et al.*, 1982). Environmental factors contributing to soil moisture stress/ water deficit predispose the plant to infection (Bhardwaj, 1995). The extent of colonization increased as osmotic water potential decreased and was most severe at -3990 J/ kg (Olaya *et al.*, 1996). The disease is most severe in a period of unusually high temperature and atleast moderately wet weather. It is more serious when soil temperature rises to 28°C or more.

MANAGEMENT

Cultural practices

Use of disease free seed is most important to reduce the havoc of this disease. Due to polyphagous nature of the pathogen, rotations are difficult for recommendation but a 3-year rotation with a cereal crop (except corn or sorghum) may help in reducing soil borne inoculum. Field sanitation measures including cutting and burning of diseased plants, may help in reducing the disease intensity during the following season. Colonization of the fungus decreased with increasing soil moisture, and addition of nitrogen to the soil completely inhibited saprophytic colonization (Dhingra and Chagas, 1981). Standardization of sowing depth may also play a key role in combating this disease as the incidence on *P. vulgaris* was minimum on sowing at a 2 cm depth

(Gonzalez and Gonzalez, 1981). Avoid drought stress, especially during periods of high temperature, good drainage, proper irrigation when needed and using a soil with abundant humus are important. Diaz and Cortinas (1988) obtained significantly higher percentage of healthy plants and yield with either 3 or 4 irrigations, water applied in each irrigation being 15 cm./ acre. Soil solarization has proved quite effective in reducing the incidence of this diseases reported by Lodha (1989) who found that sclerotial count was reduced from 350/g soil to barely 7/g soil in a period of 15 days in dry and wet polythene mulched plots compared to 335 and 279 sclerotial/g soil in their corresponding non - mulched plots. Adjustment of sowing time also helps in reducing the disease incidence (Bhardwaj, 1995). Leaf and stem fragments of lucerne, clover and wheat and cotton roots have been found to reduce the number of sclerotia in amended soils, the greatest reduction occurring with lucerne and clover after 80 days (Hakeem and Ghaffar, 1977).

Host resistance

Host resistance is the most powerful weapon to combat any plant disease because on one hand, it reduces the risks of health hazards and on the other, economy of the farmer is improved due to less expenditure on fungicides. Red and white French bean cultivars are more susceptible while black are resistant. Anthocyanin content of the cotyledons has been reported to be a direct indicator of the degree of resistance in several cultivars (Gonzalez *et al.*, 1981). Cultivar Moruna was found most resistant (Issa, 1985). In greenhouse tests with 3 cvs., the line 8437-22 showed high resistance to *M. phaseolina* while B-190 was intermediate (Echavez and Beaver, 1987). In host resistance studies by Coelho Net and Dhingra (1996), only two genotypes *viz.,* San Cristobal and EMP 86 were found resistant while IPA 1, Arona 80 and V 8025 showed intermediate reaction. Mayek Perez *et al.* (1997) obtained best performance from Selection 4 and Jamapa conditions. Six cultivars viz., Ecuador 229, Kennedy Wander 765, Mexico 235, Pinto 114, Starlight and Votomay and 6lines viz., BARC-RR-18, BelFla-RR-2, BelDak-RR-1, BelDak-RR-2, BelDak Mi-RR-4 and BelDak-RR-8 have been observed as resistant (Abdou *et al.*, 1999).

Chemical control

Charcoal rot is completely controllable by seed treatment with 2 per cent cerasan (Luttrell and Garren, 1952). Das and Sen Gupta (1963) reported mylone at 10 ppm to be fongistatic with respect to *Macrophomina phaseoli*. Papavizas (1977) after trying 13 fungicides concluded that benomyl and captan appreciably reduced sclerotial populations. Brassicol (quintozene) and captan very effectively reduced saprophytic activity (Satischandra*et al.*, 1980). Vyas *et al.* (1981) have suggested the use of systemic fungicides like bavistin, which is absorbed and translocated to the upper parts of seedlings. 40-ppm benomyl was found to inhibit the spore germination completely (Singh and Chohan, 1981). Significant control of pre and post - emergence mortality of bean plants was achieved by seed treatment with benlate with or without solvent (Sharma and Sohi, 1989). Captan, quintozene and carboxin + captan (10% +

30%) gave good control at 0.5, 0.25 and 0.5 kg /100 kg seed, respectively (Ramirez Arredono, 1991). On the other hand, Cardoso *et al.* (1997) found tolclofos methyl and quintozene to significantly control the disease under greenhouse conditions. Antibiotics like hygrostatin (Furushiro *et al.*, 1958) and tetracycline, ampicillin, griseiofulvin and agrimycin (Dubey, 1989) have been reported to inhibit fungal growth. Muthomi *et al.* (2007) reported that copper oxychloride fungicide seed treatment in combination with rhizobia inoculation was beneficial in management of root rot and enhancement of nodulation in food legumes. Treatments consisted of inoculation of legume seeds with appropriate rhizobia alone, rhizobia together with fungicide, rhizobia together with pathogen and a combination of rhizobia, fungicide and pathogen.

The efficacy of 20 organic and inorganic sodium salts, and two synthetic fungicides against eight bean root rot pathogens namely *Fusarium equiseti, F. proliferatum, F. semitectum, F. solani* f. sp. *phaseoli, F. verticillioides, Rhizoctonia solani* AG4-HG I, *Macrophomina phaseolina* and *Sclerotium rolfsii* were evaluated (Turkaan and Reper, 2015). Captan, benzoate and metabisulfite (2 %) were able to completely inhibit mycelial growth of all eight fungi. Moreover, no significant differences were observed among the inhibitory effect of these three compounds and EDTA ($P \leq 0.05$). With few exceptions, the ED_{50} values indicated captan to have a greater effect against fungi than benzoate, EDTA and metabisulfite. However, captan, benzoate and EDTA all had MIC values that varied greatly from that of metabisulfite. Whereas captan, benzoate and EDTA showed fungitoxic activity against all fungi tested at concentrations greater than 0.1 %, metabisulfite showed fungitoxic activity against all fungi tested at concentrations of 0.025–0.25 %. Soil bioassays showed 0.25 % metabisulfite to completely inhibit mycelial growth of *F. proliferatum, F. semitectum, R. solani* AG–4 HG I, *M. phaseolina* and *S. rolfsii*, but not *F. equiseti, F. solani* f. sp. *phaseoli* and *F. verticillioides*. Higher concentrations of captan and benzoate were required to achieve total inhibition in soil bioassays when compared to metabisulfite, whereas EDTA was not able to completely inhibit growth of any of the fungi tested, even at the highest concentration. Moreover, the application of 1.0–2.0 % EDTA was found to be phytotoxic to bean seeds in terms of both seed germination and root elongation, whereas 0.1–0.75 % captan, 0.1-0.75 % benzoate and 0.1 % metabisulfite did not exhibit any phytotoxicity in terms of germination; 0.5 % captan, 0.1 % benzoate and 0.1 % metabisulfite did, however, have a negative effect on root elongation.

The nanoform of a commercial fungicide trifloxystrobin 25% + tebuconazole 50% (75 WG) with broad spectrum of action for improving its antifungal activity against *M. phaseolina* were developed (Kumar *et al.*, 2016). The fungicide commercially available as trifloxystrobin 25% + tebuconazole 50% (75 WG) was converted into its nanoform using ball milling method and assessed for its efficacy against the soil borne fungal pathogen *M. phaseolina* at various concentrations, namely 5, 10, 15, and 25 ppm using poisoned food technique. Nanoform of the fungicide was characterized using Scanning Electron Microscopy (SEM) and Particle Size Analyzer (PSA). The average particle size of nano trifloxystrobin 25% + tebuconazole 50% (75 WG) was about 108 nm. Fungicidal potential of nanoform was better in comparison to the conventional ones. Nanoform of the fungicide was effective at 10 ppm and it exerted hyphal abnormality, hyphal lysis and abnormality of sclerotial formation on *M. phaseolina* when tested under *in vitro* than control.

Plants oils

Maiti *et al* (1985) have suggested activity of essential oils from *Mentha piperata* and *M. citrata* against this pathogen. Vir and Sharma (1985) obtained 100 per cent inhibition of the pathogen by oil (10%) from *Azadirachta indica* under *in vitro* conditions. Similarly, a notable decrease in sclerotial germination has been observed by Dwivedi and Dubey (1986). Chalfoun and Carvalho (1987) have advocated the use of both the industrial oil and garlic extract for inhibiting the mycelial growth of the fungus.

Biological control agents

Soil mycoflora, particularly *Actinomycetes* spp. Stimulated by addition of sodium nitrate has been proved inhibitory to the growth of the fungus (Kovoor, 1954). *Trichderma viride* and *Thielavia terricola* were found as the invader of the colony of *Macrophomina phaseolina* (Nortow, 1954). Radha (1961) demonstrated the role of a thermostable and fungicidal metabolite of *Aspergillum niger* in suppressing the growth of this fungus. Abdalla and Tayab (1976) studied the interaction between *A. flavus* and *M. phaseolina* and reported that the linear growth of the latter was decreased due to toxigenic effects of the metabolites produced by the former organism. Antifungal activity of *Nocardia salmonicolor* isolate from Jabalpur soil has been demonstrated against *M. phaseolina* (Aggarwal and Joshi, 1982). Elad *et al.* (1986) identified four isolates of *Trichoderma harzianum* that could inhibit linear growth and microsclerotia production and ultimately reducing the viable inoculum. Hajra *et al.* (1992) reported antagonistic activity of plant pathogenic bacteria like *Aeromonas* sp. and *Psedomonas* sp. against *M. phaseolina*. Mukherjee and Sen (1992) identified four potential antagonists of *M. phaseolina* such as *Aspergillus fumigatus*, *A. terrus*, *Penicillium citrinum* and *P. simplicissimum* by general screening and dual cultures. Out of 166 bacterial isolates evaluated, 19 inhibited growth of *M. phaseolina*, and when applied as seed treatments, increased plant emergence from 12 to 100 per cent (Campo *et al.*, 1994). Perdomo *et al.* (1995) observed that *Pseudomonas cepacia* strain UPR 5C consistently restricted fungal growth. *Trichoderma* isolate T-12 successfully suppresses both *Rhizoctonia* root rot and foot rot in greenhouse (Cardoso *et al.*, 1997). Fungal antagonists *Trichoderma viride, T. harzianum* and *Gliocladium virens* were effective in inhibiting the growth of *M. phaseolina* in dual culture. These antagonists alone and in combination with fungicides were effective in reducing the disease incidence and increasing seed germination and pod yield over the control. A combination of seed treatment with Vitavax + *T. viride* was most effective followed by seed treatment with thiram + *T. viride* and Captaf + *T. viride.* Seeds treated with *T. viride* alone showed superiority over seeds treated only with fungicides (Dubey, 2002).

Antagonistic activities of six *Trichoderma* isolates associated with healthy common bean seeds were evaluated against two pathogens, *Macrophimina phaseolina* and *Rhizctonia solani* through a dual culture assay (Naglaa *et al.*, 2020). Of them, *Trichoderma* isolate T22 remarkably suppressed the growth of both pathogens (72.42% and 68%, respectively), 5 days after inoculation. Both pathogens were completely overgrown by T22 mycelia, and sclerotia formation by the two pathogens was lacking compared to the control plates, after 8 days. The T22 isolate was morphologically identified as *T. atroviride*. The scanning electron microscopy (SEM) confirmed the mycoparasitism of *T. atroviride* T22 against *M. phaseolina* and *R. solani*, resulting in mycelium collapse of both pathogens (Plate 2). A total of nine unique volatile compounds of *T. atroviride* T22 filtrate were identified by GC-MS technique, representing 75.03% of the total components. Among them, 5-hydroxymethylfurfural (46.11%), and 4 H-pyran-4-one, 2,3-dihydro-3,5-dihydroxy-6-methyl- (15.66%) were detected as major components, while furfural (3.14%), hexadecanoic acid, 2-hydroxy-1-(hydroxymethyl) ethyl ester (2.81%), 1,2-cyclopentanedione (2.79%), and 2-furanmethanol (2.09%) were found to be minor compounds.

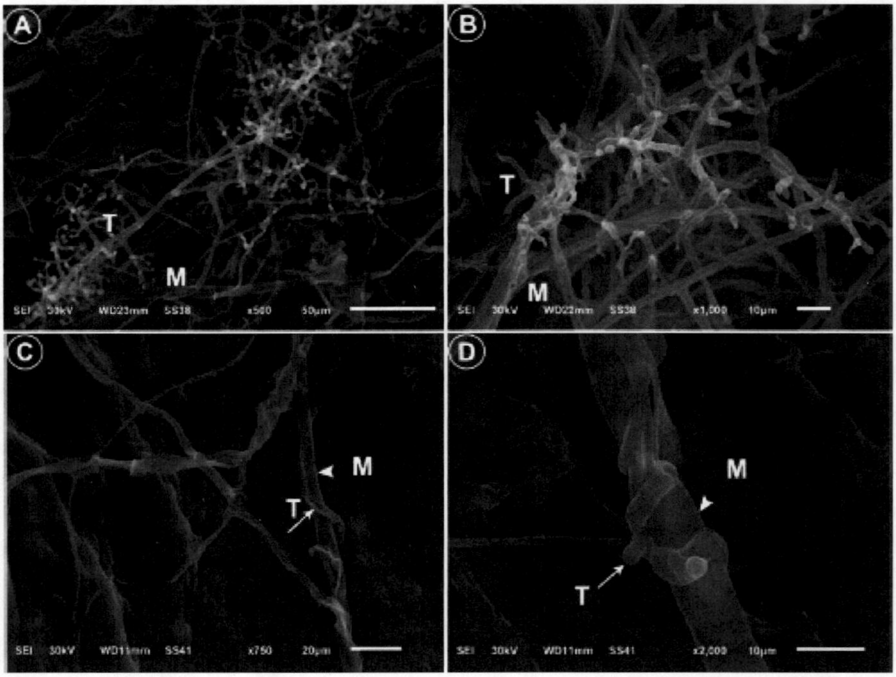

Plate. 2. SEM micrograph showing the parasitic action of *Trichoderma atroviride* T22 (T) against *Macrophomina phaseolina* (M). (A) Growth of T hyphae interacting with those of M. (B) A magnified part showing the growth of T over the hyphae of M. (C) Mycoparasitism process as parallel growth and coiling formation of T hyphae (arrow) over the hyphae of M (arrowhead). (D) A magnified part showing coiling of T hyphae (arrow) over the hyphae of M (arrowhead). Scale bar = 10 µm.

(Source: Naglaa *et al.*, 2020)

REFERENCES

Abawi G.S. and Pastor-Corrales M.A. 1990. Seed transmission and effect of fungicide seed treatments against *Macrophomina phaseolina* in dry edible beans. *Turrialba* 40:334-339.

Abdalla, M.H. and Tayeb, N.M. 1976. Studies on the interaction between *Sclerotium bataticola* and *Aspegillus flavus*. *Mycopathologia* 60(1): 27-32.

Abdou, El-S, Alla, H.M.A., El-Bana, A.A. and Galal, A.A. 1999. Response of common bean cultivars and lines and effects of inoculum density, soil moisture, soil texture and calcium sulphate on charcoal rot. *Annals Agric. Sci. (Cairo)* 44(2): 495-509.

Agarwal, G.P. and Joshi, R. 1982. A polygene antibiotic produced by soil *Nocardia. Current Science* 51(12): 613.

Agarwal, N.K. and Jain, B.L. 1978. Histopathological studies of 'Rajma' infected by *Macrophomina phaseolina. Indian J. Mycol. Pl. Pathol.*8: 59.

Anonymous. 1942. Sixteenth Annual Report of the Department of Scientific and Industrial Research, New Zealand. 33 pp.

Anonymous. 1945. Divisions of Plant Pathology and Seed Investigations. *Rep. N.Y. St. Agric. Exp. Sta.* pp. 40-52, 59-63.

Bhardwaj, C.L. 1995. Charcoal rot incidence and efficacy of seed treatment with carbendazim in French bean relative to variety and environment. *Indian J. Mycol. Pl. Pathol.* 25(3): 246-249.

Campo, R., Shroder, E., Chavez, R.E. and Zapata, M. 1994. Biological control of *Macrophomina phaseolina* (Tassi) Goid. on bean (*Paseolus vulgaris*) by rhizobacteria. *Fitopathologia Columbiana* 18(1-2): 36-41.

Cardoso, J.E., Silva, S.A.G. and Marques, E.E. 1997. Chemical and biological control of bean root rots. *Phytopatologia Brasileira* 22(1): 39-44.

Chalfoun, S.M, and Carvalho, V.D.de. 1987. Effect of industrial garlic oil extract on fungal development. *Phytopathologia Brasileira* 12(3): 234-235.

Coelho Netto, R.A. and Dhingra, O.D. 1996. Methods for evaluating bean genotype reaction to *Macrophomina phaseolina. Fitopatologia Brasileira* 21(2): 236-242.

Dhingra, O.D. and Chagas, D. 1981. Effect of soil temperature, moisture and nitrogen on competitive saprophytic ability of *Macrophomina phaseolina. Transactions British Mtcological Society* 77(1): 15-20.

Dhingra, O.D. and Sinclair, J.B. 1975. Survival of *Macrophomina phaseolina* sclerotia in soil: effects of soil moisture, carbon sources and nitrogen concentrations. *Phytopathology* 65(3): 236-240.

Diaz Franco, A. and Cortinas Escobar, H. 1988. Effect of irrigation regimes on susceptibility of bean to *Macrophomina phaseolina. J. Rio Grande Valley Hort. Soc.* 41: 47-50.

Dubey, R.C. 1989. Effect of some antibiotics on *Macrophomina phaseolina*, a charcoal rot fungus of soybean. *Hindustan Antibiotics Bull.* 31(1-2): 38-39.

Dwivedi, R.S. and Dubey, R.C. 1986. Effects of volatile and non-volatile fractions of two medicinal plants on germination of *Macrophomina phaseolina* sclerotia. *Transactions British Mycological Society* 87(2): 326-328.

Echavez-Badel, R. and Beaver, J.S. 1987. Resistance and susceptibility of bean to ashy stem blight (*Macrophomina phaseolina*) (Tassi) Goid. *J. Agric. Univ. Puerto Rico* 71(4): 403-405.

Elad, Y., Zvieli, Y. and Chet, I. 1986. Biological control of *Macrophomina phaseolina* (Tassi) Goid. by *Trichoderma harzianum*. *Crop Protection* 5(4): 288-292.

Furushiro, K., Shimizu, K., Minogata, M. and Fujisawa, T. 1958. Hygrostatin, a new antibiotic substance. *IyakuShigen ken kyushoNempo*: 24-39.

Gangopadhayay, S., Wyllie, T.D. and Teague, W.R. 1982. Effect of bulk density and moisture content of soil on the survival of *Macrophomina phaseolina*. *Plant and Soil* 68(2): 241-247.

Ghaffar, A. and Zentmyer, G.A. 1968. *Macrophomina phaseoli* on some new weed hosts in California. *Pl. Dis. Reptr.* 52(3): 223.

Goidanich, G. and Camici Leontina. 1946. A microsclerotial parasite resembling *Sclerotium* (*Rhizoctonia*) *bataticola* Taub. present in Italy. *Reprinted from Ric. Sci. Ricostruz, Roma*, xvi,II 6 pp.

Gonzalez Avila, Alardo, R. and Marrero, H. 1981. Correlation between the anthocyanin content of the hypocotyls of different types of bean and infection by *Macrophomina phaseolina*. *Ciencias de la Agricultura* 9: 79-83.

Gonzalez Avila, M. and Marrero Gonzalez, H. 1981. Effect of sowing depth on incidence of *Macrophomina phaseolina* (Tassi) Goid. on bean seedlings. *Ciencias de la Agricutura* No. 10: 3-7.

Gupta, S.K., Dohroo, N.P. and Shyam, K.R. 1991. Antagonistic studies on seed borne mycoflora of French bean (*Phaseolus vulgaris* L.). *Indian J. Pl. Pathol.* 9(1&2): 62-63.

Hajra, K.K., Khatua, D.C. and Mukherjee, N. 1992. Antagonistic bacteria against fungal pathogens. *J. Mycopath. Res.* 30(1): 65-70.

Hakeem, A. and Ghaffar, A. 1977. Reduction of the number of sclerotia of *M. phaseolina* in soil by organic amendments. *Phytopathologische Zeitschrift* 88(3): 272-275.

Hoffmaster, D. E., McLaughlin, J.H., Ray, W.W. and Chester, K.S.1943. The problems of dry rot caused by *Macrophomina phaseoli* (=*Sclerotium bataticola*). *Abst. Phytopathology* 12: 1113-1114.

Huskell, R.J. and Wood Jessie, L. 1926. Diseases of vegetable and field crops in the United States. *Plant Disease Reports, Supplement* 54: 209-333.

Issa, E. 1985. Varietal resistance in French bean (*Phaseolus vulgaris* L.) to ashy stem blight caused by *Macrophomina phaseoli* (Maubl.) Ashby. *Biologico* 51(5): 113-117.

Kaur, S., Dhillon G.S., Brar, S.K., Vallad, G.E., Chauhan, V.B. and Chand, R. 2012. Emerging phytopathogen *Macrophomina phaseolina*: biology, economic importance and current diagnostic trends. *Crit Rev Microbiol* 38:136–151.

Kovoor, A.T.A. 1954. Some factors affecting the growth of *Rhizoctonia bataticola* in the soil. *J.Madras Univ.* 24(1): 47-52.

Kumar, G. D., Natarajan, N. and Nakkeeran, S. 2016. Antifungal activity of nanofungicide Trifloxystrobin 25% + Tebuconazole 50% against *Macrophomina phaseolina. African Journal of Microbiology Research* **10(4)**:100-105.

Lodha, S. 1989. Soil solarization for the control of *Macrophomina phaseolina* in arid soils. *Indian Phytopath.* 42(1): 170-171.

Luttrell, E.S. 1946. A pycnidial strain of *Macrophomina phaseolina. Phytopathology* 11: 978-980.

Luttrell, E.S. and Garren, K.H. 1952. Blights of snap bean in Georgia. *Phytopathlogy* 42(11): 607-613.

Mackie, W.W. 1931. A new disease of maize and beans. *Abst. Phytopath.* 10: 996.

Maiti, D., Cole, C.R., and Sen, C. 1985. Antimicrobial efficacy of some essential oils. *Zeitschrift furr Pflanzenkrankheiten und Pflanzenschutz* 92(1): 64-68.

Muthomi, J.W., Otieno, P.E., Chemining wa, G.N., Nderitu J.H. and Wagacha, J.M. 2007. Effect of legume root rot pathogens and fungicide seed treatment on nodulation and biomass accumulation. *Journal of Biological Sciences, 7: 1163-1170.*

Mathur, R.S. 1965. Interaction of saprophytic and parasitic fungi in culture. *Labdev Jnl. Sci. Techol.* 3(1): 30-33.

Maublang, A. 1905. Especes Nouvelles De Champignons Inferieurs. *Soc. Mycol. de France Bull.*21: 87-94.

Mayek Perez, N., Lopez Castaneda, C. and Acosta Gallegos, J.A. 1997. Variation *in vitro* cultural characteristics of *Macrophomina phaseolina* isolates and its virulence on common bean. *Agroceincia* 31(2): 187-195.

Mayek Perez, N., Pedroza Flores, J.A., Villreal Garcia, L.A. and Valdes Lozano, C.G.S. 1997. Genetic and environmental factors related to temporal dynamics and disease effects on dry beans (*Phaseolus vulgaris* L.) at Marin, Mexico. *Revista de Mexicana Fitopatologia*13(1): 1-9.

Meyer, W.A., Sinclair, J.B. and Khare, M.N. 1973. Biology of *Macrophomina phaseoli* in soil with selective media. *Phytopathlogy* 63(5): 613-620.

Mukherjee, B. and Sen, C. 1992. *Aspergillus* and *Penicillium* spp., potential agents for biocontrol of *Macrophomina phaseolina. Indian Phytopath.* 45(1): 39-45.

Naglaa M. El-Benawy, Gamal M. Abdel-Fattah, Khalid M. Ghoneem and Yasser M. Shabana. 2020. Antimicrobial activities of *Trichoderma atroviride* against common bean seed-borne *Macrophomina phaseolina* and *Rhizoctonia solani*. *Egyptian Journal of Basic and Applied Sciences* **7(**1): 267-280.

Nortow, D.C. 1954. Antagonism in soil between *M. phaseoli* and selected soil-inhabiting organisms. *Phytopathology* 44(9): 522-524.

Olaya, G., Abawi, G.S. and Barnard, J. 1996. Influence of water potential on survival of sclerotia in soil and colonization of bean stem segments by *Macrophomina phaseolina. Pl. Dis.* 80(12): 1351-1354.

Papavizas, G.C. 1977. Some factors affecting survival of sclerotia of *Macrophomina phaseolina* in soil. *Soil Biology* 9(5): 337-341.

Perdomo, F., Echavez-Badel, R., Alameda, M. and Schroder, E.C. 1995. *In vitro* evaluation of bacteria for the biological control of *Macrophomina phaseolina. World J. Microbiol. and Biotech.* 11(2): 183-185.

Perisic, M.M. 1964. Contribution of the study of *Sclerotium bataticola* as a parasite of beans. *ZastBilja* 15(80): 411-419.

Radha, K. 1961. The genus *Rhizoctonia* in relation to soil moisture. II *Rhizoctonia bataticola*, cotton root organism. *Indian Cocon. J.* 13(4): 137-144.

Reichert, I. 1930. Palestine: root diseases caused by *Rhizoctonia bataticola. Intl. Bull. Pl. Protect.*2: 17.

Reichert, I. And Hellinger, E. 1929. A blight of beans in Palestine. Reprinted from Yedeoth, ii, 1-2: 5.

Rocha, D.C., Oliveira, M.B., de Freitas, M.A. and Petrofeza, S. 2017. Rapid detection of *Macrophomina phaseolina* in common bean seeds using a visual loop-mediated isothermal amplification assay. *Australasian Plant Pathol.* 46: 205-212.

Romirez Arredono, J.A. 1991. Chemical control of bean (*Phaseolus vulgaris* L.) root diseases with seed treatments at Costa de Hermosillo. *Revista Mexicana de Phytopathologia* 9(1): 53-56.

Sarejanni, J.A. and Cortzas, C.B. 1935. A note on the parasitism of *Macrophomina phaseolina* (Maubl.) Ashby. *Ann. Inst. Phytopath., Benaki, Greece* 3: 38-44.

Satishchandra, K.M., Hiremath, R.V. and Hegde, R.K. 1980. Effect of organic amendments and fungicides on the saprophytic activity of *Rhizoctonia bataticola* causing root rot of bean. *Indian Phytopath* 32(4): 543-546.

Sharma, S.R. and Sohi, S.H. 1989. Diseases of French bean in India and their control. *Indian Rev. Life Sci.* 9: 153-168.

Singh, R.S. and Chohan, J.S. 1981. Effect of benlate on growth and spore germination of *M. phaseolina. Pesticides* 15(7): 8.

Sohi, H.S., Sharma, S.L., Verma, B.R. and Sachdeva, K.B. 1965. Floury leaf spot of bean caused by *Ramularia phaseolina* Pertak. *Indian Phytopath.* 18: 384.

Songa, W. and Hillocks, R.J. 1996. Charcoal rot in common beans with special reference to Kenya. *Intl. J. Pest Mgt.*: 213-219.

Turkkan, M. and Erper, I. 2015. Inhibitory influence of organic and inorganic sodium salts and synthetic fungicides against bean root rot pathogens. *Gesunde Pflanzen* 2/2015: doi https://doi.org/10.1007/s10343-015-0339-z.

Vir, D. and Sharma, R.K. 1985. Efficacy of fungicide: xxxii. Evaluation of neem oil for control of plant pathogens. *Asian Farm Chemicals* 1(7/8): 23-24.

Watanabe, T. 1973. Survivability of *M. Phaseoli* (Maubl.) Ashby in naturally infested soils and longevity of the sclerotia formed *in vitro. Annals Phytopathological Soc. Japan* 39(4): 33-337.

10

ALTERNARIA LEAF AND POD SPOT

This disease is of minor importance but when mites have already injured plants or week due to malnutrition or other causes, it may cause appreciable yield losses.

GEOGRAPHICAL DISTRIBUTION

The disease was first reported from Nyasaland (Lawrence, 1949). Subsequently, it was reported from Wyoming (Schnathorst, 1954), Argentina (Muntanola, 1955), Brazil (Shands *et al.,* 1964), Costa Rica (Gonzalez, 1973), Cambridge (Russell and Brown, 1976), Chile, Mexico, Venezuela (Wellman, 1977), Romania (Crisan *et al.,* 1979), Canada (Tu, 1985) and Poland (Pieta and Labuda, 1994). In India, this disease was first time reported in 1990 from Jammu (Singh *et al.,* 1992).

LOSSES

As the disease is of minor importance, there are no reports regarding the economic losses caused by this disease alone except Abawi *et al.* (1977) who found that the disease was very severe in commercial fields in New York during 1974 and 1976 where up to 12 per cent pods were infected which were rendered unfit for processing.

SYMPTOMS

The symptoms start from the leaflets as small, irregular, reddish brown flecks with a dark brown border. As the spots enlarge, they develop as concentric rings within the affected area. Older portion of the lesion may fall out, leaving a shot hole appearance. Lesion often coalesce, form large patches and became irregular, bristle, dark olive green or black in colour due to fungal growth on either side of the leaves. In severe cases stem and petiole necrosis also occur (Gomes and Dhingra, 1983). On pods reddish to dark brown or black flecks may coalesce and produce streaks. Seeds with severely affected pods show varying degrees of greyish discolouration, which persist through processing (Tu, 1982).

PATHOGEN

Alternaria leaf and pod spot is caused by various *Alternaria* species including *A. alternata* (Fr.) Keissler, *A. brassicae* f. sp. *phaseoli* Brun., *A. fasciculata* (Cke. and Ell.) L.R. Jones and Grout, *A. tenuis* Nees and *A. brasiliensis* Queiroz, Muniz & Menezes sp. nov (Saad and Hagedorn, 1969; Zaumeyer and Thomas, 1957; Abawi, 1977;

Russell and Brown, 1977; Singh *et al.*, 1992; Queiroz *et al.*, 2001) but most common among these is *A. alternata*. Conidiophores are simple or branched, olivaceous to brown measuring 2-6 x 3.2-4 μm. Conidia are produced in long chains, obclavate, obpyriform, ellipsoidal having up to eight transverse and usually several longitudinal and oblique septa, 8-17 x 17-60 μm, with short beak 2-5 μm. Conidia measure 16-19 x 120- 240 μm. The pathogen multiplies best on potato-dextrose agar medium with 25 $\pm$ 2° C as optimum temperature. Artificial inoculation of bean plants result in typical symptoms within 5 days (Singh *et al.*, 1992).

In *A. brassicae* f. sp. *phaseoli*, the conidiophores are erect, unbranched, conidia are solitary or in short chains of 2-3, with 3-15 cross septa, and several longitudinal septa, smooth wall, long-beaked, measure 40–350 × 9–33 μm. In *A. fasciculate*, the conidiophores are short, bulged at base, solitary or in fascicles of 2-8, measure 40-90.3 × 4.7 μm, conidia catenulate 2–5, with a short beak, 5-9 cross and 1-5 longitudinal septa, constricted at septa, measure 63–105 × 10.5-22 μm. In *A. brassicicola,* the conidiophores are solitary or in fascicles of 2-5, bulged at base, sometimes branched, measure 33.6–76 × 5.3–6.5 μm, conidia catenulate up to 12, with 0-5 longitudinal septa, and 4-13 cross septa, constricted at cross septa, measure 54.6–113.5 × 11–22 μm, and beakless but pseudorostrate. The basic difference between *A. brasiliensis* and *A. brassicae* f. sp. *phaseoli* is the presence in this latter species of more cells in the body of conidia, which presents a much shorter beaked and taper gradual, broad, while in the former species the conidia beak is long, and filiform (Fig. 1). The difference between *A. brasiliensis* and both *A. fasciculata* and *A. brassicicola* is the presence of conidia usually formed in long chains, with a short beak or beakless.

Effect of proteinaceous extracts from red kidney bean cotyledons on mycelium of *Alternaria alternata* growing on potato dextrose agar showed unexpected conidia formation in response to applied crude extracts (Alizadeh *et al.*, 2011). A PDA disc method was developed to quantify conidia formed. A purified fraction retaining conidiation inducing effect (CIE) was obtained following several protein purification procedures including the last step of eluting bound proteins from an Affi-gel blue gel column. Based on MALDI (matrix assisted laser desorption/ionization) mass spectrometric analysis, a previously identified mannose-binding lectin (MBL) called PvFRIL (*Phaseolus vulgaris* fetal liver tyrosine kinase 3-receptor interacting lectin) was present in this conidiation inducing fraction. The PvFRIL was subsequently purified using a single step mannose–agarose affinity column chromatography. When the lectin was applied exogenously to *A. alternata*, increased conidiation resulted. The conidia produced in response to the MBL were similar to those induced by other methods and their germ tubes were longer after 12 h growth than those induced under white light. To our knowledge this is the first report of exogenous application of a PvFRIL or another purified protein from a plant inducing conidia formation in a fungus.

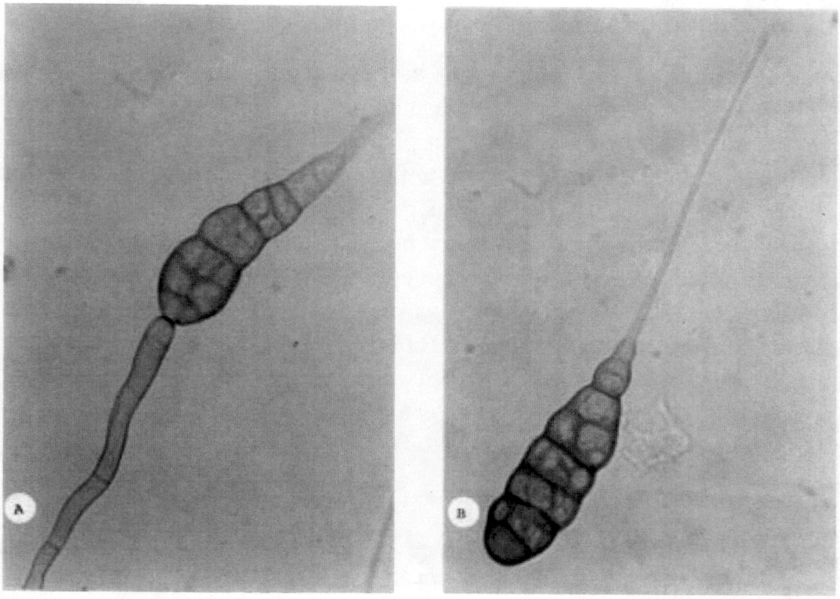

Fig.1. ***Alternaria brasiliensis*** **(A) Young conidium in the formation conidiophore (B) Conidium mature in production of filiform beak.**

VARIABILITY

Rahim *et al* (1990) observed morphological diversity among strains of Alternaria related to a) the seed from some cultivars and sample b) the external and internal localization of the fungus in the seed c) transmissibility of *Alternaria* spp. Isolated from the seed and leaves to aerial parts of bean plants through seed, and d) pathogenecity to seedlings.

HOST RANGE

Besides *Phaseolus vulgaris*, pathogen also attacks *P. coccineus* and *Lycopersicon esculentum*.

DISEASE CYCLE

The pathogen over winter either in the form of mycelium or spores on infected plant debris and in or on the seed. If the pathogen is seed borne, it may cause damping off/ stem lesions and collar rot. However, spores are produced abundantly, especially during heavy dews or frequent rains on mycelium present in debris or on infected cultivated plants and weeds and are blown to the main crop. The germinating spores may penetrate either through stomata or wounds on older plant tissue during periods of high humidity and produce new crop of conidia, which are further disseminated by moist wind and splashing rains to the neighbouring fields or areas.

EPIDEMIOLOGY

The pathogen attacks only older plant tissues and requires high degree of relative humidity (> 90%) for spore germination and infection. Pre-inoculation air temperature has no influence on disease severity but plant growth is affected at 16°C while post-infection temperature of 16-20°C is quite favourable for the severe attack of the crop (Saad and Hagedorn, 1968). Bean plants grown under N or K deficient conditions are more susceptible to the infection caused by *A. tenuis* (Saad and Hagedorn, 1969). *A. tenuis* also produces a toxin (tentoxin) in culture, which induces plant chlorosis when applied to roots (Saad *et al*., 1970). However, the fungus does not produce detectable quantities of tentoxin during natural infection of leaves or pods.

MANAGEMENT

Host resistance

Several workers have reported existence of variation among different French bean cultivars against this disease. Gonzalez (1973) observed 27R and ME X 80R as having lesser susceptibility than the black seeded varieties. Abawi *et al.* (1977) found that Roma was more severely affected by *A. alternata* than Bush Blue Lake 274 while Early Galltin Fleetwood, a late maturing cultivar was more resistant than the early Seafarer (Tu, 1982). Tu and Park (1992) screened 80 commercial cultivars for three consecutive crop seasons and reported that 50 varieties were having moderate to high degree of resistance. Out of 29 *P. vulgaris* cultivars, only Kentucky Wonder, EC 26392 and PBU-14-1 were resistant while EC 57080, Sel. 9, MB-Radish spotted, SVM-1, P-37, Sel-48-2-3 and Sel. -48-1-2 were moderately resistant (Singh and Singh, 1998).

Cultural practices

Cultural practices like late plantings, harvesting soon after maturity and rapid drying reduce the intensity of the disease (Tu, 1993). Overhead irrigation during the growing season has been reported to reduce the establishment of epiphytotic population on the leaves and consequently the incidence of infestation. Monoculture also adversely affect the plant health and proper crop rotations should be followed (Czajka and Cwazina, 1996).

Chemical Control

Various chemicals like chlorothalonil (Abawi *et al*., 1977), Vitavax, binapacryl and Quinolate(Crisan *et al*., 1979) and iprodione (Tu, 1985) have been found effective for the management of this malady. Issa (1985) suggested alternate use of fungicides for obtaining best control of this disease. He found that *A. alternata* was best controlled by chlorothalonil, captafol or mancozeb alternated with benomyl + captafol and mancozeb alternated with captafol. Seven sprays of mancozeb at 10 days interval starting from 20 days after plant emergence were fond effective in the management of this disease (Barros *et al.*, 1992). Management of the disease with Mancozeb was also reported by Singh *et al.* (2003) under field conditions in cv. PDR-14. Foliar application of iprodione just prior to plant senescence has proved quite effective against the pathogen (Tu, 1992). Combination of different synergistic fungicides like thiophanate methyl and chlorothalonil, chlorothalonil and mancozeb provided effective control of this disease (Vieira *et al.,* 1998). It was also observed that fungigation using central pivot irrigation system gave better control of *A. alternata* than conventional spraying methods (Oliveira *et al.*, 1992).

Biocontrol

A glass house experiment was chosen to study the influence of *Alternaria alternata* and *Rhodotorula mucilaginosa* on the activities and functions of mycorrhizal fungus *Glomus mosseae* during growth of *Phaseolus vulgaris* (El-Sherbeny *et al.*, 2012). Mycorrhizal plants exhibited improvements in growth compared with non-mycorrhizal ones. In addition, the paired inoculation of *G. mosseae* with *R. mucilaginosa* improved plant growth compared with either single inoculation with *G. mosseae* or paired with *A. alternata*. Phosphatases activity, dry weight, chlorophyll, carbohydrates and protein contents of mycorrhizal plants were significantly increased in the presence of *R. mucilaginosa*. It was also noticed that *R. mucilaginosa* exhibited higher mycorrhizal colonization including; higher percentage in mycorrhizal infection, more colonization intensity and higher arbuscular formation compared with single inoculation of AM fungus as well as in case of *A. alternata* all over the experimental periods. Mycorrhizal plants dependency was increased all over the experimental periods in the presence of *A .alternata* while, it was decreased in the presence of *R. mucilaginosa* during the current work. The benefits of symbiosis of AM fungus and *Phaseolus* plants was increased in the presence of *R. mucilaginosa*, while it was decreased in the presence of *A. alternata*.

REFERENCES

Abawi, G.S., Crosier, D.C. and Cobb, A.C. 1977. Pod flecking of snap beans caused by *Alternaria alternata. Pl. Dis. Reptr*. 61(11): 901-905.

Alizadeh, H., Leung, D.W.M. and Cole, A. L.J. 2011. Conidiogenic effects of mannose-binding lectins isolated from cotyledons of red kidney bean (*Phaseolus vulgaris*) on *Alternaria alternata. Phytochemistry* 72 (1): 94-99.

Crisan, A. Bedelean, C. and Jacob, R. 1979. Aspect of the biology annd control of *Alternaria* species on bean. *Babes Balyai* : 241-250.

Czajka, W. and Cwazira, B. 1996. The intensity of spot diseases occurrence on field bean grown in different crop rotation and monoculture. *Phytopathologia Palonica* 12: 35-42.

El-Sherbeny, G. A., Rabie, G. H. and Elgazar, N.S. 2012. Interaction between *Rhodotorula mucilaginosa* and *Alternaria alternata* on *Phaseolus vulgaris-Glomus mosseae* association. *Journal of Basic and Applied Mycology* 3: 27-34

Gomes, J.L.L. and Dhingra, O.D. 1983. *Alternaria alternata* - a serious pathogen of white coloured snap bean (*Phaseolus vulgaris*) seeds. *Fitopatologia Brasileira* 8(1): 173-177.

Gonzalez, L.C. 1973. Leaf spot of bean (*Phaseolus vulgaris*) caused by *Alternaria* spp. in Costa Rica. *Turrialba* 23(2): 238-239.

Issa, E. 1985. Comparative study of fungicides used singly, alternatively and in mixtures to control bean pod diseases. *Biologico* 51(6): 141-146.

Lawrence, E. 1949. Report of the acting Director of Agriculture. *Rep. Dept. Agric. Nyasald*: 23.

Muntanola, M. 1955. Cryptogamic parameters of bean (*Phaseolus* spp.) in the province of Tucuman. *Rev. Agron. Naroeste Argent*. 1(3); 283-299.

Queiroz, F., de F.S. Muniz, M. and Menezes, M. 2001. *Alternaria brasiliensis* sp. nov., a leaf pathogen on *Phaseolus vulgaris*. *Mycopathologia* 150:61-65.

Russell, P.E. and Brown, L. 1977. *Alternaria alternata* on *Phaseolus vulgaris*. *Plant Pathol.* 26(1): 47.

Saad, S. and Hagedorn, D.J. 1969. Symptomatology and epidemiology of *Alternaria* leaf spot of bean, *Phaseolus vulgaris. Phytopathology* 59: 1530-1533.

Schnathorst, W.C. 1954. Fungi in seeds and plants of certified bean varieties. *Phytopathology* 44(10): 588-592.

Shands, H., Clibas, V. and Zaumeyer, W.J. 1964. Observation on dry bean diseases in Brazil. *Pl. Dis. Reptr*. 48(10): 784-787.

Singh, A. and Singh, A.K. 1998. Response of French bean cultivars to *Alternaria alternata* (Fr.) Keissler under field conditions. *Crop Res. (Hisar)*: 15(1): 130-131.

Singh, S. R., Rai, M. and Kuwar Singh. 2003. Management of Alternaria leaf spot of French bean. *Farm Science Journal* 12(2): 172-173.

Tu, J.C. 1982. Etiology of black pod disease and seed coat discolouration of white beans. *Canadian J. Pl. Sci.* 62(2): 277-284.

Tu, J.C. 1985 a. Biology of *Alternaria alternata*, the causal agent of black pod disease of white bean in South Western Ontario. *Canadian J. Pl. Sci.* 65(4): 913-919.

Tu, J.C. 1985b. Efficacy of iprodione against *Alternaria* black pod of beans. *Canadian J. Pl. Pathol.* 5(2): 133-135.

Tu, J.C. and Park, S.J. 1992. Response of cultivars and breeding lines of *Phaseolus vulgaris* L. to black pod fungus, *Alternaria alternata* in South Western Ontario. *Canadian J. Pl. Dis. Survey* 72(1): 9-11.

Vieira, R.F., Paula Junior, T.J.de and Berger, P.G. 1998. Effect of fungicides and cartap insecticide on the control of bean foliar diseases during winter-spring season. *Summa Phytopathologica* 24(1): 17-22.

11

SCLEROTINIA WILT

Sclerotinia wilt also known as white mould (Sherbakhoff, 1917), is an important disease of this crop after a period of warm and humid weather and causes losses both under field and during transportation.

GEOGRAPHICAL DISTRIBUTION

The fungus is polyphagous in nature and has been reported from all the French bean growing areas of the world. It has been encountered in Argentina (Hauman, 1915), Bermuda (Anonymous, 1921), Germany (Jaap, 1922), Italy (Bongini, 1924), England (Biffen, 1926), Bermuda, Canada (Baribeau and Raicot, 1928), New Zealand (Brien, 1932), France (Nicolas and Aggery, 1933), Morocco (Anonymous, 1937), Tanganyika Territory (Wallace, 1944), Denmark (Buihwald, 1947), Australia (Aberdeen, 1949) and Mexico (Yerkes, 1955).

In India, the occurrence of this disease was reported for the first time from Kinnaur district of Himachal Pradesh (Gupta *et al.*, 1997). Later, Bag (2000) reported the occurrence of this disease from Arunachal Pradesh.

LOSSES

This disease causes heavy losses to the crop resulting in a major reduction in seed yield. Zaumeyer (1930) reported that this disease can cause a good deal of damage to full mature crop in Montana. On the other hand, Kerr *et al.* (1978) recorded 44% reduction in seed yield when there was 30% infection of this disease in irrigated field of bean. In Tasmania, highest disease incidence in the crop was recorded up to 86% with corresponding 64% yield losses (Wong *et al.*, 1980). In Japan, yearly losses due to this disease have been reported to be 36% (Abai, 1981). In German Democratic Republic, losses due to this disease to the tune of 42% were observed (Rudolph, 1986). White mold, considered as a serious problem with a large number of beans growing under agricultural areas and low tunnels in Ismailia governorate, Egypt (Elgorban *et al.*, 2013, Hatamleh *et al.*, 2013). Bag (2000) reported that this disease caused severe losses in total production (about 10-85% of the pods in each plant were rotten) beside reducing the market value of the produce in Arunachal Pradesh.

SYMPTOMS

Symptoms of this disease appear on all the above ground parts of the plants (Fig.1). Firstly, the symptoms appear on the stems as irregular water-soaked spots followed by branches and leaves. Spots gradually enlarge to cause soft watery rot and sticky brown exudate develops on the stem. On leaves, chlorosis followed by wilting is the main symptom. Superficial white mycelium appears on the affected plant parts in early stages in wet and cool weather especially on pods in contact with the soil (Fig. 1a). In the later stages, sclerotia are observed to be embedded in the mycelium (Fig. 1b). Sclerotia are initially light in colour, but finally turn dark and become black in colour.

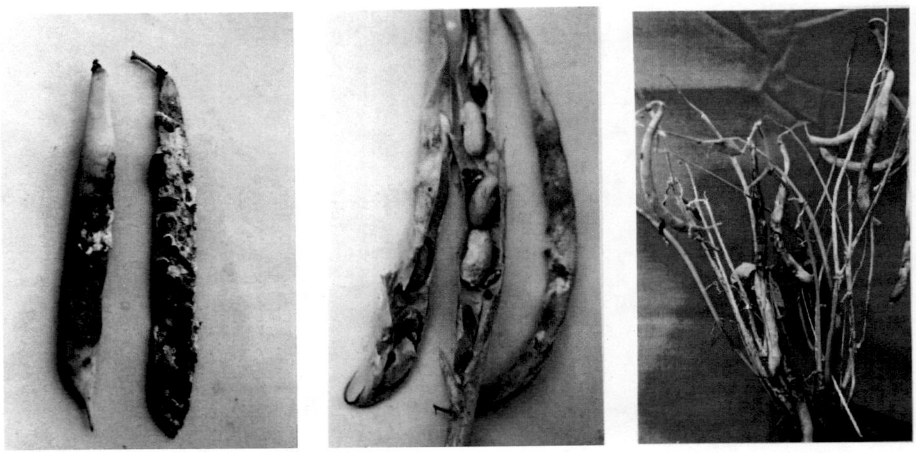

Fig. 1. (a) Symptoms of *Sclerotinia sclerotiorum* on pods (b) Sclerotia of *S. sclerotiorum* inside pods c) Completely infected bean plant with *S. sclerotiorum*

THE PATHOGEN

Till 1924, the Americans used to refer the fungus as *Sclerotinia libertiana* Faik., but in the same year, Wakefield pointed out that in Europe the fungus was named as *Sclerotinia sclerotiorum*. Since then, the name in vogue is *Sclerotinia sclerotiorum* (Lib.) Mycelium is septate with inter and intracellular ramification of hyphae between the cuticle (Lumsden and Wergin, 1980). The mycelium produces resting bodies called sclerotia having a high degree of resistance to drought and cold. Under moist conditions, sclerotium gives rise to apothecium. The apothecia are smooth, pale and funnel shaped and 4-8mm in size. Asci are 125-130 x 8-10μm in dimensions (Saccardo, 1911) and are cylindrical. Spores contained in an ascus measured 9-13 x 4-6.6 μm with ellipsoid shape. Abawi *et al*. (1975) reported blossoms as the energy source of the fungus with the result germination of ascospores within 6hr. of inoculation.

HOST RANGE

The pathogen has a very wide host range from field crops to vegetable crops infecting approximately > 400 species and > 200 genera of higher plants including clover, sunflower, lemon, green apricot and apple (Figueirêdo *et al.*, 2010). Young (1936) reported a long list in Montana, which included squash, pumpkin, peas, carrot, celery, lettuce, potato, Shasta daisy (*Chrysanthemum maximum*), zinnia, and white and sweet clovers. The attack of this pathogen on hosts like bean (*P. vulgaris*), cabbage, vegetable marrow, hemp, certain vines, plums, almonds, figs and citrons, Passion fruit was amongst the principal hosts in Queensland (Blackford, 1944). Weed hosts like *Amaranthus viridis, Artemisia scupoonia, Brassica arvensis, Datura stramonium, Heliotropium ellipticum, Holoptelea integrifolia, Phyllanthus fraternus, Trianthema portulacastrum, Tridax procumbens, Vernonia cinerea* and *Withania somnifera* were also reported as the host of this fungus (Ghsolia *et al.*, 2004).

DISEASE CYCLE AND EPIDEMIOLOGY

Infection arises from contaminated soil, rotten plant debris and sclerotia decayed seed (Hangerfords and Pitts, 1953). The sclerotia serve as the chief source of primary inoculum germinating after several years of dry storage or winter. Maximum sclerotial germination was observed after a period of three years when buried at a depth of 10cm (Cook *et al.*, 1975). The fungus can also survive in the form of dormant mycelium on seed testa and cotyledons (Tu, 1988). He also observed four types of infections:

1. Infection from ascospores on flower petals, which lodge on the stem.
2. Infection resulting from ascospore contaminated leaves in contact with moist soil.
3. Infection associated with injuries and necrotic tissues.
4. Leaves encountering sclerotia on the soil surface.

Type of soil also have a profound impact on the survival of sclerotia as these survived for 2 years in sandy soils, but their mortality occurred after six months in wet peat soil (Chambers and Hardie, 1964). High temperature (20^{0}C) favours apothecial production (Terui and Harada, 1966). According to Weiss *et al.* (1980), 10-25°C is the favourable air temperature for disease development. Ascospores do not survive when exposed to 25°C temperature or more and RH > 35% (Caesar and Pearson, 1983). Bag (2000) also observed that during the epidemic, average rainfall, RH and temperature were 18.27 mm, 63-93% and 21.75 – 25.50 ° C, respectively. The severity of this disease was more when cropping sequence as pea (rabi season) – French bean (kharif/ summer season) – carrot/ pea (rabi season) is followed as all these crops are the host of the pathogen and facilitates its perpetuation from one season to another (Bag, 2000).

FORECASTING

Forecasting methods developed for any disease not only help in planning the management strategies, but also prevent indiscriminate use of fungicides. Hunter *et al.* (1984) have proposed a forecasting method based on soil matric potential. According to them, for areas with a history of white mould, it is proposed that the disease be anticipated whenever the average soil moisture potential is more than or equal to –30 kPa for a period of 1-2 week just before or during flowering.

MANAGEMENT

Cultural practices

Like resistant varieties, cultural practices also play an important role in creating conditions which are unfavourable for infection thus preventing environmental pollution and health hazards by reducing indiscriminate use of chemicals. Moore (1949) found that flooding of fields with marl muck or sandy soils destroyed the sclerotia within 23-45 days irrespective of the depth of burial. Germination of the sclerotia can be controlled by trenching to a depth of 18 inch (Chambers and Hardie, 1960). They have further suggested wider spacing of plants and rows and dusting with quintozene @ 30 pounds/ acre. Trellis grown plants give a higher seed yield with a very less disease incidence (Coyne *et al.*, 1974). Disease development was significantly less in near isogenic determinate than in indeterminate lines regardless of row spacing (Steadman *et al.*, 1973). On the other hand, Gaxiola (1979) found that on increasing separation between the rows from 50 to 95 cm, yield rose from 796 to 1133 kg/ha. Heavy irrigation increases disease incidence as over irrigation particularly at flowering and pod formation stages may create conditions which may increase the incidence of the disease (Weiss *et al.*, 1980). Presence of extended periods of water droplets on the surface of plant parts may increase the likelihood of infection. Keeping the plants off the soil surface with wire fencing tunnels significantly reduce the infection and increase seed yield (Fuller *et al.*, 1984). Huang and Kemp (1989) have suggested growing cultivars with a strong basal stem and narrow upright growth. The incidence of white mould disease is reduced in dense, erect canopies (Saindon *et al.*, 1995). Crop rotation also plays a significant role in reducing the overwintering sclerotia in the soil. The efficacy of crop rotation with paddy or maize have been reported effective (Gupta *et al.*, 1987).

The incorporation of organic matter in the soil has been used as a viable option for the control of fungal diseases, especially those whose causal agents give rise to sclerotia that allow the pathogen to survive under a large range of adverse environmental conditions, which restricts chemical control. Viana *et al.* (2000) evaluated the effect of different sources of organic matter on the control of *S. sclerotiorum* and obtained best effects on plant survival with liquid Swine manure, *Bacillus subtilis* (10^8cfu/ ml) suspension and Iprodione (0.1% a.i.) and suggested that incorporation of organic matter in the substrate is a viable alternative. Plantation of biofumigant cover crops eg. mustard species, which release compounds in the soil that inhibit *Seclerotinia* growth and green manuring of these crops not only increase the organic matter in the soil but also promote soil microflora by acting as a food source. Due to the large population of soil microflora, there will be a competition for food, space and oxygen and soil microbes can out compete soil pathogens like *Sclerotinia* and reduce their numbers in soil (Villata and Pung, 2010).

Host resistance

Resistant varieties of a host against particular disease if available, not only ensure a pesticide free produce, but also make production of a crop more economical. As for as this disease is concerned, different workers have reported several varieties /lines having varying degree of resistance. Varieties having tolerance to the disease include Black Turtle, Soup. Sanilar, Capitol, Aurora, NY-69-6207-2 (Anderson *et al.*, 1974) and Ex Rico-23 (Tu and Beversdorf, 1974). Low sugar and high cellulose contents also confer resistance to this disease as found in varieties Biala Wyborowa, Weijsba, Sexobel, IGE 1/79 and IGE 2/79 (Pieta, 1990). Middleton *et al.* (1995) identified a series of resistant sources viz., PER 257, Bat 447-IC, Bac 17, Xan 170, A 480, etc. A high level of resistance was shown by lines Rabia de Gato, A 55, 491-Cal-71, REC-6, BAT 1290, DRK 2, AFR 91 and SUG 10 (Mushi and Ngulu, 1995).

Abawi *et al.* (1978) have reported a single dominant gene governing resistance against this disease and they found a white seeded selection of runner bean having a high level of resistance.

Chemical control

Various chemical compounds have been suggested for the control of this disease. Hartzfeld (1957) obtained good control with Terraclor (PCNB as dust, EC, or WP). Pegg (1962) suggested post-harvest dip in Allisan @ 1-2 pounds/gal. Dichloran applied as dust in furrows also proved effective. In 1965, Pegg advocated 2 pounds Allisan at weekly intervals starting from blossom fall to first harvest stage. Two sprays of DuPont 1999 @1 pound/acre pre- and post-blossom gave best results (Natti, 1967). One spray of Ronilan at flowering has been recommended by Baraer (1979). One spray of benomyl @ 0.5 pounds a.i./50-gal water/ acre, applied 3-5 days before full bloom is also effective (Hunter *et al.*, 1978).

Abai (1981) obtained good control with dichlozoline (20 WP) @ 100L/ha applied thrice at 10 days intervals starting from 3-5 days after flowering. On the other hand, Tu (1983) has recommended iprodione 10 μg/ml. Disease can also be managed effectively with benomyl +chlorothalonil (0.15+0.7 kg a.i./ha), benomyl + mancozeb (0.2 + 1.6 kg a.i./ha), benomyl + captan (0.25 + 1.2 kg a.i./ha) and chlorothalonil 1.0 kg a.i./ha (Oliveira, 1995). Vitavax and Bavistin reduced disease incidence and increased the yield under field conditions (Palat *et al.*, 2004). Efficacy of foliar sprays of various fungicides like azoxystrobin, carbendazim, thiophanate-methyl, boscalid, iprodione, prothioconazole, tebuconazole, trifloxystrobin and vinclozolin in the management of Sclerotinia diseases have been reported (Bradley *et al.*, 2006). Rotation of chemicals from different groups is necessary, as relying on the one chemical group may increase the risk of development of resistance in the pathogen.

Systemic acquired resistance

Plants have passive defensive systems that can be triggered upon pathogen attack (Dann and Deverall, 1996). Various chemicals and environmental stimuli may act on plants to induce systemic acquired resistance (SAR) as trigger to subsequent pathogen attack. This technique is cheaper and less hazardous to environment. There is no involvement of direct effect on pathogen rather than acquiring immunity against the trespasser. SAR chemicals like isotonic acid or salicylic acid (100 ppm) as seed treatment resulted in significant less seed mortality, lesion length, disease intensity and disease incidence of Sclerotinia stem rot of mustard (Mehta *et al.*, 2010).

Biological control agents

Biological agents if available, could also help a great deal in preventing environmental pollution due to pesticide residues as well as provide a healthy produce to the consumers. *Trichoderma lignorum* has been found antagonistic to *S. sclerotiorum* (Fedorinchik, 1956). Trutman *et al.* (1982) have reported good sclerotial control by hyperparasite *Coniothyrium minitans*. They also found good control by *Trichoderma koningii* in the temperature range of 20-35°C. Pre-sowing treatment has been found most effective. Foliar application of *C. minitans* also controls the infection of this pathogen on beans (Bremer *et al.*, 2000). Nearly hundred per cent destruction of sclerotia of this pathogen by *C. minitans* occurred at 30°C temperature, 30% soil moisture and 10^6 spores/ ml of bioagent within 4-week period (Tu, 1999). He further suggested that field application of *C. minitans* should be made as early as possible, immediately after the bean harvest in early to mid-September when temperature is warm and soil is moist due to autumn rains. Seed treatment with talc-based formulation of *T. harzianum* + CMC + sublethal dose of carbendazim (0.05 and 0.10%) resulted in maximum reduction in disease incidence and increased dry weight of roots, shoots and yield of crop (Nath *et al.*, 2004).

A combination of seed treatment with *T. harzianum* + soil amendment with FYM effectively managed the disease and increased plant growth and yield under field conditions (Das and Das, 2005). Dutta *et al.* (2008) reported *S. sclerotiorum* growth reduction of 64.93% in dual culture by *Pseudomonas fluorescens* followed by *Bacillus subtilis* (62.86%) and *T. harzianum* (59.08%). They further showed fungal antagonist initially caused coiling and then lysed the pathogen hyphae and the bacterial antagonist ceased pathogen growth after coming in contact with the pathogen.

The effects of Arbuscular Mycorrhizal Fungi (AMF) *Glomus mosseae*, *G. fasciculatum* and *Rhizobium leguminosarum* biovar *phaseoli* were examined on the patho-system of *S. sclerotiorum* and common bean (Aysan and Demir, 2009). The colonization and nodulation of two biological control agents exhibited differences because of reciprocal interactions of these items as well as the effect of the *S. sclerotiorum*. Nodulation of *R. leguminosarum* biovar *phaseoli* particularly decreased in triple inoculation. In addition, colonization of AMF significantly decreased in treatment of *S. sclerotiorum* +AMF than control AMF. Treatments of single inoculations of AMF and *R. leguminosarum* biovar *phaseoli* isolates reduced disease severity by 10.3-24.1%. It was determined that single biological control agents' inoculations were more effective than dual inoculations (AMF+ *R. leguminosarum* biovar *phaseoli*). When the morphological parameters of common bean were considered, all the morphological parameter's values were decreased in treatments which present pathogen isolate. Besides this, all biological control agents increased total contents of P and N in treated plants compared to the controls.

Seed treatment with *T. harzianum* was found has most effective in improving seed germination (18.43%), reducing infection (90.46%) and increasing yield (69.51%) over control plot under field conditions followed by *B. subtilis* and *P. fluorescence* as seed treatment. Application of *Streptomyces* sp. (isolate 422) applied @15g/ kg soil as wheat bran inoculum significantly reduced the disease incidence (Mehta *et al.*, 2012).

REFERENCES

Abai, J. 1981. Studies on the epidemiology and control of *Sclerotinia sclerotiorum* (Lib.) de Bary. *Report Hokkaido Prefectural Agric. Exp. Stn.* No. 36, 83p.

Abawi, G.S., Potlatch, F.J. and Molin, W.J. 1975. Infection of beans by ascospores of *Whetzelinia sclerotiorum. Phytopathology* 65(6): 673-678.

Abawi, G.S., Provvidenti, R., Crosier, D.C., and Hunter, J.E. 1978. Inheritance of resistance to white mould disease in *Phaseolus coccineus*. *J. Heredity* 69(3): 200-202.

Aberdeen, J.E.C. 1949. Report on nestiness in beans. *Banenaoul Sydney* 138: 20.

Anderson, F.N., Steadman, J.R., Coyne, D.P. and Schwartz, H.F. 1974. Tolerance to white mould in *Phaseolus vulgaris*, dry edible bean types. *Pl. Dis Reptr.* 58(9): 782-784.

Anonymous. 1921. Report of the pathologist for the period June 10th to Dec. 31st. *Bermuda Dept. Agri. Rep.*: 30-64.

Anonymous. 1937. Memoranda of the plant protection service of Morocco. 62pp.

Aysan, E. and Demir, S. 2009. Using Arbuscular Mycorrhizal Fungi and *Rhizobium leguminosarum* biovar *phaseoli* against *Sclerotinia sclerotiorum* (Lib.) de Bary in the common bean (*Phaseolus vulgaris* L.). *Plant Pathology Journal* 8(2): 74-78.

Bag, T.K. 2000. An outbreak of watery pod rot of French bean in the hills of Arunachal Pradesh. *J. Mycol. Pl. Pathol.* 30(1): 130.

Baraer, J. 1979. Protection against *Botrytis cinerea* and *Sclerotinia sclerotiorum* of vegetable crops for canning. *Revue Horticole* No. 198: 53-54.

Baribeau, B. and Raicot, H.N. 1928. Studies on diseases caused by sclerotia-producing fungi in Quebec, Canada. *Dept. Air. Dominion Bot. Rep*. 1927-28: 220-222.

Bedlan, G. 1984. *Pflanzenarzt* 37(5): 67-69.

Biffen, R.H. 1926. Annual report for 1926 of the Botanist. *Royal Agric. Soc. England J.* 87: 340-351.

Blackford, F.W. 1944. Sclerotinia or cottony rot. *Queensland Agric. J.* 4(3): 161-163.

Bongini, V. 1924. La"Sclerotinia". *Negri Ortaggi D efera delle Piantecontro Mal. Ed iParassiti.* 1: 39-42.

Bradley, C.A., Lamey, H.A., Endres, G.J., Hensen, R.A., Mekay, K.R., Helvorson, M., Le Gare, D.G. and Porter, P.M. 2006. Efficacy of fungicides for the control of Sclerotinia stem rot of canola. *Plant Dis*. 90: 129-134.

Brien, R.M. 1932. Host range of *Sclerotinia sclerotiorum* in New Zealand. *New Zealand J. Agric.* 44: 127-1.

Bremer, E., Huang, H. C., Selinger, L. J. and Davis, J.S. (2000). Competence of *Coniothyrium minitans* in preventing infection of bean leaves by *Sclerotinia sclerotiorum*. *Plant Pathology Bulletin* **9**, 69-74.

Buihwald, N.F. 1947. Sclerotiniaceae Daniae. *EnFlonstisk-systematiskoversigt over Del Danmark Fundne Knoldbaewgersvampe* 3: 235-330.

Caesar, A.J. and Pearson, R.C. 1983. Environmental factors affecting survival of ascospores of *Sclerotinia sclerotiorum*. *Phytopathology* 73(7): 1024-1030.

Chambers, S.C. and Hardie, M. 1960. *Sclerotinia* rot of beans. *J. Agric. W. Aust.* 1(11): 977-983.

Chambers, S.C. and Hardie, M. 1964. *Sclerotinia* disease of vegetables-survival of the fungus n soil. *J. Agr. W. Aust.* 5(3): 169-170.

Cook, G.E., Steadman, J.R. and Boosalis. 1975. Survival of *Whetzelinia sclerotiorum* and initial infection of dry edible beans in western Nebraska. *Phytopathology* 65(3): 250-255.

Coyne, D.P., Steadman, J.R. and Anderson, F.N. 1974. Effect of modified plant architecture of Great Northern dry bean vars. on white mould severity and components of yield. *Pl. Dis. Reptr.* 58(4): 379-382.

Dann, E.K. and Deverall, B.J. 1996. 2,6-dichloro-isonicotinic acid (INA) induces resistance in green beans to the rust pathogen, *Uromyces appendiculatus* under field conditions. *Australas. Plant Pathol.* 25: 199-204.

Das, M. G. and Das, B. C. 2005. Potentiality of antagonists in reducing white rot disease of French bean in amended soil. *Crop Research Hisar* 29(3): 503-508.

Dutta, P., Das, B. C. and Islam, M. 2008. Eco-friendly strategies for management of Sclerotinia rot of French bean. *Journal of Biological Control* 22(2): 405-410.

Elgorban A., Al-Sum B.A., Elsheshtawi M., Bahkali A.H. 2013. Factors affecting on *Sclerotinia sclerotiorum* isolated from beans growing in Ismailia, Egypt. *Life Sci. J.* 10(4):1278–1283.

Fedorinchik, N.S. 1956. The role of antagonists in controlling infection sources of causal agents of soil borne diseases. *Trud. vs esoyuz Inst. Zashch. Rast*. 3: 69-73.

Figueirêdo G., Figueirêdo L., Cavalcanti F., Santos A., Costa A., Oliveira N. 2010. Biological and chemical control of *Sclerotinia sclerotiorum* using *Trichoderma* spp. and *Ulocladium atrum* and pathogenicity to bean plants. *Braz. Arch. Biol. Technol.* 53(1):1–9.

Fuller, P.A., Steadman, J.R. and Coyne, D.P.1984. Enhancement of white mould avoidance and yield in dry beans by canopy elevation. *HortScience* 19(1): 78-79.

Gaxiola, L.S. 1977. Effect of different distances between furrows on the incidence of *Whetzelinia sclerotiorum* causing white mould of bean. *Agricultura Tecnicaen Mexico* 4(1): 73-76.

Ghasolia, R.P., Shivpuri, A. and Bhargava, A.K. 2004. *Sclerotinia* rot of Indian mustard (*Brassica juncea*) in Rajasthan. *Indian Phytopathol* 57:76-79.

Gupta, S.K., Agarwala, R.K. and Sharma, S.L. 1987. Effect of crop rotation on the management of stalk rot (*Sclerotinia sclerotiorum*) of cauliflower. *Indian J. Plant Pathol.* 5(1): 79-80.

Gupta, S.K., Shyam, K.R. and Dohroo, N.P.1997. *Sclerotinia* wilt of French bean. *Indian Phytopath.* 50(4): 593.

Hangerford, C.W. and Pitts, R. 1953. The *Sclerotinia* disease of beans in Idaho. *Phytopathology* 43(9): 519-521.

Hartzfield, F.G. 1957. Terraclor, a new soil fungicide. *Agric. Chem.* 21(7): 31-33.

Hatamleh A., El-Sheshtawi M., Elgorban A., Bahkali A., Al-Sum B. 2013. Pathogenicity of *Sclerotinia sclerotiorum* to bean (*Phaseolus vulgaris*, L.) Cultivars. *J. Pure Appl. Microbiol.* 7(4):3275–3279.

Hauman, M.L. 1915. Les Parasites Vegetaux Des Plantes Cltivees en Argentine et dans Les regions Limitrophs. *Buenos Aires Mus. Nac. De Hist. Nat. An.* 26: 163-225.

Huang, H.C. and Kemp, G.A. 1989. Growth habit of dry beans (*Phaseolus vulgaris*) and incidence of white mould (*Sclerotinia sclerotiorum*). *Pl. Prot. Bull. Taiwan* 31(3): 304-309.

Hunter, J.E., Abawi, G.S. and Crosier, D.C. 1978. Effects of timings, coverage and spray oil on control of white mould of snap beans with benomyl. *Pl. Dis. Reptr.* 62(7): 633-637.

Hunter, J.E., Pearson, R.C., Smith, C.A. and Palumbo, D.R. 1984. Relationship between soil moisture and occurrence of *Sclerotinia sclerotiorum* and white mould disease on snap beans. *Protection Ecol.* 7(4): 269-280.

Jaap, O. 1922. Weitere beitrage zur pilzflora von triglitz in der prignitz. *Bot. Ver. Der Brandenb. Verhandl.* 64: 1-60.

Kerr, E.P., Steadman, J.R. and Nelson, L.A. 1978. Estimation of white mould disease reduction of yield and yield components of dry edible beans. *Crop Sci.* 18(2): 275-279.

Lumsden, R.D. and Wergin, W.P. 1980. Scanning electron microscopy of infection of bean by *Sclerotinia*. *Mycologia* 72(6): 1200-1209.

Mehta, N., Hieu, N.T. and Sangwan, M.S. 2010. Influence of systemic acquired resistance chemicals on the management of white stem rot of mustard caused by *Sclerotinia sclerotiorum* (Lib.) de Bary. *Pl. Dis. Res.* 25: 171-173.

Mehta, N., Hieu, N.T. and Sangwan, M.S. 2012. Efficacy of various antagonistic isolates and species of *Trichoderma* against *Sclerotinia sclerotiorum* causing white stem rot of mustard. *J. Mycol. Pl. Pathol.* 42: 244-250.

Middleton, K.J., Tatnell, J. and Redden, R.J. 1995. Germplasm screening of *Phaseolus vulgaris* for resistance to *Sclerotinia sclerotiorum*. *Aust. Pl. Pathol.* 24(2): 118-125.

Moore, W.D. 1949. Flooding as a means of destroying the sclerotia of *Sclerotinia sclerotiorum*. *Phytopathology* 39(11): 920-927.

Mushi, C.S. and Ngulu, F.S. 1995. Screening for white mould resistance in common bean in Tanzania. *Landbouwversiteit Wageningen*: 262-267.

Nath, S., Das, B. C. and Sarmah, D. K. 2004. Management of white mold (*Sclerotinia sclerotiorum*) of French bean (*Phaseolus vulgaris*) with carrier-based formulation of *Trichoderma harzianum*. *Indian Journal of Agricultural Sciences* 74(3): 168-170.

Natti, J.J. 1967. Bean white mould control with foliar sprays. *Farm Res.* 33(4): 10-11,16.

Nicolas, G. and Aggery, B. 1933. Notes Mycologiques et Phytopathologiques. *Soc. d Hist. Nat. Bul.* 65 : 506-507.

Oliveira, S.F.H., Recco, C.A., Sugahara, E. and Oliveira, D.A. 1995. Comparative evaluation of fungigation and conventional spray of fungicides to bean *Sclerotinia sclerotiorum* control. *Summa Phytopathologica* 21(3/4): 249-252.

Palat, R., Narain, U. and Singh, P. N. 2004. Management of Sclerotinia stem rot of French bean through fungicides and biopesticides. *Annals of Plant Protection Sciences* 12(2): 447.

Pegg, J.J. 1967. Bean *Sclerotinia* control. *Qd. Agric. J.* 91(5): 262-264.

Pegg, K.G. 1962. Control of *Sclerotinia* rot of French bean. *Qd. J. Agric. Sci.* 19(5): 561-564.

Pieta, D. 1990. Susceptibility of French bean to fungal infection-to chemical composition of leaves and root exudates. *Hodowla Roslin, Aklimatyzacja Nasiennictwo* 34(3-4): 53-54.

Rudolph, M. 1986. *Sclerotinia sclerotiorum* (Lib.) de Bary on *Phaseolus vulgaris* L. *Nachrichtenblatt Fur den Pflanzenschutz in der DDR* 40(9): 195-196.

Saccardo, P.A. 1911. Notae Mycologicae. *Ann. Mycol.* 9: 249-257.

Saindon, G., Huang, H.C. and Kozub, G.C. 1995. White mould avoidance and agronomic attributes of upright common beans grown at multiple planting densities in narrow rows. *J. Amer. Soc. Hort. Sci.* 120(5): 843-847.

Sherbakoff, C.D. 1917. Some important diseases of truck crops in Florida. *Fla. Agri. Expt. Sta. Bull.* 139: 193-277.

Steadman, J.R., Coyne, D.P. and Cook, G.E. 1973. Reduction of severity of white mould disease on Great Northern bean by wider row spacing and determinate plant growth habit. *Pl. Dis. Reptr.* 57(12): 1070- 1071.

Terui, M. and Harada, Y. 1966. Effect of low temperature treatment of sclerotia on apothecial production in *Sclerotinia sclerotiorum*. *Bull. Fac. Hirosaki Univ.* 12: 24-30.

Trutman, P., Keane, P.J. and Merriman, P.R. 1982. Biological control of *S. sclerotiorum* on aerial parts of plants by the hyper parasite *Coniothyrium minitans. Trans. British Mycol. Soc.* 78(3): 521-529.

Tu, J.C. 1983. Efficacy of iprodione against *Alternaria* black pod and white mould of white beans. *Can. J. Pl. Pathol.* 5(2): 133-135.

Tu, J.C. 1988. The role of white mould infected white bean (*Phaseolus vulgaris* L.) seeds in the dissemination of *Sclerotinia sclerotiorum. J. Phytopathology.* 121(1): 40-50.

Tu, J.C. and Beversdorf, W.D. 1982. Tolerance to white mould (*Sclerotinia sclerotiorum*) in Ex Rico 23, a new white bean. *Can. J. Pl. Sci.* 62(1): 65-69.

Tu, J.C. 1999. Effect of temperature and moisture on the destruction of sclerotia of *Sclerotinia sclerotiorum* by *Coniothyrium minitans.* In Proc. 51st Int. Symp. on Crop Prot. Gent. Belgium, 4thMay, 1999 Part II. *Mededelingen – Faculteit Landbouw kundigeentoegepaste Biologische Wetenschappen, Universiteit Gent* 64 (3b): 411-417.

Viana, F.M.P., Kobori, R.F., Bettiol, W. and AthaydeSobrinho, C. 2000. Control of damping-off in bean plant caused by *Sclerotinia sclerotiorum* by the incorporation of organic matter in the substrate. *Summa Phytopathologica* 26(1): 94-97.

Villata, O. and Pung, H. 2010. Managing *Sclerotinia* diseases in vegetables-new management strategies for lettuce drop and white mould of beans. Department of Primary Industries. The State of Victoria. 4p.

Wallace, M.M. 1944. Sclerotinia disease of beans and other crops. *East African Agri. J.* 9: 171-172.

Weiss, A., Hipps, L.E., Blad, B.L. and Steadman, J.R. 1980. Comparison of within canopy microclimate and white mould disease (*Sclerotinia sclerotiorum*) development in dry edible beans as influenced by canopy structure and irrigation. *Agric. Meteorol.* 22(1): 11-21.

Weiss, A., Kerr, E.D. and Steadman, J.R. 1980. Temperature and moisture influences on development of white mould disease (*Sclerotinia sclerotiorum*) on Great northern beans. *Pl. Dis.* 64(8): 757-759.

Wong, J.A., Cox, J.I. and Maynard, J.R. 1980. White mould in green beans: research progress. *J. Agric. Tasmania* 51(4): 108-111.

Yerkes, W.D.Jr. 1955. Sclerotinia wilt of beans in Mexico. *Plant Disease Reporter* 39: 47.

Young, P.A. 1936. *Sclerotinia* rot of squash and pumpkin. *Phytopathology* 22(2): 184-190.

Zaumeyer, W.J. 1930. Bean diseases in western United States in 1930. *Pl. Dis. Reptr.* 14(22): 228-239.

12

CROWN ROT

Crown rot or Southern blight which is also referred to as Southern wilt of French bean, is a widespread disease throughout the world and causes colossal yield losses to this crop.

GEOGRAPHICAL DISTRIBUTION

Southern blight, southern wilt, or crown rot caused by *Sclerotium rolfsii* Sacc. was first recognized as a distinct disease by Rolfs (1893) for whom Saccardo named the species in 1911. The disease was first reported on beans from Florida, U.S.A. in 1916 (Peltier, 1916) and thereafter, it was reported from New South Wales (Birmingham, 1924), Beltsville (Cooley, 1938), Brazil (Viegas, 1945), Georgia (Luttrell and Garren, 1952), Colombia (Cardona - Alvarez, 1954), Uruguay (Pastirino, 1964), Venezuela (Casanova and Polanco, 1966), Mexico (Crispin and Campos, 1976), Costa Rico (Echandi, 1976) and Chile (Montealegre and Esterio, 1989).

In India, Roy (1977) observed a serious disease of this crop exhibiting collar, stem and root symptoms due to this disease in Assam. Subsequently it was reported from Manipur (Singh and Singh, 1990) and Bihar (Sinha *et al.,* 1993). Crown rot phase of this disease on French bean was observed during 1998 in Solan area of Himachal Pradesh (Gupta and Shyam, 1998). Sudden wilt and leaf spots were observed during early growth to pre-flowering stage in the Mysore and Mandya districts of Karnataka State (Mahadevakumar *et al.*, 2015). This was the first report on the occurrence of *S. rolfsii* causing southern blight and leaf spot of common bean in India.

LOSSES

Losses from this disease vary greatly in different localities and from year to year. Direct estimation of losses caused by this pathogen alone is not available (Zaumeyer and Thomas, 1957). Crop losses due to this disease start from 12 to 13 days after emergence and continue till crop maturity. Wilhem (1942) reported up to 5 per cent losses due to this disease. Being an aerobic pathogen, it causes up to 90 per cent crop losses in loose soils (Borzini and Picco, 1949). In field trials with several host crops, Niemann *et al.* (1967) observed that bean (*Phaseolus vulgaris*) was one of the most susceptible hosts and losses in yields ranged from 50.7 to 61 per cent. In India, Gupta and Shyam (1998) reported about 25 per cent incidence of the disease during 1996 crop season in the fields around Solan, Himachal Pradesh.

SYMPTOMS

The symptoms of the disease as described by Gupta and Sharma (2004) are as follows:

The pathogen attacks the crop both in vegetative as well as in reproductive phase. The initial symptom of the disease appears as dark brown, water soaked lesion on the stem at collar region just below the soil line (Plate 1 a). The lesions enlarge rapidly and girdle the stem. The infection of the underground part of the stem and tap root extends downward and destroy the cortex. As the disease progress, water-soaked spots also appear on the upper branches of the plants. At the base of the stem and on the ground around the plant a moldy growth of white mycelium intermixed with many sclerotial bodies are produced at all growth stages of the plants (Plate 1b). The earliest symptoms on leaves of the affected plants are slight yellowing of lower leaves. As the disease further progress, leaves on the upper branches of the plants become yellow and drop off eventually (Plate 1c, d). Sometimes the under surface of the affected leaves is also covered with moldy growth of white mycelium and intermixed with sclerotial bodies. Pods which are in close proximity of soil also get infected. Coarse fan-shaped mycelial strands along with sclerotial bodies radiate over infected pods (Plate 1e). Developing seed in the infected pod may also get infected.

THE PATHOGEN

Sclerotium rolfsii Sacc. consists of coarse hyphae and the size of cells ranges from 150-250 µm x 2-9 um (Zaumeyer and Thomas, 1957). Higgins (1922) reported that the hyphae of this pathogen are characterized by the presence of clamp connections and a peculiar method of branching, which are valuable aids in ascertaining its taxonomic position. The sclerotia (0.5-1.5 mm in dia.) at first white but becoming brown with age are more or less smooth and glossy and somewhat resemble mustard seed.

Goto (1930) first described the teliomorphic stage of *S. rolfsii* who provisionally placed it in *Corticium centrifugum* (Lev.) Bres. Curzi (1931) later proposed that *C. rolfsii* should be designated as the basidial state of *S. rolfsii*. Barrett (1934) observed the basidial stage of the fungus in United States. West (1947) transferred it to *Pellicularia rolfsii* and subsequently placed in *Botryobasidium* by Venkatarayan (1950). Talbot in 1973 suggested that the basidial state of *S. rolfsii* belonged to the genus *Athelia,* and acting upon his suggestion, Tu and Kimbrough (1978) made the combination *Athelia rolfsii* (Curzi) Tu and Kimbrough. The teleomorph of *S. rolfsii* has been reported infrequently from nature, but there are some reports of hymenial formation in culture. The hyphae from which basidia developed initially appeared slender with terminal pro-basidial swelling. As the terminal portions enlarged, clavate basidia were discernable. Sterigmata developed as the basidia matured and basidiospores subsequently formed.

Basidiospores are resupinate, effused, in patches, usually 1-2 cm in diameter, sometimes up to 6 cm, pellicular, usually 50-100 µm thick and loosely adherent to the substrate and separable as a layer, sometimes up to 3-5 mm thick depending on the substrate. Subicular hyphae small, 3-5µm diameters loosely arranged, simple septate with rare clamp connections, hyaline binucleate. Hymenophore coriaceous-membranous, merulioid, but becoming smooth on drying, whitish to creamy-white.

Cystidia are lacking. The basidia arise in clusters, relatively short, clavate, mostly 4-spored without a basal clamp, (11.5-) 15.8-19(-26) x (4-) 5-6(-6.5) µm, length of sub-basidial stalk or branches varying with isolate, substrate and environmental conditions. The spores are hyaline, thin walled, smooth measuring 5.5-7x3-5 µm, subglobose to ellipsoid and binucleate (Punja *et al.*, 1982).

The genome size of *Athelia rolfsii* is approximately 73 Mb having 8919 contigs (Iquebal *et al.,* 2017). Annotation analysis revealed 16830 genes which are involved in fungicide resistance, virulence and pathogenicity along with putative effector and lethal genes. Secretome analysis revealed CAZY genes representing 1085 enzymatic genes, glycoside hydrolases, carbohydrate esterases, carbohydrate-binding modules, auxillary activities, glycosyl transferases and polysaccharide lyases.

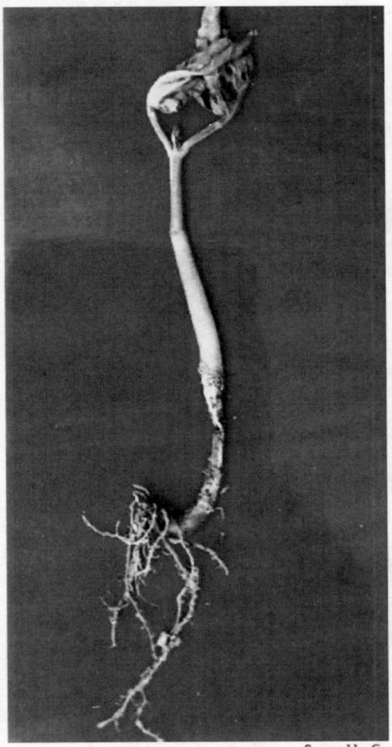

Plate 1a. Symptoms on seedling

Plate 1b. Symptoms on grown up plants

Plate 1c. Symptoms on leaf-dorsal view

Plate 1d. Symptoms on leaf-ventral view

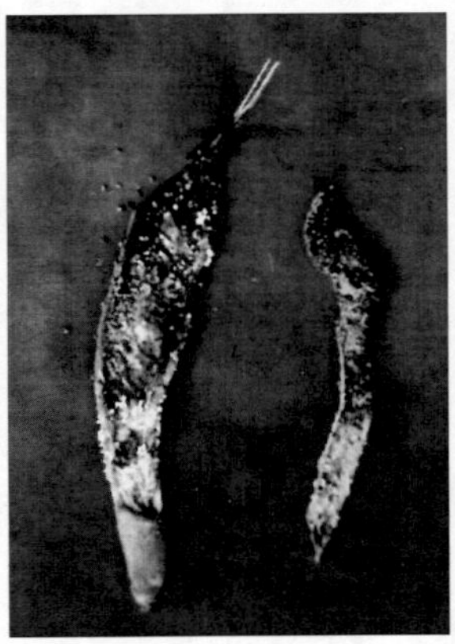

Plate 1e. Symptoms on pods

Plate 1. Symptoms of crwon rot of French bean

VARIABILITY

Goto (1934) divided different strains of *S. rolfsii* into 4 groups. Group 1 included strains which infected potato, barley, sugarcane, cantaloupe and *Setaria italica* while, pepper (*Capsicum annum*), beans (*Phaseolus vulgaris*), iris, peach and *Cedrus deodara* were infected by group II strains from United States, whereas group III comprised strains of the fungus from *Garcinia spicata*, clover (*Trifolium repens*), *Allium fistulosum* and *Petasites japonicus*, all at Taihokn. Group IV was represented by strains from potato (India) and species of Cyperaceae, beans, *Lochnera rosea*, *Mentha arvensis* var. *piperascens*, groundnut, *Cymbidium longisepalum* and *Jasminum sambac* all at Taikohn. Eight isolates of *S. rolfsii* studied by Fajardo and Mendoza (1935) fell into three groups as shown by differences in growth rate, type of colony, time required for sclerotial formation, and number, size and colour of the sclerotia. Isolates I to IV from tomato, bean (*Phaseolus vulgaris*), groundnut and pineapple were apparently identical and commonest strain found locally producing round to ovoid sclerotia with light to dark brown colour. On the other hand, Epps *et al.* (1951) reported four strains of the fungus that could be readily distinguished by their mutual aversion in agar plate culture. They differed slightly but consistently in virulence as measured by the rapidity with which young plants were killed. One strain was morphologically distinct and readily identified by its larger sclerotia. While most cultures are homothallic, heterothallism do occur (Cooper, 1961). Strains of the fungus do differ in their tolerance to Fytolan (50% copper oxychloride). A parent strain of *S. rolfsii* grew at less than 7000 ppm of the solution (Kapoor, 1962).

In all, 348 isolates of *S. rolfsii* were obtained from bean fields in seven agroecological zones of Uganda and morphological and pathogenic characterization was carried out (Paparu *et al.*, 2020). It was found that mycelial growth rate and the number of sclerotia produced on artificial media varied among agroecological zones but not within a zone. The five bean varieties tested were found to be susceptible to *S. rolfsii*, including varieties MLB49-89A and RWR719 that are resistant to Fusarium and Pythium root rots, respectively. Pre-emergence damping-off ranged between 0 and 100%, and disease severity index ranged between 4.4 and 100%. The widespread and high levels of *S. rolfsii* virulence on varieties of common bean indicate that management intervention is urgently required to help reduce losses incurred by the pathogen.

HOST RANGE

Sclerotium rolfsii Sacc. infects more than 500 species of plants belonging to 100 families and causes economically important diseases in warm moist climates throughout the world (Aycock, 1966). Various workers from different parts of the world carried out infection experiments and clearly demonstrated that it was parasitic on several hosts widely separated in taxonomic relationship (Fulton, 1908, Wolf, 1914, Harter, 1916 and Taubenhaus, 1919). Different hosts includes rhubarb, decaying fallen leaves of antirrhinum, French bean, delphinium, imported yams *(Dioscorea* spp.), bananas, sword bean (*Canavalia ensiformis*), *Zinnia elegans,* pepper, iris, cantaloupe, peach, *Cedrus deodara*, *Garcinia spicata*, clover (*Trifolium repens*), *Allium fistulosum, Mentha arvensis,*var. *piperascens,* groundnut, *Jamonium sambac,* beet, carnation, carrot, cauliflower, convolvulus, cotton, cowpea, chrysanthemum, eggplant, fig, grasses, hydrangea, rock melon, (*Cucumis melo*), squash, sweet potato, tobacco, tomato, water melon and citrus (Birmingham, 1925; Briant and Martyn, 1929; Ramakrishnan, 1930; Goto, 1934; Magee, 1947). Among fifteen vegetables screened for host range of this pathogen under field conditions, only onion was found as a non-host and rest were moderately susceptible to highly susceptible (Niemann *et al.*, 1967). Roy (1977) also observed this fungus in Assam, India, on vegetables like chilli (*Capsicum annuum* L.), Frenchbean (*Phaseolus vulgaris* L.), carrot (*Daucus carota* L. cv. *Nantes*), Pea (*Pisum sativum* L.), cauliflower (*Brassica oleracea* var. *botrytis* L.), cabbage (*Brassica oleracea* var. *capitata*), potato (*Solanum tuberosum*) and arum (*Colocasia* sp.).

DISEASE CYCLE

The pathogen is both soil and seed borne (Bolkan *et al.*, 1976) in nature. The inoculum of *S. rolfsii* consists of sclerotia, mycelium and basidiospores. The role of basidiospores in the life cycle of this pathogen is of minor importance. Sclerotial germination is induced by volatiles that emanate from crop residues in the soil (Linderman and Gilbert, 1975). Disease development is affected by high temperature and moisture, which favour sclerotial germination and optimum mycelial growth. The pathogen penetrates host tissue mechanically and gives rise to coenocytic intra and intercellular hyphae (Milthrope, 1941). Before penetration can occur, appreciable mycelial growth must take place on the plant surface (Abeygunawarena and Wood, 1957; Coley-Smith and Cooke, 1971). Death of the cytoplasm precedes the invading hyphae by one cell layer. Dissolution of the middle lamellae occurs shortly after entry. Higgins (1927) believed that the death of cells is caused by the toxic action of oxalic acid and that oxalic acid is secreted by the pathogen hyphae. The pathogen also produces protopectinase and pectinase that cause cell disintegration in bean hypocotyls (Zaumeyer and Thomas, 1957). Baterman (1969) reported the production of cellulase, and Van Etten and Baterman (1969) detected enzymes, which readily degraded pectic galactan, galactomannan. These substances may also play some role in infection caused by *S. rolfsii*. After appearance of symptoms, mycelium and sclerotia are produced on the infected tissue in favourable weather conditions. Dispersal of mycelium and sclerotia may occur by contaminated irrigation water, soil, adhering to cultural tools and feet of animals or seed. The sclerotia of this pathogen can pass through the digestive tract of animals without losing viability and therefore, be transported relatively long distances by animals which feed on infected host materials (Leach and Davey, 1942), serve as the source of primary inoculum in the ensuing season.

EPIDEMIOLOGY

Sclerotia are the chief source of survival for this pathogen. The pathogen prefers a warm climate and considerable soil and air moisture. The fungus is not well adapted to low temperature conditions, but in culture it may grow at temperatures between 13- 37° C with an optimum temperature for mycelial growth is 28-32°C and the corresponding water potential is –2.3 bars (Yu *et al.*, 1996). Sclerotia germinate at 10- 35° C, and the fungus requires relative humidity > 99 per cent. Sclerotial germination decreases with increased soil depth due to reduced aeration (Abeygunawarena and Wood, 1957). Germination occurs at a pH range of 2.6- 7.7 with 2.6- 4.4 as optimum (Coley-Smith and Cooke, 1971). Sclerotial germination is induced by volatiles that emanate from crop residues in the soil (Linderman and Gilbert, 1975). Beans are infected at 8-37°C (Lauritzen *et al.*, 1933). Gupta *et al.* (2002) reported that the disease was serious at 25-30° C and 50% soil temperature and moisture, respectively. Sclerotial formation is favoured by light and normal aeration (Perez Sendin *et al.*, 1986).

MANAGEMENT

Cultural practices

Changes in time of planting as a control measure is generally suggested for the management of soil borne diseases. Crops such as corn or wheat may result in less disease incidence in subsequent years by lowering inoculum levels. As this disease does not affect most of the cucurbitaceous crops, crop rotation with such crops may be helpful in minimizing losses to the bean crop (Palti, 1954). Cover crops, weeds and refuse should be buried significantly deeply so that dead leaves do not provide a substratum for the growth of the pathogen (Boyle, 1961; Garren, 1961). Eradication of susceptible weed hosts and destruction of infected host residues by burning or deep ploughing reduce soil populations of *S. rolfsii*. Selecting fields also can reduce inoculum levels with low soil acidity and good drainage. Wider plant spacing and application of lime to increase soil pH can also be utilized in reducing the infection. Messiaen *et al.* (1976) reported mycelial growth inhibition of the pathogen by soluble nitrogen compounds as urea, ammonium nitrate and ammonium sulphate. Soil amendment with these materials either as a fertilizer or pre-plant treatment can reduce *S. rolfsii* infection (Leach and Davey, 1942; Henis and Chet, 1968)). Reynolds (1970) reported that soil amendment with coconut mulch reduced infection and increased yields considerably. Esqueda Valle and Zenteno Zaveda (1991) have suggested mulching with black or transparent polythene of 200 µm thickness that resulted in 71 to 100% control of *S. rolfsii*. Soil flooding or submergence reduces viability of the sclerotia and reduce seedling blight (Sariah and Tanaka, 1995). Montealegre (1996) have advocated 40 days of soil solarization at 5 or 10 cm soil depth for effective control of this pathogen.

Host resistance

Borzini and Picco (1949) reported variation in susceptibility and resistance among five varieties and Regima altadi Monsummano was the most resistant. Plant resistance has been identified and includes the cvs. Ibibe Redondo as resistant (Cardona Alvarez, 1954) while Mexico 348-2 and Blanco, were moderately tolerant to *S. rolfsii* (Vieira, 1967). Echevarria *et al.* (1982) reported existence of significant differences in level of resistance among different varieties screened while Hernandez *et al.* (1988) found none of the cultivar as resistant against this disease. Additional research is required to identify more sources of resistance and/or tolerance to the fungus.

Chemical control

Soil or foliar application of 2,4-dichlorobenzyl tributyl phosphonium chloride or cycocel have proved quite effective in reducing seedling infection (Tahori *et al.*, 1965). In sandy loam soils, addition of ammonium hydroxide (0.2%) raises the soil pH to toxic level with complete loss of sclerotial germinability (Henis and Chet, 1967). Fungicides like benodanil and carboxin also inhibit sclerotial germination (Maiti and Choudhuri, 1975; Menten, 1980). Seed treatment with carbendazim (0.2%), triadimenol (0.5%) provided good control of this pathogen (Asghari and Mayee, 1991; Sharma *et al.*, 1994). Several systemic and non-systemic fungicides like carbedazim, flutolanil, diniconazole, tebuconazole, propiconazole, hexaconazole, epoxiconazole, carboxin and captaf provided good control of this pathogen when used as soil drenches (Hagan *et al.*, 1991; Tiwari, 1995; Das *et al.*, 1997). *In vitro* evaluation of fungicides against the pathogen showed that penconazole, hexaconazole, propineb and mancozeb completely inhibited the mycelial growth of the pathogen (Gupta and Sharma, 2004). Maximum inhibition of sclerotial germination was also observed in penconazole. Under field conditions propineb was the most effective in reducing disease incidence and in producing maximum green pod yield while other systemic fungicides like carbendazim, benomyl and thiophanate methyl were found ineffective against this disease (Huang *et al.*, 1996, Das *et al.*, 1997, Gupta and Sharma, 2004). Gupta and Upmanyu (2005) collected 25 isolates of the pathogen from different states of the country for *in vitro* screening against carbendazim (1000 ppm) and reported that 68 per cent isolates were insensitive while 32 per cent showed only slight inhibition. Hexaconazole (0.1%) and difenoconazole (0.1%) and kresoxim methyl (0.1%) proved most effective in containing the incidence of this disease under laboratory conditions.

The efficacy of five each systemic fungicides viz., Bavistin (Carbendazim), Vitavax (carboxin + thiram), Tilt (propiconazole), Contaf (hexaconazole), SAAF (carbendazim 12% + mancozeb), non-systemic fungicides Zineb (indofil Z-78), Blitox-50 (copper oxychloride), Antracol (propineb), Indofil M-45(Mancozeb) and Kavach (chlorothalonil) were tested against *S. rolfsii* causing collar rot of French bean (Kumar *et al.*, 2018). Among the systemic fungicides propiconazole, Vitavax, hexaconazole and SAAF each at 0.025%, 0.05% and 0.01% concentration completely inhibited the mycelial growth of *S. rolfsii.* Among the five non-systemic fungicides tested zineb, propineb, mancozeb and Kavach were found to be more effective showing 100% inhibition of mycelial growth at all concentration (0.1, 0.2 and 0.3%).

Biological control agents

Use of biocontrol agents is an ecofriendly method of plant disease control and it should be advocated wherever it is applicable. *Trichoderma viride, Penicillium* spp. and certain bacteria affect natural functioning of sclerotia (Curl and Hansen, 1964). *Penicillium* spp. also inhibits the production of mycelium and sclerotia (Diaz Polanco and Luis Castro, 1977). *T. harzianum* has been found to increase the yield of bean besides invading the mycelium of the fungus under conditions adverse to growth of the pathogen (Elad *et al.*, 1980). Severity of this disease can be reduced by as much as 35 per cent by using TN-21 isolate of *Trichoderma* (Silveira *et al.*, 1994). Sanchez *et al.* (1994) have advocated seed treatment of bean with *Psuedomonas cepacia* to protect against infection by root rot pathogens. Pellets of Th V isolate of *T. harzianum* when added to fumigated soil, suppressed plant damage by 83 per cent (Montealegre and Larenas, 1995). Roberti *et al* (1996) investigated the activity of *Trichoderma* species against bean root rot (*S. rolfsii*) and applied as seed treatment and reported that the *Trichoderma* strains reduced bean root rot effectively. Further under laboratory conditions, they observed that non-volatile metabolites released by *T. viride* 117 in agar completely inhibited growth of *S. rolfsii*, while those released by *T. harzianum* 74 and *Trichoderma* sp. 6A only reduced its growth partially. In another study, *Gliocladium virens* followed by *Trichoderma viride, T. hamatum, T. harzianum* and *Chaetomium globosum* proved effective to cause inhibition in mycelia growth (Plate 2) and also resulted in reduction in disease incidence in pot experiments (Sharma, 2001).

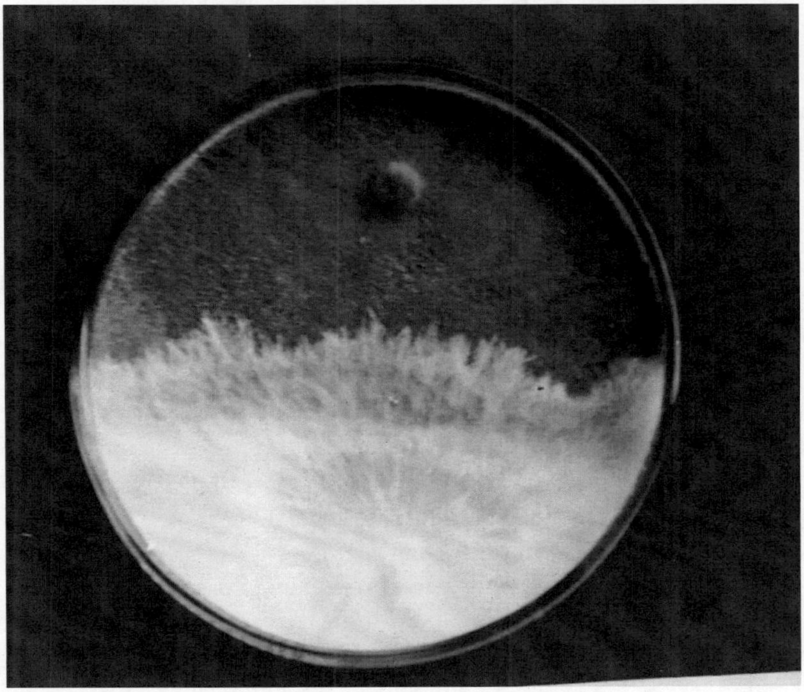

Plate 2. *In vitro* inhibition of *Sclerotium rolfsii* by *Gliocladium virens*

The role of four bioagents, i.e. *Bacillus subtilis, Pseudomonas fluorescens*, yeast (*Saccharomyces cerevisiae*) and *Trichoderma viride*, in controlling damping-off disease of bean (*Phaseolus vulgaris*) caused by *S. rolfsii* was evaluated under greenhouse and field conditions (Eid, 2014). The greenhouse experiment indicated that all the tested bioagents significantly reduced the incidence of the disease compared to control treatment. The most effective treatments were *B. subtilis, T. viride* and *P. fluorescens* which reduced disease incidence by more than 83.7 and 74.5% for pre- and post-emergence damping-off, respectively, while increased the plant survival by 90.3, 86.1 and 87.6%, respectively, compared to 26.3% in untreated plants. These treatments also increased the dry and fresh weights of bean shoot and root as well as resulted in considerable increases in the activities of peroxidase, polyphenoloxidase and chitinase by 260.0, 109.0 and 218.3%, respectively. Under field conditions, all the tested bioagents significantly reduced disease incidence with considerable increases in plant survival and higher seed yield during the two seasons (February of 2009 and 2010) of study. During the first season, *B. subtilis T. viride, S. cerevisiae* and *P.*

fluorescens were found to be the most effective bioagents which reduced the disease incidence with more than 61.3 and 41.3% than the control for pre- and postemergence damping-off, respectively. The corresponding percentages of survived plants were 78.2, 79.0, 75.2 and 76.8%, respectively, viz. 38.5% for the control. On the other hand, the most effective treatments for increasing seed yield were *S. cerevisiae* followed by *P. fluorescens*, being 894.95 and 748.1 kg/feddan viz. 269.2 kg/feddan for the control. The other two bioagents showed moderate effect. The same trend was obtained during the second season.

Sixty-five *Trichoderma* spp. isolates were tested against the mycelia growth of *S. rolfsii* and the following ones were selected *in vitro* for the *in vivo* tests: 5, 11, 12, 15, 102, 103, 127, 136, 137, 1525 (*T. longibrachiatum*), 1637 (*T. reesei*), 1642, 1643 (*T. harzianum*), 1649 (*T. harzianum*), 1700 (*T. asperellum*) and EST 5 (Pacheco *et al.*, 2016). The most promising isolates were identified by Sequencing of the internal transcribed spacer regions ITS1, ITS2, and the 5.8s rRNA genomic region, using the ITS5 and ITS4 primers and compared with sequences in the National Center for Biotechnology Information (NCBI) database. These selected isolates 1649 (*T. harzianum*), 1525 (*T. longibrachiatum*) and 1637 (*T. reesei*) were tested for evaluation of sclerotial germination inhibition under laboratory conditions, and to evaluate the effects of these on disease of bean plants under greenhouse conditions. The *Trichoderma* isolates 1649, 1525 and 1637 were more efficient in reducing sclerotial germination. In addition to 1649, 1525 and 1637, the isolates 5, 12, 102 and 1525 (*T. longibrachiatum*) significantly reduced de amount of diseased bean plants under greenhouse conditions.

Application of the bio-inoculants *viz.,* arbuscular mycorrhizal fungi *(Claroideoglomus etunicatum), Saccharomyces cerevisiae,* or *Trichoderma viride* solely or in different combinations to bean plants, enhanced plants resistance against infection with *Sclerotium rolfsii* and also improves plant nutritive status (Abdelhafez *et al.*, 2021). All tested bioinoculants significantly raised the activities of plant defense enzymes i.e. chitinase, peroxidase and polyphenoloxidase as compared to non-inoculated control. Likewise, pre-, post- and plant survival percentages significantly increased due to these bio-inoculations. Increased survival percentages were attributed to the concurrent increases in uptake of N, P and Zn nutrients by plants treated with bioinoculants. In this concern, plant nutrients uptake was higher in combined than single bio-inoculant treatments. Moreover, the uptake values of plant nutrients owing to the combined bio-inoculants were higher than the corresponding ones achieved due to fungicide treatment.

Integrated disease management

A combination of heat treatment with *T. harzianum* controlled *S. rolfsii* in beans by 90 to 100 per cent (Elad *et al.*, 1980). Similarly, combined use of *T. viride* and fungicide quintozene gives best control of *S. rolfsii* on beans (Olivos and Mont, 1993).

REFERENCES

Abdelhafez, A. A., Eid, K. E., El-Abeid, S. E., Abbas, M. H. H., Ahmed, N., Mansour, R. R.M.E., Zou, G., Iqbal, J., Fahad, S., Elkelsih, A., Alamri, S., Siddiqui, M. H. and Mohamed, I. 2021. Application of soil biofertilizers to a clayey soil contaminated with *Sclerotium rolfsii* can promote production, protection and nutritive status of *Phaseolus vulgaris*. *Chemosphere* 271: 129321.

Abeynawarena, D.V.W. and Wood, R.S.K. 1957. Factors affecting germination of sclerotia and mycelial growth of *S. rolfsii* Sacc. *Trans. Brit. Mycol. Soc.* 40: 221-231.

Asghari, M.R. and Mayee, C.D. 1991. Comparative efficiency of management practices on stem and pod rot of groundnut. *Indian Phytopath.* 44: 328-332.

Aycock, R. 1966. Stem rot and other diseases caused by *Sclerotium rolfsii*. *NC Agric. Exp. Sta. Tech. Bull.* 174. 202 pp.

Barrett, J.T. 1934. Observations on the basidial stage of *Sclerotium rolfsii*. *Phytopathology*. 24: 1137-1138.

Bateman, D.F. 1969. Some characteristics of the cellulase system produced by *Sclerotium rolfsii* Sacc. *Phytopathology* 59: 37-42.

Birmingham, W.A. 1924. A fungus and some of its host plants (*Sclerotium rolfsii*) Sacc. *Agric. Gaz. New South Wales* 35(6): 441-444.

Bolkan, H.A., Silva, A.R. and Cupertino. 1976. Fungi associated with soybean and bean seeds and their control in central Brazil. *Pl. Dis. Reptr*. 60: 545-549.

Borzini, G. and Picco, D. 1949.Verification of the relative resistance of bean races to *S. rolfsii* and a *Rhizoctonia* sp. *Notiz. Malatt. Piante* 2: 25-28.

Boyle, L.W. 1961. The ecology of *S. rolfsii* with emphasis on the role of saprophytic media. *Ga Exp. Sta.*: 117-119.

Briant, A.K. and Martyn, F.B. 1929. Diseases of cover crops. *Trop. Agric.* 6(9): 258-260.

Cardona Alavarez, C. 1954. Fungal root rot of the bean (*Phaseolus vulgaris*) in the Valley of Medellin-Columbia. *Rev. Fsac. Agron. Medellin* 15(46): 137-209.

Casanova, J.R. and Polanco. 1966. Lab enfermedades fungosasmas importantes de la caraota (*Phaseolus vulgaris*) en la zona central de Venezuela. *Agron. Trop. (Maracay)*. 16: 129-139.

Coley-Smith, J.R. and Cooke, R.C. 1971. Survival and germination of fungal sclerotia. *Ann. Rev. Phytopath*. 9: 65-92.

Cooley, J.S. 1938. Susceptibility of crop plants and weeds to *Sclerotium rolfsii*. *Phytopathology* 28(8): 594-595.

Cooper, W.E. 1961.Strains of, resistance to, and antagonists to *Sclerotium rolfsii* *N.Carol. State College, Raligh* pp.113-116.

Crispin, A. and Campos, J. 1976. Bean diseases of importance in Mexico in 1975. *Pl. Dis. Reptr*. 60: 534-535.

Curl, E.A. and Hansen, J.D. 1964. The microflora of natural sclerotia of *Sclerotium rolfsii* and some effects on the pathogen. *PL. Dis. Reptr.* 48(6): 446-450.

Curzi, M. 1931. STUDI SULO "*Sclerotium rolfsii*". *Bot. R. Staz. Pat. Veg. (n.s.)*. 11: 306-375, illus.

Das, S.R., Pani, B.K. and Panda, S.N. 1997. Evaluation of fungicides against *Sclerotium rolfsii* causing collar rot of tuberose. *Orissa J. Hort.* 25(1): 6-48.

Diaz Polanco, C. and Luis Castro, J. 1977. Studies on biological control of *Sclerotium rolfsii Agronomioa Tropical* 27(5): 539-547.

Echandi, E. 1976. Principles enfermedades de hongo del frifol (*Phaseolus vulgaris*) en los tropicsamericanosy su distribution en diferentes zone sdologicas. *Fitopatologia Brsiliera* 1: 171:177.

Echevarria, E., Gonzalez, A.M. and Marrero, H. 1982. Two methods of controlling the fungus causing southern blight of bean. *Gencias de la Agricultura* 12: 11-16.

Eid, K. 2014. Biological control of bean damping-off caused by *Sclerotium rolfsii. Egyptian Journal of Phytopathology* 42 (1): 179-191.

Elad, Y., Chet, Y. and Kartan, J. 1980. *Trichoderma harzianum*: a biocontrol agent effective against *S. rolfsii. Phytopathology* 70(2): 119-121.

Elad, Y., Kartan, J. and Chet, I. 1980. Physical, biological and chemical control integrated for soil borne diseases in potatoes. *Phytopathology* 70(5): 418-422.

Epps, W.M., Patterson, J.C. and Freeman, I.E. 1951. Physiology and parasitism of *Sclerotium rolfsii. Phytopathology* 41(3): 245-256.

Esqueda Valle, M. and Zenteno Zevada, M. 1991. The effect of polythene mulch on the mycoflora associated with bean. *Serei Botanica* 61(1): 11-19.

Fajardo, T.G. and Mendoza, J.M. 1935. Studies on the *Sclerotium rolfsii* Sacc. Attacking tomato, peanuts and other plans in the Philippines. *Philipp. J. Agric.* 4(4): 387-424.

Fulton, H.R. 1908. Diseases of pepper and beans. *La. Agr. Expt. Sta. Bul.* 101:21.

Garren, K.H. 1961. Control of *Sclerotium rolfsii* through cultural practices. *Va Agric. Exp. Sta.*, Holland: 120-124.

Goto, K. 1930. On the perfect stage of *Sclerotium rolfsii* Sacc. produced on culture media. Preliminary report. *J. Soc. Trop. Agric*. 2: 165-175.

Goto, K. 1934. *Sclerotium rolfsii* Sacc. in perfect stage. Some correlation between sporation and cultural characteristics. *Trans. Nat. Hist. Soc. Formosa* 22: 37-43, 75-90.

Gupta, S.K. and Shyam, K.R. 1998. A new record of crown rot of French bean from Himachal Pradesh. *Pl. Dis. Res.* 13(2): 168.

Gupta, S.K. and Sharma, A. 2004. Symptomatology and management of crown rot of (*Sclerotium rolfsii*) of French bean. *J. Mycol. Pl. Pathol*. 34(3): 820- 823.

Gupta, S.K., Sharma, A., Shyam, K.R. and Sharma, J.C. 2002. Role of soil temperature and moisture on the development of crown rot (*Sclerotium rolfsii*) of French bean. *Pl. Dis. Res.* 17(2): 366-368.

Gupta, S.K. and Upmanyu, S. 2005. Management of *Sclerotium rolfsii*, causal agent of crown rot of French bean, having decreased sensitivity to carbendazim. IInd Global Conference Plant Health- Global Wealth, organized by Indian Society of Mycology and Plant Pathology at MPUAT, Udaipur w.e.f. Nov. 25-29, 2005 (Abstr.).

Hagan, A.K., Weeks, J.R. and Bowen, K. 1991. Effects of application timing and method of control of southern stem rot of peanut with foliar applied fungicides. *Peanut Sci*. 18:47-50.

Harter, L.L. 1916. Storage rots of economic aroids. *J. Agri. Res.* 6:549-571.

Henis, Y. and Chet, I. 1967. Mode of action of ammonia on *Sclerotium rolfsii*. *Phytopathology* 57(4): 425-427.

Henis, Y. and Chet, I. 1968. The effect of nitrogenous amendments on the germinability of sclerotia of *Scleroium rolfsii* and on their accompanying microflora. *Phytopathology* 58: 209-211.

Hernandez,G.A. and Gracia, C. 1988. Screening of bean varieties for resistance to *Sclerotium rolfsii*. *Gencias de la Agricultura* 34-35:15-17.

Higgins, B.B. 1922. Notes on morphology and systematic relationship of *Sclerotium rolfsii* Sacc. *J. Elisha Mitchell Sci. Soc.* 37(3-4): 167-172.

Higgins, B.B. 1927. Physiology and parasitism of *Sclerotium rolfsii* Sacc. *Phytopathology* 17: 414-418.

Huang, D., Chen, J. and Zhu, Y. F. 1996. Southern Sclerotium blight of water caltrop and its control. *Plant Protection* 22(4): 32-34.

Iquebal, M.A., Tomar, R.S., Parakhia, M.V., Singla, D., Jaiswal, S., Rathod, V. M., Padhiyar, S. M., Kumar, N., Rai, A. and Kuma, D. 2017. Draft whole genome sequence of groundnut stem rot fungus *Athelia rolfsii* revealing genetic architect of its pathogenicity and virulence. *Scientific Reports* **7:** 5299 1-10.

Kapoor, S. 1962. Copper resistant strain of *Sclerotium rolfsii*. *Curr. Sci.* 31(7): 277-278.

Kumar, A., Kudada, N. and Saurabh, A. 2018. *In-vitro* efficacy of fungicides, bioagents and plant leaf extracts against *Sclerotium rolfsii* causing collar rot disease of French bean (*Phaseolus vulgaris* L.). *Ecology, Environment and Conservation* 24 (4): 1759-1767.

Lauritzen, T.I., Harter, L.L. and Whitney, W.A. 1933. Environmental factors in relation to snap bean diseases occurring in shipment. *Phytopathology* 23(5): 411-445.

Leach, L.D. and Davey, A.E. 1942. Reducing southern Sclerotium rot of sugarbeets with nitrogenous fertilizers. *J.Agr. Res*. 64: 1-18.

Linderman, R.G. and Gilbert, R.G. 1975. Influence of volatiles of plant origin on soil borneplant pathogens. *In, Bruehl, G.W. ed. Biology and control of soil borne plant pathogens*. Pp. 90-99. APS. Saint Paul Minnesota.

Luttrell, E.S. and Garren, K.H. 1952. Blights of snap bean in Georgia. *Phytopathology* 42(11): 607-613.

Magee, C.J. 1947. Slerotium stem rot caused by *Sclerotium rolfsii Agric Gaz. NSW* 58(5): 265-267.

Mahadevakumar S., Tejaswini, G.S., Janardhana, G. R. and Yadav V. 2015. First report of *Sclerotium rolfsii* causing southern blight and leaf spot on common bean (*Phaseolus vulgaris*) in India. *Plant Disease* 99(9):1280.

Maiti, S. and Chaudhuri, S. 1975. Effect of some systemic fungicides on sclerotial germination and growth of *Sclerotium rolfsii in vitro*. *Univ. Kalyani, WB, India* 82(4): 233-235.

Menten, J.O.M. 1980. Sensitivity *in vitro* of *Sclerotium rolfsii* Sacc. to some fungicides. *Revista de Agricultura* 55(3): 175-186.

Messiaen, C.M., Mampouya, P.C. and Belliard Alowzd, L. 1976. Effect of soluble nitrogen compounds on mycelial growth of *Selrotium rolfsii* Sacc. in soil. *Annales de Phytopathologie* 8(1): 17-23.

Milthrope, F.L. 1941. Studies on *Corticium rolfsii* (Sacc.) Cruzi (*Sclerotium rolfsii* Sacc.) I. Cultural characters and perfect stage. II. Mechanism of parasitism. *Linn. Soc. N.S. Wales Proc*. 66: 65-75.

Montealegre, A. and Esterio, G.1989. Presence of *Sclerotium rolfsii* Sacc. In bean (*Phaseolus vulgaris* L.), located in the V Region, Chile. *Agricultura Tecnica* 49(1): 66-68.

Montealegre, J.R. and Larenas, C. 1995. Use of *Trichoderma harzianum* as biological control of *Sclerotium rolfsii* in bean. *Fitopatologia* 30(3): 160-166.

Montealegre, J.R., Rojas, M.A., Varnero, M.T. and Abalay, E. 1996. Effect of soil solarisation on the control of *Sclerotium rolfsii* and nematode in the Metropolitan Region of Chile. *Fitopatologia* 3(1): 70-83.

Niemann, E., Scharif, G. and Mirkamali, A. 1967. Field susceptibility of some cultivated plants to *Sclerotium rolfsii*. *Entomologia Phytopath. Appl.* 26: 22-26.

Olivos, M.I. and Mont, R. 1993. Use of fungicides and *Trichoderma viride* for the control of *Sclerotium rolfsii*. *Fitopatologia* 28(1): 16-21.

Pacheco, K. R., Viscardi, B. S. M., de Vasconcelos, T. M. M., Moreira, G. A. M., de Vale, H. M. M. and Blum, L. E. B. 2016. Efficacy of *Trichoderma asperellum, T. harzianum, T. longibrachiatum* and *T. reesei* against *Sclerotium rolfsii*. *Bioscience Journal* 32 (2): 412-421.

Palti, J. 1954. Root diseases and crop rotation of farm and garden crops. *Adv. Leaft. Agric. Consultants* 7: 8 pp.

Paparu, P., Acur, A., Kato, F., Acam, C., Nakibuule, J., Nkuboye, A., Musoke, S. and Mukankusi, C. 2020. Morphological and pathogenic characterization of *Sclerotium rolfsii*, the causal agent of southern blight disease on common bean in Uganda. *Plant Disease* 104(8):2130-2137.

Pastirino, A. 1964. A polyphagous fungus: *Sclerotium rolfsii*. *Boln Fac. Agron. Univ. Montevideo* 750: 1-6.

Peltier, G.L. 1916. A serious disease of cultivated perennials caused by *Sclerotium rolfsii*. *Ill. Agric. Expt. Sta. Gr.* 187: 4.

Perez Sendin, M. de Los A., Gomez Izaguirre, G. and Gonzalez Barrios, J. 1986. Biological aspects of *Sclerotium rolfsii* in Cuba. *Ciencias de la Agricutura No.* 27: 12-17.

Punja, Z.K., Grogan, R.G. and Adams, G.C. jr. 1982. Influence of nutrition, environment and the isolate on basidiocarp formation, development and structure in *Athelia (Sclerotium) rolfsii*. *Mycologia* 74(6): 917-926.

Ramakrishnan, T.S. 1930. A wilt of Zinnia caused by *Sclerotium rolfsii*. *Madras Agric J.* 9 pp.

Reynolds, S.G. 1970. The effect of mulches on southern blight (*Sclerotium rolfsii*) in dwarf bean (*Phaseolus vulgaris*). *Trop. Agr*. 47: 137-144.

Roberti, R., Aari, P. and Pisi, A. 1996. Biological control of soil borne *Sclerotium rolfsii* infection by treatment of bean seeds with species of *Trichoderma. Petria* 6: 105-106.

Rolfs, P.H. 1893. The tomato and some of its diseases. *Fla. Agr. Exp. Sta. Bul.* 21: 38 pp.

Roy, A.K. 1977. Attack of *Sclerotium rolfsii* on some plants and capability of the fungus to cause storage rot. *Indian Phytopath.* 30: 425-426.

Saccardo, P.A. 1911. *Notae Mycologicae*. *Ann. Mycol.*9: 249-257.

Sanchez, A., Echavez Badel, R. and Schroder, E.C. 1994. *Pseudomonas cepacia*, a potential biofungicide for root rot pathogens of beans. *J. Agric. Univ. Puerto Rico* 78(1-2): 55-57.

Sariah, M. and Tanaka, K. 1995. Effect of flooding on the viability of sclerotia of *Sclerotium rolfsii* in Malaysian soil. *Intl. J.Pest Mgt.* 41(2): 97-99.

Sharma, A. 2001. Studies on crown rot of French bean caused by *Sclerotium rolfsii* (Sacc.). *M. Sc. Thesis. Dr. YashwantSingh Parmar University of Horticulture and Forestry, nauni, Solan.* 55p.

Sharma, B.K., Sugha, S.K. and Tyagi, P.D. 1994. Evaluation of fungicides for the control of chickpea collar rot. *Pl. Dis. Res*. 9: 139-141.

Silveira, N.S.S., Michereffi, S.J. Menezes, M. and Campos Takaki, G.M. 1994. Potential of *Trichoderma* spp. isolates on the control of *S. rolfsii* on beans. *Summa Phytopathologica* 20(1): 22-25.

Singh, B.K. and Singh, N.I. 1990. Seed and collar rot diseases of broad bean, a new record from Manipur, India. *New Agriculturist* 1(1): 93-94.

Sinha, B.K., Sinha, R.K.P., Roy, B. and Sinha, B.B.P. 1993. Two new diseases of rajmash (*Phaseolus vulgaris*). *J. Applied Biol.* 3(1/2): 102-103.

Tahori, A.S., Zeidler, G. and Halevy, A.H. 1965. Effect of some plant growth compounds on fungal diseases. *Pl. Dis. Reptr.* 49(9): 775-777.

Talbot, P.H.B. 1973. Aphyllophorales I: General characteristics: thelephoroid and cupuloid families. Pp. 327-349. In: *The Fungi*, Vol. IVB. Eds. G.C. Einsworth, F.K. Sparrow and A.S. Sussman, Academic Press, New York.

Taubenhaus, J.J. 1919. Recent studies of *Sclerotium rolfsii* Sacc. *J. Agri. Res.* 18:127-138.

Tiwari, R.K.S. 1995. Comparative evaluation of three systemic fungicides against *Sclerotium rolfsii* causing root rot in gram and sunflower. *Indian J. Mycol. Pl. Pathol.* 25: 243-245.

Tu, C. C. and Kimbrough, J.W. 1978. Systematics and phylogeny of fungi in the *Rhizoctonia* complex. *Bot. Gaz. (Crawfordsville)* 139: 454-466.

Van Etten, H.D. and Bateman, D.F. 1969. Enzymatic degradation of galactan, galactomannan, and xylan by *Sclerotium rolfsii. Phtopathology* 59: 968-972.

Venkatarayan, S.V. 1950. Notes on some species of *Corticium* and *Pellicularia. Indian Phytopath.* 3: 81-86.

Viegas, A.P. 1945. Some cassava fungi. I. II. *Bragantia. S. Paulo* 3(1): 1-19; 2: 21-29.

Vieira, C. 1967. *O feijoeirocomun: cultura, doencas e melhoramento, Imprensa.*

Weber, G.F. 1931. Blight of carrots caused by *S. rolfsii* with geographic distribution and host range of the fungus. *Phytopathology* 21(12): 1129-1140.

West, E. 1947. *Sclerotium rolfsii* Sacc. And its perfect stage on climbing fig. *Phytopathology* 37: 67-69.

Wilhelm, A.F. 1942. *Botany-Rep Ga Exp. Sta.* 68-75.

Wolf, F.A. 1914. Leaf spot and some fruit rots of peanut. *Aca. Agr. Exp. Sta. Bul.* 180:127-150.

Yu, R.C., Chen, L.C., Chen, T.Z., Wang, C.H. and Young, H.C. 1996. Studies on the physiological characterization of *Sclerotium rolfsii. Pl. Pathol. Bull.* 5(2): 91-99.

Zaumeyer, W.J. and Thomas, H.R. 1957. A monographic study of bean diseases and methods for their control. *U.S.D.A. Tech. Bull. No.* 868: 255.

13

FUSARIUM ROOT ROT

Fusarium root rot probably occurs to some degree in most bean fields throughout the world. The disease usually causes little damage if the plants are not stressed by other environmental factors that restrict root growth.

GEOGRAPHICAL DISTRIBUTION

It was originally described in 1916 from USA (Burkholder,1916). Since that time the disease has been reported from Uganda (Small, 1925), Barco de Avila (Benlloch and Del Canizo, 1926), California (Harter, 1929), New South Wales (Anonymous, 1941), Southern Idaho (Meiners, 1950), Norway (Roll Hansen, 1950), Canada (Conners *et al.*, 1950), Peru (Dongo and Osores, 1961),Hawaii (Trujillo, 1962), England and united States (Armstrong and Armstrong, 1964), Bazil (Reibiro and Ferraz, 1984).

LOSSES

Under severe disease pressure ranged from 6-53 per cent depending upon the cultivar. Sppill and Hall (1982) have reported a reduction of 29.1 per cent in seed weight per plant due to infection of *F. solani* f. sp. *phaseoli*.

SYMPTOMS

The symptoms initiates as reddish lesions or streaks on the hypocotyls. Similar symtoms also appear initially on primary root of one to two weeks old seedlings. As infection progresses, the lesions coalesce, become brown and may extend to the soil surface but rarely beyond. The lesions have no definite margins and may increase longitudinally. The primary and lateral roots are frequently attacked by the fungus and killed (Plate 1). When the primary root is killed, the lower stem may become pithy or hollow. There is no pronounced host wilt, though plant growth may be retarded and exhibit leaf yellowing and premature defoliation. Lateral roots often develop above the initial lesions and support plant growth so that some yield still is produced if surface soil moisture is adequate, although pod number per plant and seed size may be reduced.

THE PATHOGEN

The causal pathogen is *Fusarium solani* (Mart.) Appel and Wollenw f. sp. *phaseoli* (Burk.) Snyder and Hansen. In culture the aerial mycelium is sparse and usually greyish-white. The fungus mostly produces macroconidia which are hyaline and fusiform with a pointed, slightly-beaked apical cell. Size of macroconidia and number of septa vary according to culture medium used and incubation conditions. Size generally range in length from 44 to 55 μm in width from 5.1 to 5.3 μm and are 3 or 4 septate but rarely five septate. Microconidia are rare and develop from sparsely-branched conidiophores. Microconidia are broad, oval and may have one septation. Chlamydospores are 6-16 μm in diameter and form terminally on short lateral branches or intercalary. They form singly, in pairs or occassionally in short chains and may be round, subglobular or pear shaped.

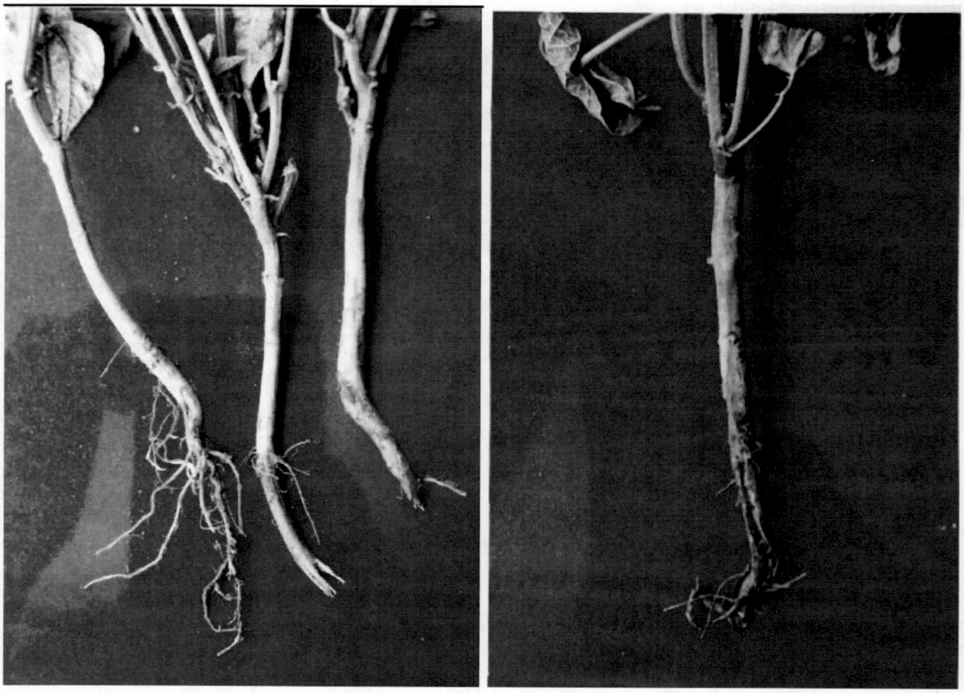

Plate 1. Symptoms of Fusarium root rot

HOST RANGE

The host range of *F. solani* f. sp. *phaseoli* comprises *Phaseolus vulgaris*, *P. limensis*, *P. coccineus*, *P. angularis*, *P. acutifolius* var. *latifolius*, *P. acontifolius*, *P. lunatus*, *Pisum sativum*, *Pueraria thunbergiana*, *Onobrychus vicifolia* and *Vigna unguiculata*.

DISEASE CYCLE

The pathogen does not survive in seeds and hence is a not seed-borne. It has a low mobility and exists in naturally infested soil as chlamydospores associated with or embedded in tissue or humus particles (Nash *et al.*, 1961). Chlamydospores provide primary inoculum for infection under field conditions and are stimulated to germinate by seed and hypocotyl exudates.

EPIDEMIOLOGY

Plant susceptibility is influenced by environmental conditions such as soil compaction, temperature and pH. Tu and Tan (1991) demonstrated that root rot severity significantly increased in compacted soil and this effect was comparatively less in clay loam than sandy loam soils. In culture, the growth of the fungus is rapid at 30-32°C but it cause more damage at 22-23°C than at 30°C (Sharma and Sohi, 1989). Infection is reported to be favoured by acidic soil or by soils fertilized with NH_4-N, and may be suppressed by soils fertilized with NO_3-N (Smiley, 1975). Guerra and Andersson (1985) also reported the increased lesion size of *Fusarium* affected *P. vulgaris* in Fe and Bo deficient soils. Root rot damage is further aggrevated during periods of high soil moisture when the oxygen diffusion rate is low (Miller and Burke, 1977). Letourneau and Msuku (1992) reported that root rot infection by *F. solani* f.sp. *phaseoli* is enhanced in the association of beanfly *(Ophiomyia puparia)*.

The pathogen is primarily disseminated as chlamydospores or conidia through drainage and irrigation water, soil adhering to agricultural tools and animals, been straw, manure and possibly in soil or as spores washed by rain or floods. Once the fungus introduced in new area, it may survive indefinitely as a soil saprophyte on organic matter. Trichomes on bean roots may provide an entry point for infection (Mulligan *et al.*, 1990).

MANAGEMENT

Cultural practices

Long term crop rotation with non-susceptible plants reduces soil population of *F. solani*. However, this method is of seldom practical utility. Field sanitation is very important. Deep subsoiling and narrow row spacing are also helpful in reducing the disease severity. Okumura *et al.* (1994) demonstrated the role of preceding crops like red clover and lucerne in reducing soil borne inoculum of *Fusarium*. Soil amendment with crop residues may improve biological control potential of soil microbial population. Maier (1961) demonstrated the effect of incorporation of barley straw into soil infested with *F. solani* in reducing the disease incidence. Mutitu *et al.* (1989) reported root rot control under glasshouse conditions with soil amendment of coffee hulls and farmyard manure. However, coffee hulls affected the host by causing N deficiency symptoms.

Host resistance

Aim is lost when unfavourable features are linked. It has been discovered that cultivars with purple hypocotyls and black seed coats are more resistant to this disease than cultivars with other colours.

Despite the fact that numerous root rot resistant cultivars have been reported from various regions of the world, the undesirable traits were found to be linked with the disease resistance. Cultivars having purple hypocotyls and black seed coats were more resistant compared than the cultivars having hypocotyls and seed coats of other colours (Statler, 1970). Boomstra (1975) evaluated 700 bean accessions against this disease and found 18 plant introductions of Mexican origin as resistant. Cvs./ lines viz., Wade, Harvester (Munoz and Perez, 1975), USWA-20 were found to be resistant across the world. Susceptibility to *F. solani* f. sp. *phaseoli* was found to be dominant over resistance (Silbernagel *et al.*, 1998). It is considered that recurrent selection is more suitable for developing resistance in *P. vulgaris* than back crossing or pedigree selection (Boomstra and Bliss, 1977). The genotype VAX 3, which is a dwarf and red coloured variety, has a high level of resistance against Fusarium root rot disease and could be used in breeding programmes of bean (Bilgi *et al.*, 2008). Two inbred back cross line populations (IBL) were developed from crosses between a Mesoamerican black bean with an Andean kidney bean and an Andean cranberry bean and nine QTLs were identified in two IBL explaining 5–53% of the phenotypic variation (Roman-Aviles and Kelly, 2005). QTL mapping study of Middle American recombinant inbred lines developed from landrace Puebla 152 and the commercial black cultivar 'Zorro' identified one QTL on Pv05 linked to resistance against Fusarium root rot (Nakedde *et al.*, 2016). Hagerty *et al.* (2015) used snap bean RIL population to locate a QTL associated with Fusarium root rot resistance on Pv03. 310 genotypes were assessed for resistance against root rot in the fields of Minnesota, out of which only five genotypes were found resistant to Fusarium root rot (Vasquez-Guzman, 2016).

Chemical control

Various chemicals such as Nabam, Formaldehyde, Thiram, PCNB, benomyl, Difolatan, Busan and thiabendazole have been reported to reduce root rot. Thiabendazole can be used alone or in combination with other fungicides as a dry seed treatment or a soil drench to achieve effective control (Mussa, 1986). However, most chemical treatments are not completely effective, since the lateral root development receives little or no benefit from the fungicide (Burke and Barker, 1966).

Herbicides viz. treflan or trifluralin and basagran or bentazon and the pesticide Metasystox or oxdemeton methyl stimulated the growth of *F. solani* and aggravate root rot problems while Epitam inhibited (Anicuta, 1985).

Biopesticides

Garlic extract prepared as a juice from crushed cloves inhibited the growth of *F. solani* f. sp. *phaseoli* under *in vitro* conditions and seed treatment with either the extract from crushed garlic cloves or with an aqueous preparation of commercial lyophilized garlic powder provided sufficient control of foot rot under *in vivo* conditions (Russell and Mussa, 1977).

Biological control agents

Buonassisi *et al.* (1986) demonstrated the role of seed inoculation with root nodulating *Rhizobium* strains in controlling *Fusarium* root rot and were also antagonistic to the mycelium of *Fusarium solani* f.sp. *phaseoli*. Seed inoculation of bean seed with

Pseudomonas cepacia protected the plants from this disease (Sanchez *et al*., 1994). *Gliocladium virens* and *Bacillus subtillis* @ 101 to 104 propagules/cm^3, respectively, satisfactorily controlled root rot under glasshouse conditions (Tu, 1991). Efficacy of T-12 isolate of *Trichoderma* in suppressing this disease under glasshouse conditions has also been reported (Cardoso *et al*., 1997). Plants inoculated with *Rhizobium leguminosarum* alone or in combination with *Glomus musseae* showed more tolerance to the root rot pathogen (Dar *et al*., 1997, Beigh *et al.,* 1998). *T. harzianum* (6.7 - 45%) and *T. viride* (8.1 –30%) provided efficient management of Fusarium root rot (Ellil *et al*., 1998).

The efficacy of arbuscular mycorrhizal fungus (AMF) *Glomus intraradices* against *F. solani* f. sp. *phaseoli* in bean plants and the surrounding mycorrhizosphere soil was evaluated and it was found that infected plants previously colonized by *G. intraradices* remained healthy whereas non-mycorrhizal bean plants infected with the pathogen had distinctive symptoms of Fusarium root rot (Filion *et al*., 2003). The amount of genomic DNA of *F. solani* f. sp. *phaseoli* was significantly reduced in mycorrhizal bean plants. The presence of *G. intraradices* in the mycorrhizosphere was not significantly modified, although the mycorrhizal colonization of roots was slightly increased in the presence of the pathogen. The results suggest that the reduced presence of Fusarium as well as root rot symptoms are caused by biotic and/or abiotic modifications of the mycorrhizosphere as a result of colonization with *G. intraradices*.

Siderophore producing *Pseudomonas flourescens* GL 20 has also been found to be effective for the management of Fusarium root rot by inhibiting the growth of fungal hyphae and germination of fungal spores (Lim *et al*., 1999). *B. subtilis* and *T. harzianum* effectively antagonized *F. solani* f. sp. *phaseoli* (Abeysinghe, 2007). Further, the efficacy of the *B. subtilis* and *T. harzianum* were tested against Fusarium root rot disease in bean plants in pots under greenhouse and it was found that seed treatment with *B. subtilis* and *T. harzianum* significantly reduced the disease in bean seedlings compared to the untreated control. Reduction in disease was more in plants treated with *T. harzianum* compared to *B. subtilis* treated plants. Additionally, enhanced root growth was also found in *T. harzianum* treated plants which suggested that there were biotic modifications of the mycorrhizosphere due to colonization with *T. harzianum*.

Application of biocontrol agents namely *Trichoderma harzianum* T 22, *Bacillus subtilis* QST 713 and *Pseudomonas chlororaphis* as seed treatment provided effective management of Fusarium root rot disease (Gilardi *et al*., 2008). A proteomic approach was used to analyse the ability of *T. harzianum* to promote growth in common bean and to modulate metabolism and defense response in bean plants against phytopathogenic fungi *Rhizoctonia solani* and *Fusarium solani* (Pereira *et al*., 2014). *T. harzianum* augmented the growth of common bean plants by increasing the root/foliar areas and size. Modulation in the expression of defense-related genes namely Glu1, pod3 and lox1 in bean roots was also observed in plants treated with *T. harzianum* and infected with the pathogenic fungi. Proteomic maps of roots and leaves of plants challenged or unchallenged by *T. harzianum* and phytopathogenic fungi highlighted the change in the protein profiling. Reference gels presented differences in spot distribution (absence/presence) and relative volumes of common spots (up or down-regulation). MALDI-TOF mass spectrometry identified the differential spots. 48 identified spots (19 for leaves and 29 for roots) were grouped into protein functional classes. 33%, 22% and 11% of the identified proteins of leaves were categorized in three groups i.e. metabolism, defense response and oxidative stress response, respectively. Similarly, 17.2%, 24.1% and 10.3% of the identified proteins of roots were categorized in metabolism, defense response and oxidative stress response group, respectively.

REFERENCES

Abeysinghe, S. 2007. Biological control of *Fusarium solani* f. sp. *phaseoli* the causal agent of root rot of bean using *Bacillus subtilis* CA32 and *Trichoderma harzianum* RU01. *Ruhuna Journal of Science* 2:82-88.

Anicuta, P. 1985. Effect of some herbicides used in bean cultivation on *Fusarium oxysporum* f. sp. *phaseoli* and *Fusarium solani* f. sp. *phaseoli*. *Probleme de Protectia Plantelor* 13(4): 347-357.

Anonymous 1941. Plant Diseases. Notes contributed by the biological branch. *Agric. Gaz. NSW.* Pp. 632-636.

Armstrong, G.M. and Armstrong, J.K. 1964. Pathogenicity of isolates of bean wilt *Fusarium* from England and the United States. *Pl. Dis. Reptr*. 48: 846-847.

Beigh, G.M., Dar, G.M., Bhat, B.A. and Zargar, M.Y. 1998. Biological control of Fusarium root rot in common bean with *Rhizobium leguminosarum. J. Hill. Res.* 11(2): 160-170.

Benlloch, M. and Del. Canizo, J. 1926. Diseases of French beans in Barco de Avila (Fusariose). *Bull. De La Estacion de Patol. Vegetal*. 1(1): 1-7.

Bilgi, V. N., Bradley, C. A., Khot, S. D., Grafton, K. F., and Rasmussen, J. B. 2008. Response of dry bean genotypes to *Fusarium* root rot, caused by *Fusarium solani* f. sp. *phaseoli*, under field and controlled conditions. *Plant Dis.* 92: 1197–1200.

Boomstra, A.G. 1975. Breeding for resistance to Fusarium root rot in beans (*Phaseolus vulgaris*). *Dissert. Abst. Intt. –B* 36(3): 1002 B.

Buonassisi, A.J., Copeman, R.J., Pepin, H.S. and Easton, G.W. 1986. Effect of *Rhizobium* spp. on *Fusarium solani* f.sp. *phaseoli. Can. J. Plant Pathol.* 8(2): 140-146.

Burke, D.W. and Barker, A.W. 1916. Importance of lateral roots in the Fusarium root rot of beans. *Phytopathology* 56: 292-294.

Burkholder, W.H. 1916. Some root diseases of the bean. *Phytopathology* 6: 104.

Cardosa, J.E., Silva, S.A.G. and Marques, E.E. 1997. Chemical and biological control of bean root rots. *Fitopatol. Brasiliera* 22(1): 39-44.

Dar, G.M., Zargar, M.Y. and Beigh, G.M. 1997. Biocontrol of Fusarium root rot in the common bean (*Phaseolus vulgaris* L.) by using symbiotic *Glomus mosseae* and *Rhizobium leguminosarum. Microbiol. Ecol.* 34(1): 74-80.

Dongo, D.S. and Osores, D.A. 1961. *Fusarium* spp. in the cultivation of bean in Peru. *Turrialba.* 11(4): 161-162.

Ellil, A.H.A.A., Awad, N.G.H. and El Haleem, S.T.A. 1998. Biocontrol of vegetable root rot disease by *Trichoderma harzianum* and *T. viride*, role of sugars, proteins and amino acids in host resistance. *African Journal of Mycology and Biotechnology* 6(2): 25-41.

Filion, M., St-Arnaud, M. and Jabaji-Hare, S. H. 2003. Quantification of *Fusarium solani* f. sp. *phaseoli* in mycorrhizal bean plants and surrounding mycorrhizosphere soil using real-time polymerase chain reaction and direct isolations on selective media. *Phytopathology* 93:229-235.

Hagerty, C. H., Cuesta-Marcos, A., Cregan, P. B., Song, Q., McClean, P. and Noffsinger, S. 2015. Mapping *Fusarium solani* and *Aphanomyces euteiches* root rot resistance and root architecture quantitative trait loci in common bean. *Crop Sci.* 55:1969-1977.

Harter, L.L. 1929. A *Fusarium* disease of beans. *Phytopathology* 19(1): 84.

Letourneau, D.K. and Msuku, W.A.B. 1992. Enhanced *Fusarium solani* f.sp. *phaseoli* infection by bean fly in Malawi. *Plant Dis*. 76: 1253-1255.

Lim, H.S., Lee, J.M. and Kim, S.D. 1999. Role of siderophore in biological control of *Fusarium solani* by *Pseudomonas fluorescens* GL 20. *Bull. Instt. Comprehens. Agric. Sci., Kinki Univ. No.7*: 47-58.

Maier, C.R. 1961. Selective effect of barley residue on fungi of the pinto bean spot complex. *Pl. Dis. Reptr.* 45: 808: 811.

Meiners, J.P. 1950. Bean diseases in Southern Idaho in 1949. *Pl. Dis. Reptr*. 39: 14.

Mulligan, M.F., Safir, G.R. and Klomparen, K.L. 1990. Association of *Fusarium solani* f.sp. *phaseoli* with trichomes of *Phaseolus vulgaris. Mycological Res*. 94(3): 409-411.

Munoz de Con, L. and Perez Martinez, J.J. 1975. Evaluation of varieties and genetic material in French bean. *Revista de Agric., Cuba* 8(2): 6-11.

Mussa, A.E.A. 1986. The control of *Fusarium solani* f.sp. *phaseoli* by fungicide mixtures. *J. Phytopathol*. 117(2): 173-180.

Nakedde, T., Ibarra-Perez, F. J., Mukankusi, C., Waines, J. G., and Kelly, J. D. 2016. Mapping of QTL associated with *Fusarium* root rot resistance and root architecture traits in black beans. *Euphytica* 212: 51-63.

Nash, S.M., Christou, T. and Snyder, W.C. 1961. Existence of *Fusarium solani* f. *phaseoli* as chlamydospores in soil. *Phytopathology* 51: 308-312.

Okumura, M., Higashida, S., Yamagami, M. and Shimono, K. 1994. Effects of different preceding crops on Fusarium root rot of kidney bean. *Jpn. J. Soil Sci. Plant Nut.* 65(3): 274-281.

Pereira, J.L., Queiroz, R.M.L., Charneau, S.O., Felix, C.R., Ricart, C.A.O., et al. 2014. Analysis of *Phaseolus vulgaris* response to its association with *Trichoderma harzianum* (ALL-42) in the presence or absence of the phytopathogenic fungi *Rhizoctonia solani* and *Fusarium solani. PLoS One* 9(5): e98234.

Rava, C.A., Sartorato, A. and Costa, J.G. C. da. 1996. Reaction of French bean genotypes to *Fusarium oxysporum* f.sp. *phaseoli* in the greenhouse. *Fitopatologia Brasileria* 21(2): 296-300.

Roman-Aviles, B., and Kelly, J. D. 2005. Identification of quantitative trait loci conditioning resistance to *Fusarium* root rot in common bean. *Crop Sci.* 45:1881-1890.

Russell, P.E. and Mussa, A.E.A. 1977. The use of garlic (*Allium sativum*) extracts to control foot rot of *Phaseolus vulgaris* caused by *Fusarium solani* f.sp. *phaseoli. Ann. Appl. Biol.* 86(3): 369-372.

Sanchez, A., Echavez Badel, R. and Schroder, E.C. 1994. *Pseudomonas cepacia,* a potential biofungicide for root rot pathogens of beans. *J. Agric. Univ. Puerto Rico* 78(1-2): 55-57.

Sharma, S.R. and Sohi, H.S. 1989. Diseases of French bean in India and their control. *Indian Rev. Life Sci.* 9: 153-168.

Sibernagel, M.J., Hang, A.N. and Miklas, P.N. 1998. Registration of USWA-20 virus and root rot resistant pinto dry bean germplasm. *Crop Sci.* 38(3): 899.

Sippell, D.W. and Hall, R. 1987. Effects of *Fusarium solani phaseoli, Pythium ultimum* and *Fusarium oxysporum* on yield components of whilte bean. *Can. J. Plant Pathol.* 4(1): 54-58.

Small, W. 1925. Notes on species of *Fusarium* and *Sclerotium* in Uganda. *Kew Bull. Mix. Inform*. 3: 118-126.

Trujillo, E.E. 1962. Occurrence of the bean root rot fungus, *Fusarium solani* f.sp. *phaseoli* in Hawaii. *Pl. Dis. Reptr*. 46: 819.

Tu, J.C. 1991. Comparison of the efficacy of *Gliocldium virens* and *Bacillus subtilis* in the control of seed rots and root rots of navy bean. *Rijksuniver-satie-Gent*. 56: 229-234.

Tu, J.C. and Tan, C.S. 1991. Effect of soil compaction on growth, yield and root rots of white beans in clay loam and sandy loam soil. *Soil Biol. Biochem*. 23(3): 233-238.

Vasquez-Guzman, J. 2016. *Identifying dry bean genotypes and genomic regions associated with root rot resistance with emphasis on Fusarium solani f. sp. phaseoli.* Ph.D. thesis, North Dakota State University, Fargo ND.

14

FUSARIUM YELLOWS

This is an important disease in those areas where the crop is grown in warm and dry conditions.

GEOGRAPHICAL DISTRIBUTION

The occurrence of the disease was first recorded in California during 1928 (Harter, 1929) on dry beans. Several years later, the disease was observed in Colorado, Idaho, and Montana. Since then, the disease has been reported from California (Kendrick and Snyder, 1942), Peru (Dongo and Osores, 1961), Costa Rica (Echandi, 1967), Brazil (Cardoso *et al.*, 1966), Spain (Tuset Barrachina, 1973), Canada (Thomas and Wood, 1981), Italy (Aloj *et al.*, 1983), Kenya (Mititu *et al.*, 1988), Yugoslavia (Balaz, 1989), Colorado (Schwartz *et al.*, 1989), China (Zhang, 1992), Rawanda and Democratic Republic of Congo (Buruchara *et al.*, 1999) and Central Africa (Buruchara and Camacho, 2008).

In India, the occurrence of this disease was first reported from Solan, Himachal Pradesh during May-June 1991(Gupta *et al.*, 1993).

LOSSES

Though the disease is of minor importance but in favourable weather cause significant losses by killing the plants prematurely. Up to 50% damage in two large plantings of pink beans, one of them covering 50 acres in California have been reported (Kendrick and Snyder, 1942). Dongo and Osores (1961) noticed 60-70% crop damage in the Central Coastal region of Peru due to attack of *Fusarium oxysporum* in association with *F. equisetti*. In Colorado, cultivar U.I. 114 was severely affected by this disease and caused >75% defoliation within 75 days of planting (Schwartz *et al.*, 1989).

SYMPTOMS

The initial symptoms of the disease appear as slight yellowing and drooping of the lower leaves, followed by those of the next leaves in order up the stem. As the disease become more severe, the leaves become increasingly yellow and finally bright yellow (Plate 1a). Reddish brown lesions also appear on subterranean stem and tap root of the plants later becoming dark brown to black (Plate 1b). The affected plants ultimately dried showing wilt type symptoms.

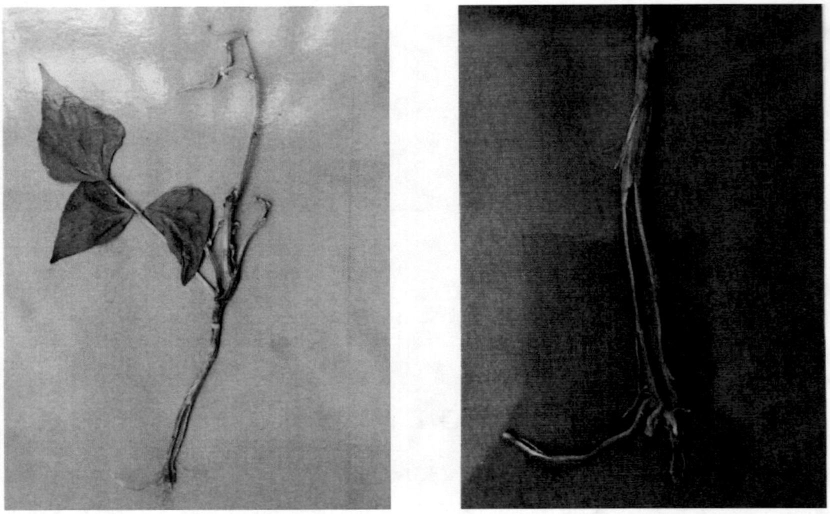

Plate 1. (a) Symptoms of Fusarium yellows (b) Browning of vascular bundles in *Fusarium* infected stem

THE PATHOGEN

The disease is caused by *Fusarium oxysporum* Schlecht. f. sp. *phaseoli* Kendrick and Snyder. The fungus produces hyaline, non-septate chlamydospores measuring 6-15 x 2-4 µm in size. Elongated macroconidia are curved with two to three septa and measure 25-35 x 3-6 µm in size (Weber, 1973).

VARIABILITY

The isolates of *F. oxysporum* f.sp. *phaseoli* isolated from runner bean and French bean did not show any difference in their pathogenicity on resistant and susceptible cultivars indicating the absence of pathogenic races among the pathogen isolates (Armstrong and Armstrong, 1964). However, Aloj *et al.* (1987) isolated a new pathotype in Italy, which differ in its pathogenicity from the known races to some differential cultivars of *P. vulgaris*. Virulence of this pathotype was higher at 26° C than at 20° C. Differences in aggressiveness of different isolates of the pathogen on *P. vulgaris* cultivars has been reported from different parts of the world (Piza, 1993, Nascimento *et al.,* 1995). Among 30 isolates of *F. oxysporum* f.sp. *phaseoli* collected from different geographical areas (Brazil, Czech Republic, Slovakia, Columbia, Japan, Greece, Italy, Poland and USA) only 12 were virulent on differential set of 9 *P. vulgaris* cultivars and comprised of 5 races (Zoina *et al.*, 1995). Two races were described for the first time for this forma species. The Brazilial isolates Fop 2, 3 and 27 were in race 2, the Coloumbian isolates Fop 6, 7, 8 and 28 in race 3 and the Greek isolates Fop 31 and 32 in race 5. Five Fop isolates from Spain and one from Colorado, USA, were tested for pathogenicity on 2 American lines Pinto U.I. 114, considered as a universal susceptible check and Flor de Mayo, a Mexican landrace of which two of the Spanish isolates showed resistant reactions on both lines at all inoculum concentrations (Velasquez-Valle *et al.*, 1997). Susceptible reaction to rest of the three Spanish isolates and the Colorado isolate were observed on U.I. 114 regardless of inoculum concentration, though severity ratings increased with increased inoculum concentration.

F. oxysporum f. sp. *phaseoli* strains isolated from runner bean plants showing Fusarium wilt symptoms were characterized (de Vega-Bartol *et al.*, 2011). The analysis of the genetic diversity of these strains and the comparison with strains formerly isolated from diseased common bean plants grown in the same region of Spain indicated a close genetic similarity among them. Pathogenicity assays carried out on runner bean plants showed virulence differences that allowed the classification of these strains into three groups: super virulent, highly virulent, and weakly virulent. However, all the analyzed strains behaved as highly virulent when inoculated on common bean plants, indicating that virulence is specific of the host-pathogen interaction. The number of copies and expression of the gene encoding the transcription factor ftf1, which has been shown to be specific of virulent *F. oxysporum* strains and highly up regulated during plant infection was also analyzed. *In planta* real-time quantitative polymerase chain reaction expression analysis showed that expression of ftf1 was correlated with the degree of virulence. The comparative analysis of the polymorphic copies of ftf1 detected in the strains here characterized and those detected in the genome sequence of *F. oxysporum* f. sp. *lycopersici* strain 4287 indicates that some of the copies are likely nonfunctional.

High variability in pathogenicity of *F. oxysporum* f. sp. *phaseoli* was observed and 27 different physiological races of the pathogen were proposed (Henrique *et al.*, 2015). The transcriptome and metabolome responses in common bean roots in response to *F. oxysporum* f. sp. *phaseoli* revealed significant changes in structural defense i.e., cell-wall loosening and weakening characterized by hyper accumulation of cell-wall loosening and degradation related transcripts (Chen *et a.*, 2019). The levels of pathogenesis related genes were significantly higher upon *F. oxysporum* f. sp. *phaseoli* inoculation. There was involvement of glycosylphosphatidylinositol- anchored proteins (GPI-APs) in signal transduction in response to *F. oxysporum* f. sp. *phaseoli* infection. The pathogen induced energy metabolism and nitrogen mobilization in infected common bean roots as compared to control. Importantly, the flavonoid biosynthesis pathway was the most significantly enriched pathway in response to *F. oxysporum* f. sp. *phaseoli* infection as revealed by the combined transcriptome and metabolome analysis.

HOST RANGE

Besides *P. vulgaris*, the host range of the pathogen includes *Cajanus cajan, Dolichos lablab, Canavalia ensiformis, Mucuna aterrima, Crotolaria spectabilis, P. lunatus, Glycine max, Crotolaria juncea* etc. (Coelho Netto and Dhingra, 1999).

DISEASE CYCLE

The pathogen is seedborne and the fungus is probably being transported by the spores adhering to the seed (Kendrick and Snyder, 1942). They further reported that the spores were deposited on the surface of the seed during harvesting. The pathogen also survives in soil as chlamydospores. In favourable weather, the spores already adhering to the seed surface germinate and cause infection. Similarly, the chlamydospores also germinate and cause infection.

EPIDEMIOLOGY

The pathogen is both seed and soil borne. Dry hot season is quite favourable for development of the pathogen (Conners *et al.*, 1957). Acidic soils with lower pH are suppressive to this disease as these conditions inhibit spore germination (Kobayashi and Komada, 1995).

MANAGEMENT

Cultural practices

Since the pathogen is seed borne, use of healthy seed is advocated for the control of this disease. The pathogen is also soil borne and once introduced in a field, very difficult to eradicate. Follow long crop rotation with non-susceptible crop to keep the pathogen population below economic threshold level. Soil amendment with oil cakes, green manure and other decomposable plant residues may enhance natural biological control by resident soil microorganisms. Mitutu *et al.* (1988) demonstrated that out of four amendments tried in greenhouse against this disease, FYM was fairly effective and coffee hulls also gave good control when applied at very high levels in the ratio of 2:1 w/w (soil: hulls).

Host resistance

Bastos *et al.* (1974) observed that wax bean (*P. vulgaris*) cvs. like Cherokee Wax, Snico, Contender and Processor and dry bean cvs. viz., Manteigao Preto, Pintado, Manteigao Lustroso and Manteigao 41 were résistant to this disease. Cvs. / lines viz., White Select, Wiejska, Saxobel, IGE 1/ 79 and IGE 2/79 (Pieta, 1987), black bean cvs. Negrito 897, Rico 1735 and Milionario 1732 (Zambolim *et al.*, 1987), IAPAR-14, 16, 20, 31 and 44, FT Taruma, Goiano Precoca, Rio Tibagi, G 4000, EMP 81, IAPAR BAC 87-4, IAPAR BAC 87-9 (Miranda *et al.*, 1994), IAPAR 44, Milionario 1732, PT Trauma, Serrano, Sao Jose and Rico 1735 (Rava *et al.*, 1996), Carioca MG, Ouro Negro, Diaool Calima, CNF 05, Jalo, Negro Argel, ESAL 574, LM 30406, CI 11, P 78, CIAT 430, CI 164-1, Jaime 208 and CI 164-4 (Rocha Junior *et al.*, 1998), Xan 160, PI 150414, A 417, PI 175829 purple, Xan 161, A 420, PI 16317 and PI 175829 white (Maringoni and Lauretti, 1999), PWR 950 and G 685 (Buruchara and Camacho, 2008) have been reported resistant to this disease. Dongo and Muller (1969) reported that identified resistant varieties to the disease were generally red-seeded and produce many strong lateral roots after infection. Seventy-three bean genotypes were screened for resistance to the disease using artificial inoculation and cultivar RWR 950 and G 685 were found to resistant whereas cultivars G 2333 and MLB-48-89 A were susceptible (Buruchara and Camacho, 2008). Of the 29 climbing bean genotypes evaluated, 19 were resistant, including 11 of the 15 pre-release or released cultivars. Of the 44 bush bean cultivars evaluated, 28 were resistant, five were intermediate and 11 were susceptible.

Genotypes AL 9021332, CF 128, CP 9310635, TB 9401, and UFSC-01 were resistant whereas genotypes FT 991159, cv. Uirapuru and FT 84113 were highly susceptible (de Borba *et al.*, 2011). Resistance classification of bean genotypes following root-dip inoculation in the greenhouse was tightly associated with the reactions in infected adult bean plants assessed on the 90th day. Disease severity rate in seedlings was positively correlated (0.93, $P \leq 0.01$) with Fusarium wilt under outdoor conditions. Fungus more efficiently colonized the susceptible aerial tissues of cv. Uirapuru, resulting in earlier and stronger disease symptoms.

Chemical control

Use of healthy seed and their treatment with fungicides have been advocated for the control of this disease. Seed treatment with Semesan or Ceresan (Kendrick and Snyder, 1942) and carbendazim (Zhang, 1992) provided effective control of this disease. Several herbicides and insecticides have also been tried for the control of this disease (Anicuta, 1985).

Biological control agents

Seven out of eleven mutants obtained from *Streptomyces corchorus* were antagonistic to *F. oxysporum* f.sp. *phaseoli*, mutant 155 being the most effective (El-Abyad *et al.*, 1993). They further observed that seed soaking in 80 per cent solution of the strain 155 prior to sowing was the most effective treatment with disease incidence being reduced by 83.4% in addition to improving plant growth. Reddy *et al.* (1994) also obtained significant control of this disease by nine different strains of bacteria. *Trichoderma* isolates viz., TN-28, TN-59 and TN-15 were also found antagonistic to *F. oxysporum* f. sp. *phaseoli* (Reis *et al.*, 1995). The *Trichoderma* isolates significantly ($p < 0.01$) reduced the mycelial growth of *F. oxysporum* f. sp. *phaseoli* (Otadoh *et al.*, 2011). The principal mechanisms of niche competition, mycoparasitism, and antibiosis were observed in growth of the pathogen mycelium in the presence of *Trichoderma* spp., through development of inhibition zones. There was coiling of hyphae around the pathogen mycelium coupled by lysising of cell wall *Trichoderma* spp., where *T. reesei* and *T. koningii* were the most effective isolates.

Five *Trichoderma harzianum* isolates (CEN287, CEN288, CEN289, CEN290, and CEN316) were evaluated against *F. oxysporum* f. sp. *phaseoli* and tested for their potential use as seed treatment for the management of Fusarium yellows in bean (Carvalho *et al.*, 2014). It was found that all but one *Trichoderma* isolate (CEN290) inhibited mycelial growth of pathogen. SEM analysis revealed that only one *Trichoderma* isolate (CEN287) showed parasitic interaction with *F. oxysporum* f. sp. *phaseoli*. Two isolates (CEN287 and CEN316) significantly reduced the incidence of Fusarium yellow disease of bean and enhanced seed germination, though less effective than the fungicide mixture (carboxin + thiram). In another study, among all the evaluated bacterial endophytes, two endophytic bacteria belonging to the species *Pseudomonas brassicacearum* and *Bacillus siamensis*, were the best controllers of Fusarium root rot in common bean, caused by *F. oxysporum* f. sp. *phaseoli* (Barquero *et al*., 2016). Further, common bean plants inoculated with the pathogen *F. oxysporum*, plus an autochthonous *Rhizobium leguminosarum* strain and the *B. siamensis* isolate, showed a very weak pathogen attack.

Four fungal bio-control agents, *viz., Trichoderma harzianum, T. viride, Aspergillus* spp. and *Trichothecium* spp. were evaluated under *in vitro* conditions through dual culture technique against *F. oxyporum* f. sp. *phaseoli,* the causative agent of vascular wilt of beans and minimum colony diameter (1.6 cm) and maximum inhibition per cent over control (77.14%) were observed in *T. harzianum* which was statistically at par with *T. viride*, *Aspergillus* spp. and *Trichothecium* spp. (Rasool *et al.,* 2019).

REFERENCES

Aloj, B., Marziano, F., Zoina, A. and Noviello, C. 1983. Vascular *Fusarium* disease of bean in Italy. *Informatore Fitopatologico* 33(11): 63-66.

Aloj, B., Marziano, F., Zoina, A. and Noviello, C. 1987. Observation of a new physiological race of *Fusarium oxysporum* f.sp. *phaseoli. Annali della Facolta di Scienze Agrarie della Universita degli Studi di Napoli* 21: 51-54.

Anicuta, P. 1985. Effect of some herbicides used in bean cultivation on *Fusarium oxysporum* f.sp. *phaseoli* and *Fusarium solani* f.sp. *phaseoli*. *Probleme de Protectia Plantelor* 13(4): 347-357.

Armstrong, G.M. and Armstrong, J.K. 1964. Pathogenicity of isolates of bean wilt *Fusarium* from England and the United States. *Pl. Dis. Reptr*. 48: 846-847.

Balaz, F., 1989. *Fusarium oxysporum* f.sp. *phaseoli*, causal agent of Fusarium wilt of snap bean in Yugoslavia. *Zastita Bilja* 40(1): 47-55.

Barquero, M., Terrón A., Velázquez, E. and González-Andrés, F. 2016. Biocontrol of *Fusarium oxysporum* f.sp. *phaseoli* and *Phytophthora capsici* with autochthonous endophytes in common bean and pepper in Castilla y León (Spain). *In:* González-Andrés F., James E. (eds) *Biological Nitrogen Fixation and Beneficial Plant-Microbe Interaction*. Springer, Cham.

Bastos Cruz, B.P., Teranishi, J., Issa, E., Bernardi, J.B. and Anuda, H.V.de. 1974. Resistance of wax bean cultivars to Fusarium wilt. *Biologico* 40(1): 25-32.

Buruchara, R.A. and Camacho, L. 2008. Common bean reaction to *Fusarium oxysporum* f.sp. *phaseoli*, the cause of severe vascular wilt in Central Africa. *J. Phytopathol.* 148(1): 39-45.

Buruchara, R.A., Pastor-Corrales, M.A. and Scheidegger, U. 1999. Fusarium wilt disease caused by *Fusarium oxysporum* f.sp. *phaseoli* on a common bean cultivar G 2333, in Rawanda and the Democratic Republic of Congo. *Plant Dis*. 83: 397.

Cardoso, C.O.N., Kimati, H. and Fernandes, N.C. 1966. Note on the occurrence of *Fusarium oxysporum* f. sp. *phaseoli* causing vascular wilt in runner bean. *Anais Sup. Agric. 'Luiz Queiroz'* 23: 273-276.

Carvalho, D., Junior, M.L., Martins, I., Inglis, P., Mello, S.C., and Goiás, S.A. 2014. Biological control of *Fusarium oxysporum* f. sp. *phaseoli* by *Trichoderma harzianum* and its use for common bean seed treatment. *Tropical Plant Pathology,* 39: 384-391.

Chen, L., Wu, Q., He, W., He, T., Wu, Q. and Miao, Y. 2019. Combined De Novo Transcriptome and Metabolome Analysis of Common Bean Response to *Fusarium oxysporum* f. sp. *phaseoli* Infection. *Int. J. Mol. Sci.*, 20: 6278.

Coelho, Netto, R.A. and Dhingra, O.D. 1999. Hosts of *Fusarium oxysporum* f. sp. *phaseoli. Fitopatol. Brasileira* 24(2): 156-159.

Conners, L. L., Shoemaker, R. A. and Creelman, D. W. 1957. *Thirty-sixth Annual Report of The Canadian Plant Disease Survey,* 1956. 29, 134pp.

de Borba, M. C., Garcés-Fiallos, F. R. and Stadnik, M. J. 2011. Reactions of black bean seedlings and adult plants to infection by *Fusarium oxysporum* f. sp. *phaseoli. Crop Protection* 96: 221-227.

de Vega-Bartol JJ, Martín-Dominguez R, Ramos B, García-Sánchez MA, Díaz-Mínguez JM. 2011. New virulence groups in *Fusarium oxysporum* f. sp. *phaseoli*: the expression of the gene coding for the transcription factor ftf1 correlates with virulence. *Phytopathology* 101(4):470-479.

Dongo, D.S. and Osores, D.A. 1961. *Fusarium* spp. in the cultivation of bean in Peru. *Turrialba* 11: 161-162.

Dongo, S.L. and Muller, L.E. 1969. Pathogenicity studies of *Fusarium oxysporum* f. *phaseoli* in beans II. Varietal trials. *Turrialba* 19: 82-90.

Echandi, E. 1967. Bean yellows caused by *Fusarium oxysporum* f.sp. *phaseoli. Turrialba* 17(4): 409-410.

El-Abyad, M.S., El-Sayed, M.A., El-Shanshoury, R. and El-Batanouny, N.H. 1993. Inhibitory effects of UV mutants of *Streptomyces chorchomsii* and *S. spiroverticiltatus* on bean and banana wilt pathogen. *Can. J. Bot.* 71(8): 1080-1086.

Gupta, S. K, Dohroo, N.P. and Shyam, K. R. 1993. Fusarium root rot of *Phaseolus vulgaris. Indian Phytopathology* 46: 415.

Harter, L. L. 1929. A *Fusarium* disease of beans. *Phytopathology* 19: 84.

Henrique, F. H., Carbonell, S. A. M., Ito, M. F., Gonçalves, J. G. R., Sasseron, G. R. and Chiorato, A. F. 2015. Classification of physiological races of *Fusarium oxysporum* f. sp. *phaseoli* in common bean. *Bragantia Campinas* 74(1): 84-92.

Kendrick, J.B. and Snyder, W.C. 1942. Fusarium yellows of beans. *Phytopathology* 32: 1010-1014.

Kobayashi, N. and Komada, H. 1995. Screening of suppressive soils to Fusarium wilt from Kanto, Tozan and Tokai areas in Japan, and analysis of their suppressiveness. *Soil Microorgan.* 45: 21-32.

Maringoni, A.C. and Lauretti, R.L.B. 1999. Response of dry bean genotypes to *Fusarium oxysporum* f.sp. *phaseoli*, *Macrophomina phaseolina* and *Xanthomonas campestris* pv. *phaseoli*. *Pesquisa Agopecuaria Brasileira* 34(4): 535-542.

Miranda, P., Costa, A.F.da, Pimentel, M. de L. and Silva, J.C. da. 1994. Common bean line selection for resistance against *Fusarium oxysporum* f. sp. *phaseoli* under field conditions. *Summa Phytopathol.* 20(6): 28-29.

Mititu, E.W., Mukunya, D.M. and Keya, S.O. 1988. Biological control of Fusarium yellows of beans caused by *Fusarium oxysporum* f.sp. *phaseoli* Kendrick and Snyder using organic amendments locally available in Kenya. *Acta Hortic.* 218: 267-274.

Nacimento, S.R.C., Maringoni, A.C. and Kurozawa, C. 1995. Reaction of varieties and lines of dry beans to *Fusarium oxysporum* f.sp. *phaseoli*. *Fitopatol. Brasileira* 20(3): 458-463.

Otadoh, J., Okoth, S., Ochanda, J. and Kahindi, J. 2011. Assessment of *Trichoderma* isolates for virulence efficacy on *Fusarium oxysporum* f. sp. *phaseoli*. *Tropical and Subtropical Agroecosystem* 13: 99-107.

Pieta, D. 1987. Studies on the susceptibility of different bean varieties to infection by fungi of the genus *Fusarium*. *Roczniki Nauk Rolniczych* 16(2): 69-79.

Piza, S.M. de T. 1993. Pathogenicity of *Fusarium oxysporum* f.sp. *phaseoli* and reaction of bean germplasm (*Phaseolus vulgaris*). *Summa Phytopathol.* 19(3/4): 165-167.

Rasool, B., Summuna, B., Shah, T. A. and Umbarkar, P. 2019. *In-Vitro* evaluation of bio-control agents against Alternaria leaf blight of sunflower and vascular wilt of beans. *International Journal of Current Microbiology and Applied Sciences* 8(8): 326-335.

Rava, C.A., Sartorato, A. and Costa, J.G.C. da. 1996. Reaction of French bean genotypes to *Fusarium oxysporum* f. sp. *phaseoli* in the greenhouse. *Fitopatol. Brasileira* 21(2): 296-300.

Reddy, M.S., Hynes, R. K. and Lazarovits, G. 1994. Relationship between *in vitro* growth inhibition of pathogens and suppression of pre-emergence damping-off and post-emergence root rot of white bean seedlings in greenhouse by bacteria. *Can. J. Microbiol.* 40(2): 113-119.

Reis, A., Oliveira, S.M.A. de, Menzes, M. and Mariano, R. de L.R. 1995. Potential of *Trichoderma* isolates on biocontrol of bean Fusarium wilt. *Summa Phytopathol.* 20: 16-20.

Rocha Junior, W.C., Santos, J.B. and Mendes-Costa, M.C. 1998. Reaction of common bean lines and cultivars to *Fusarium oxysporum* f.sp. *phaseoli*. *Fitopatol. Brasileira* 23 (3): 407-409.

Schwartz, H.F., McMillan, M.S. and Silbernagel, M.J. 1989. Occurrence of Fusarium wilt of beans in Colorado. *Plant Disease* 73(6): 518.

Thomas, C.A. and Wood, R.K.S. 1981. Studies on respiration in *Fusarium* infected beans. *Can. J. Plant Sci.* 59(2): 129-131.

Tuset Barrachini, J.J. 1973. Studies on wilt and desiccation of herbaceous plants. I A *Fusarium* species pathogenic to bean in eastern Spain. *Anales del Instituta Nacional de Investigaciones Agrarias Proteccion Vegetal No*. 3: 73-93.

Velasquez-Valla, R., Schwartz, H.F. and Diaz-Minguez, J. 1997. Pathogenicity of *Fusarium oxysporum* f.sp. *phaseoli* isolates from Spain in *Phaseolus vulgaris. Plant Dis.* 81: 312.

Weber, G.F. 1973. *Bacterial and fungal diseaes of plants in tropics*. University of Florida Press, Gainesville, pp.49-67.

Zambolim, L., Veira, C., Araujo, C.A. de A., Chagas, J.M. and Silva, C.C. da. 1987. Occurrence of Fusarium wilt on bean in the Zona da Mata of Minas Gerois. *Fitopatol. Brasileira* 12(3): 287-288.

Zhang, C.C. 1992. The pathogen and control of kidney bean wilt disease. *Acta Phytophylacica Simica* 19(2): 107-110.

Zoina, A., Woo, S.L., Sorbo, G. del., Linto, M., Nanni, B., Scala, F., Noviello, C. and Del Shorbo, G. 1995. Physiological races and vegetatively compatible groups (VCG) in *Fusarium oxysporum* f.sp. *phaseoli. Annali della Facolta di Scienze Agrarie della Universita degli Studi di Napoli Portici* 21(5): 51-54.

15

CERCOSPORA LEAF SPOT

Among various foliar diseases, Cercospora leaf spots are most common in India although they are not so damaging as other foliar diseases. Two main species of *Cercospora* reported world over are *C. canescens* and *C. cruenta. C. canescens* is generally found on higher elevation i.e. more than 2400 above mean sea level (Bisht *et al.*, 1997).

GEOGRAPHICAL DISTRIBUTION

The occurrence of this disease was first described in 1882 from USA (Ellis and Martin, 1882). Subsequently, it was reported from many bean growing areas of the world like Mauritius (Anonymous, 1924), Sierra Leone (Deighton, 1928), China (Teng, 1932), Philippines (Silayan, 1937), New Caledonia (Bugnicourt *et al.*, 1951), Brazil (Shands *et al.*, 1964), Colombia, Puerto Rico, Trinidad, Jamaica, Venezuela, Argentina (Wellman, 1977) and Iran (Bakshi, 2018).

In India, the occurrence of different *Cercospora* species on French bean viz., *C. cruenta* Sacc. (Butler, 1918; Agarwal *et al.*, 1959), *C. dolichos* Ell. and Everh. (Anonymous, 1950) and *C. canescens* Ell. & Mart. (Agarwal and Hasija, 1964) has been reported.

LOSSES

This is a common foliar disease and not so damaging but sometimes cause significant damage if the attack is early (Sharma and Sohi, 1989). In Sierra Leone, significant losses due to this disease have been reported (Deighton, 1928). Premature defoliation due to this disease in Colombia has also been reported (Orozeo, 1958).

SYMPTOMS

Cercospora cruenta

Spots on leaves are first small and brown or red, then grey or brown in the center with a prominent reddish-purple border. The spots are usually angular and restricted by the veins on the leaf lamina. Sometimes the central portion of the spot drops down leaving hole in the center. Affected leaves fall down prematurely. Blemishes may occur on stems and pods, and the fungi can become seed borne.

Cercospora canescens

Spots on the leaves are circular to angular, dark brown with lighter coloured center surrounded by yellow halo (Plate 1). The spots often coalesce and result in blight symptoms. Later sporulation of the fungus can be seen in the grey portion as a dark growth. Pods may also be affected (Plate 1).

THE PATHOGEN (S)

Several species of *Cercospora* are responsible for this disease, however, *C. canescens* Ellis and Martin and *C. cruenta* Saccardo are reported world over. The fungus is placed in the class Coelomycetes of sub-divison Deutromycotina. The fungus is both inter and intra-cellular in the host tissue. The conidiophores emerge through stomata which are brown, septate with knee-bends marking the insertion of fallen spores. Conidia are hyaline with varying numbers of septations. Spores may be club, curved or straight shaped. *C. cruenta* spores measure 50-150 x 6-9µm while *C. canescens* spores measure 50-100 x 3-4.5 µm.

HOST RANGE

Host range of both these species include *Vigna sinensis*, *V. catjang*, *Dolichos lablab*, *Phaseolus mungo* var. *radiatus*, *P. aureus*, *P. radiatus*, *P. acontifolius*, *P. vulgaris* and *P. lunatus*.

DISEASE CYCLE AND EPIDEMIOLOGY

The fungus survives, for one season in diseased crop refuse and are carried long distances on or in the seed. The spores are splashed or carried on tools or clothing from one part of the field to another. In some areas there are wild hosts on which they can remain alive from season to season. It has been reported that the fungus can live as a saprophyte on crop debris and produce conidia. The conidia are wind blown to green leaves where they cause infection.

Plate 1. Symptoms of Cercospora leaf spot on French bean leaves

MANAGEMENT

Cultural practices

Long crop rotations with non-leguminous crops, well-drained soils and planting healthy seed are usually recommended for the control of this disease. Mixed cropping is also recommended as these spots cause more damage alone than in mixed cropping with maize (Candal Neto *et al.*, 1987).

Host resistance

Though there are variations in the reaction of different varieties of crop plants to the disease, little has been done to develop resistant varieties. Silayan (1937) reported cv. Chinese Wax and Black bean as the promising pole type bean resistant to *C. cruenta* while V.P. 147, Porto-Alegre-Vagem, Roxa and Rico-23 were resistant to *C. vanderysti* (Vieira and Shands, 1965). Field and greenhouse resistance has also been reported in cvs. Ricopardo, 37-R, Quinheutas Vagens 1792, Cornel 49-242, Carioca and Caraota 260 (Asmus and Dhingra, 1983).

Chemical control

The use of chemicals in the management of this disease depends on the severity of the disease in the previous years. If the disease has been serious in previous years, spraying with Bodeaux mixture (4:4:50) or Blitox (0.25%) or zineb or maneb or thiram (0.25%) at 15 day interval starting one month after planting is desirable (Sharma and Sohi, 1989).

REFERENCES

Agarwal, G.P. and Hasija, S.K. 1964. Fungi causing plant diseases at Jabalpur (M.P.). *X. Mycopath. et Mycol. appl.* 23: 314-320.

Agarwal, G.P., Nema, K.G. and BeliRam, R. 1959. Fungi causing plant diseases at Jabalpur (M.P.) I. *Proc. Nat. Acad. Sci. India* 29: 310-315.

Anonymous, 1924. Botanical Division. Investigations of plant diseases. *Annu. Rep. Dept. Agric. Mauritius for the year 1923* pp. 10-11.

Anonymous, 1950. List of common names of Indian plant diseases. *Indian J. Agric. Sci.* 20: 107-142.

Asmus, G.L. and Dhingra, O.D. 1983. Evaluation of bean cultivars and germplasm for resistance to *Cercospora vandeysti. Fitopatol. Brasileira* 8(1): 97-101.

Bakshi, M. 2018. Identification of Cercospora spp. associated with emerging Cercospora leaf spot disease of *Phaseolus vulgaris* L. in Guilan province based on multigene phylogeny. *In 23th Iranian Plant Protection Congress* at Iran.

Bisht, G.S., Harish Singh, Singh, H. and Dubey, R.C. 1997. Fungal diseases of useful plants in the Garhwal Himalaya and their management. *Himalayan Microbial Diversity Part 1*: 255-271.

Bugnicourt, F., Cohic, F. and Dadant, R. 1951. A catalogue of the animal and plant parasite of the cultivated plants of New Caledonia (RAM 1954 p.615).

Butler, E.J. 1918. *Fungi and Diseases in Plants*. Thacker Spink & Co., Culcutta.

Candal Neto, J.P., Vieira, C., Cardoso, A.A. and Galvao, J.D. 1987. Intercropping of beans and maize. Effects of cultivar and plant density of maize. *Revista Ceres* 34: 196, 619-630.

Deighton, F.C. 1928. Mycological section. *Annu. Rep. Lands and Forests Dept. Sierra Leone for the year 1927*. pp. 13-17.

Ellis, D.E. and Martin, G.B. 1882. General notes: Botany. New species of North American fungi. *Amer. Nat.* 16: 1001-1004.

Orozeo S., S.H. 1958. Mancha del Cercospora en frijol. *Tests de Ing. Agric. Univ. Nac. De Colombia. Palmira, Colombia*. 47 p.

Sharma, S.R. and Sohi, S.H. 1989. Diseases of French bean in India and their control. *Indian Rev. Life Sci.* 9: 153-168.

Shands, H., Viera, C. and Zaumeyer, W.J. 1964. Observations on dry bean diseases in Brazil. *Pl. Dis. Reptr*. 48: 784-787.

Silayan, H.S. 1937. *Annual Report of the Director of Plant Industry, Commonwealth of the Philippines, for the fiscal year ending Dec. 31, 1936*, 178pp.

Teng, S.C. 1932. Some fungi from Canton. *Contrib. Lab. Sci. Soc. China* 8(2): 121-128.

Veira, C. and Shands, H.L. 1965. Mancha gris, nova doenca do Feijoeiro no Brasil. *Revista Agric. Piracicaba* 40(1): 3-5.

Wellman, F.L. 1977. *Dictionary of tropical American crops and their diseases*. pp. 312-321. The Scareerow Press Inc. Metuchen, New Jersey, 495pp.

16

PHYLLOSTICTA LEAF SPOT

The disease is of minor importance but sometimes in favourable weather cause significant yield losses.

GEOGRAPHICAL DISTRIBUTION

The disease is not serious but wide worldover. The occurrence of the disease has been reported from Serbia (Ranojevic, 1910), Denmark (Lind, 1913), Japan (Ideta, 1914), Union of South Africa (Doidge, 1924), Bulgaria (Ivanoff, 1926), Canada (Drayton, 1926), Brazil (Muller, 1935), Belgium (Marchal, 1936), Mauritius (Anonymous, 1961), United States (Zaumeyer and Thomas, 1957, Goth and Zaumeyer, 1963), Costa Rica, Nicaragua, El Salvador, Guatemala, Peru, Argentina, Puerto Rico (Wellman, 1977).

In India, the occurrence of the disease was first reported from Dharwad, Karnataka (Rangaswami *et al.*, 1970). Subsequently it was reported from Himachal Pradesh (Gupta and Saharan, 1973).

LOSSES

Since this disease is of minor importance, there are not many reports regarding the economic losses due to this disease. In New Jersey, many snap bean fields particularly of the cv. 'Tender Crop' where 50 – 70 per cent plants were severely infected by this disease (Goth and Zaumeyer, 1963). In India, *Phyllosticta phaseolina* has been reported to cause about 12% seed rotting in bean (Gupta and Saharan, 1973).

SYMPTOMS

Symptoms are largely restricted to mature leaves, but other parts are also affected. The spots irregularly scattered over the leaf lamina, are 2-10 mm in diameter with light centers and deep rusty brown margins. Lesions on petioles and stems are narrower at first but later may extend around the petioles and halfway around the stems. Flower buds usually turn brown and fall off soon after infection. On pods, lesions usually remain small (< 3mm in dia.) with dark centers and reddish margins. With the advancement of the season, the disease becomes more severe and dark-coloured lenticular pycnidia scattered over the centers of the spots appear.

THE PATHOGEN

The fungus responsible for this disease is *Phyllosticta phaseolina* Sacc. (Saccardo, 1878). Pycnidia of the fungus are 70-90 μm in diameter. Pycnidia are filled with hyaline, one - celled, mostly straight, ovoid to oblong pycniospores that exude in long gelatinous tendrils and measure 4 - 6 x 2.0 - 2.5μm in size.

HOST RANGE

Besides *Phaseolus vulgaris*, the pathogen also attacks *P. lunatus* f. *macrocarpus, P. polystachios, Strophostyles helvola* and *Vigna cylindrica* (Zaumeyer and Thomas, 1957).

DISEASE CYCLE

The pathogen is both seed (Gupta and Saharan, 1973) and soil borne. In soil the pathogen survives in infected plant debris. In favourable weather conditions, the pathogen produce pycniospores which are blown by wind and cause infection. Seed borne inoculum also causes pre-emergence damping-off of seedlings (Gupta and Saharan, 1973). The damped-off seedlings served as substrate for multiplication of the pathogen and provide primary inoculum for secondary infection in the field.

EPIDEMIOLOGY

The fungus perennates both in infected plant debris in soil and in infected seed. The disease is favoured by high moisture and moderate temperature conditions (Goth and Zaumeyer, 1963, Shands *et al.*, 1964, Viera, 1967).

MANAGEMENT

Cultural practices like use of healthy seed, long crop rotation with non-leguminous crops and destruction of infected plant debris (Chupp and Sherf, 1960) should be followed to keep the disease in check. If sources of resistance are available should be used in breeding programmes for developing resistant varieties. Fungicides such as mancozeb, carbendazim and EBI fungicides either alone or in combination can be applied at fortnightly interval for the control of this disease.

REFERENCES

Anonymous 1961. Plant Pathology. *Rep. Dep. Agric. Mauritius*: 41-49.

Doidge, E.M. 1924. A preliminary check list of plant diseases occurring in South Africa. *So. Africa Bot. Surv. Mem*. 6: 56.

Drayton, F.L. 1926. A summary of the prevalence of plant diseases in the Dominion of Canada 1920-1924. *Canada Department of Agriculture Bulletin*. 71: 88.

Goth, R.W. and Zaumeyer, W.J. 1963. Occurrrence of Phyllosticta leaf spot in beans. *Pl. Dis. Reptr.* 47(12): 1079.

Gupta, V.K. and Saharan, G.S. 1973. Seed rot and root rot complex of beans (*Phaseolus vulgaris* L.). *Biologia Plantarum* 15(2): 123-125.

Ideta, A. 1914. *Handbook of the plant diseases of Japan*. 6th Ed., Tokyo, Japan, 553pp.

Ivanoff, B. 1926. Cryptogamic parasites of cultivated plants recorded in the coarse of the last five years (1921-1925). *Sofia Period. Bull.* 7(3): 14-17.

Lind, J. 1913. Danish Fungi as Represented in the Herbarium of E. Rostrup. Copenhagen, 648pp.

Marchal, E. 1936. Observations and researches carried out at State Phytopathological Station during the year 1935. *Bull. Inst. Agron. Gumbloux* 5(2): 105-111.

Muller, A.S. 1935. Lista Preliminar Das Doencas Cryptogamicas de Plantas Cultivadas em Minas Geras, Brasil. *Sec. da Agric. Zootech. E Vet. Bol.* 8: 67-77.

Rangaswami, G., Seshadri, V.S. and Lucy Channamma, K.A. 1970. *Fungi of South India*. Univ. Agric. Sci, Bangalore, 193pp.

Ranojevic, N. 1910. Zweiter Beitrag zur Pilzflora serbeins. *Ann. Mycol.* 8: 347-402.

Saccardo, P.A. 1878. Fungi Venti Novi Vel Critici. Vel. Mycologiae Venetae Addendi, *Phyllosticta phaseolina* Sacc. *Michelia* 1: 149.

Shands, H., Vieira, C. and Zaumeyer, W.J. 1964. Observations on dry bean diseases in Brazil. *Pl. Dis. Reptr*. 48: 784-787.

Vieira, C. 1967. O Feijoeiro – Comum. Cultura, Doenease Melhoramento. Pp. 84-124. Imprensa Universitaria, Vicosa, Brazil.

Wellman, F.L. 1977. *Dictionery of Tropical American Crops and Their Diseases*. pp. 312-321. The Scarecrow Press Inc. Metuchen. New Jersey, 495pp.

Zaumeyer, W.J. and Thomas, H.R. 1957. A monographic study of bean diseases and methods for their control. *U.S. Dept. Agric. Tech. Bull. No*. 868, 255pp.

17

CORYNESPORA LEAF SPOT

This is a minor disease of this crop and does not cause economic losses.

GEOGRAPHICAL DISTRIBUTION

There are few reports of the disease occurrence from different parts of the world. *Corynespora cassiicola* was isolated from the roots of mature field bean (*Phaseolus vulgaris*) (Seaman *et al.*, 1965). *C. cassiicola* had been reported as seed-borne in cultivar INTA Rojo of common bean in Nicaragua (Marcenaro and Valkonen, 2016).

In India, the occurrence of the disease has been reported from Assam (Barman and Roy, 1978) and Punjab (Saini *et al.*, 1983) states only.

LOSSES

The disease appears on leaves as spots which ultimately reduce the net photosynthetic area leading to yield losses. However, no estimates of yield losses due to this disease have been reported.

SYMPTOMS

The symptoms of the disease appear as light brown, circular to ovoid, distinct spots with concentric rings which measure 2-14 mm in diameter. The spots are prominent on both the sides of the leaves, scattered all over the surface and ultimately become necrotic (Saini *et al.*, 1983).

THE PATHOGEN

The fungus responsible for this disease is *Corynespora cassiicola* (Berk. & Curt.) Wei. The fungus is characterized by conidiophores which are thick walled, olivaceous, elongated, unbranched, coming out in groups by breaking the epidermis, transversely 1-4 septate, hypophyllous and measure 107 –197 x 4-6 µm in size (Fig. 1). Conidia are elongated, thick walled, pale brown, obclavate, slightly curved, smooth, transversely 9-18 septate, attached to conidiophore with a hyaline isthmus, tip obtuse and measure 86-237 x 12-16 µm in size (Fig. 1).

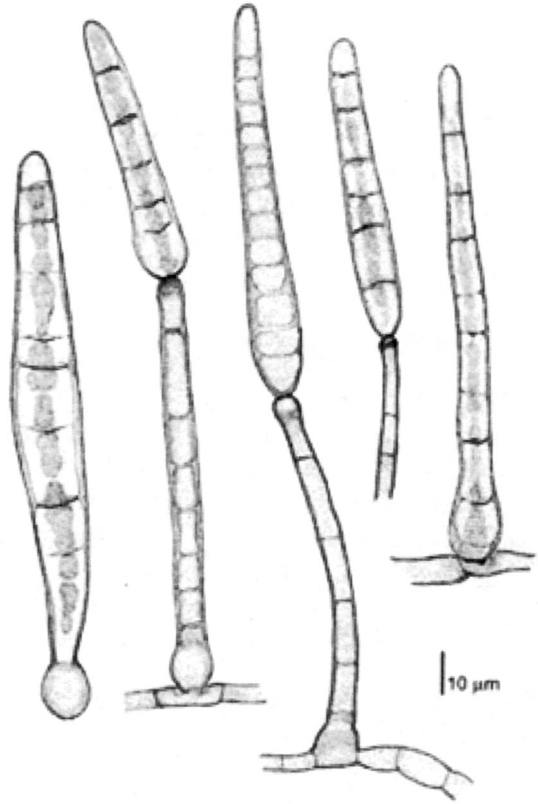

Fig. 1. Conidiophores and conidia of *Corynespora cassiicola*

(Source: http://www.mycobank.org/MycoTaxo.aspx?Link=T&Rec=296024)

Whole genome sequences of one soybean isolate and four cotton isolates of *Corynespora cassiicola* from Tennessee, USA were used to develop single nucleotide polymorphism (SNP) markers for cotton isolates (Shrestha *et al.*, 2017). It was found that cotton isolates had little diversity at the genome level and very little differentiation from the soybean isolate. Analysis of 75 isolates from cotton and soybean, using targeted-sequencing of 22 polymorphic SNP sites, revealed eight multi-locus genotypes and it appears a single clonal lineage predominates across the southeastern region. However, cotton and soybean genome sequences were significantly different from the rubber isolate of *C. cassiicola*.

MANAGEMENT

Information on the management of this disease is lacking in literature, however, it can be kept in check by collection and destruction of infected plant refuse, crop rotation and application of fungicides.

REFERENCES

Barman, B. and Roy, A.K. 1978. Corynespora leaf spot of French bean and tomato. *Sci. and Cult*. 44: 411.

Marcenaro, D., and Valkonen, J. P. 2016. Seedborne pathogenic fungi in common bean (*Phaseolus vulgaris* cv. INTA Rojo) in Nicaragua. *PloS one*, 11(12): e0168662

Saini, S.S., Suppal, H.S. and Ramzan, M. 1983. A new Corynespora leaf spot disease of *Phaseolus vulgaris*. *Indian Phytopath*. 36: 179.

Seaman, W. L., Shoemaker, R. A. and Peterson, E. A. 1965. Pathogenicity of *Corynespora cassiicola* on soybean. *Canadian Journal of Botany*. **43**(11): 1461-1469.

Shrestha, S.K., Lamour, K., Young-Kelly, H. 2017. Genome sequences and SNP analyses of *Corynespora cassiicola* from cotton and soybean in the southeastern United States reveal limited diversity. *PLoS ONE* 12(9): e0184908.

18

COTTONY LEAK

This is also a minor disease but in favourable weather conditions it causes significant yield reductions. This disease is also known as Pythium root rot.

GEOGRAPHICAL DISTRIBUTION

The occurrence of the disease caused by *Pythium debaryanum* has been reported by Bhargawa and Gupta (1957). *P. aphanidermatum* has also been reported to cause aerial infection on the pods and occurrence of this phase has been reported from Meghalaya (Maiti and Sahambi, 1979). However, the occurrence of *P. paroecandrum* on this crop has been reported from South India (Balakrishnan, 1948). Recently, morphological and molecular studies revealed that *Pythium and Phytophthora* spp. were responsible to cause cottony leak disease of common or snap bean (Olson *et al.*, 2016). Further, it was also observed that among various *Pythium* species, *P. ultimum* and *P. aphanidermatum* were most predominant and virulent on this crop.

LOSSES

The disease infects pods and rendering them unmarketable thus resulting in complete loss. Losses in seed yield of white bean due to Pythium root rot were estimated to be 44% in 1976 and 39% in 1977 (Chew and Hall, 1984). Maiti and Sahambi (1979) reported 2-6% average incidence of this disease in Meghalaya but in some cases, it was as high as 10%.

SYMPTOMS

Pythium debaryanum and *Pythium ultimum*

The symptoms include the production of large number of water-soaked lesions which enlarge rapidly. The fungus grows profusely over the surface with abundant cottony mycelium and soon dirty fluid is seen oozing out of the infected pods and this stage is known as cottony leak.

Pythium aphanidermatum

Initial symptom of the disease appears in the form of water-soaked lesions on young pods, up to 6 inches above the ground level. Under favourable weather conditions, the lesions enlarge very rapidly ultimately girdle the entire pod. The diseased tissue become rotten, but no mycelial growth occurs on the surface of the infected pods.

THE PATHOGEN (S)

The fungi responsible for this disease are *Pythium debaryanum* Hasse (Bhargava and Gupta, 1957), *P. aphanidermatum* (Edson) Comb. Nov. (Maiti and Sahambi, 1979), *P. ultimum* and *Phytophthora drechsleri* (Olson *et al.*, 2016). The sporangia of *P. aphanidermatum* are in the form of stout swollen hyphae or lobulated, bud like outgrowth, ranging 410-530 µm in length and 15-25 µm in width (Fig. 1). These structures are very rich in protoplast. At the time of germination, vesicle is formed at the apex of the outgrowth (Fig. 1).

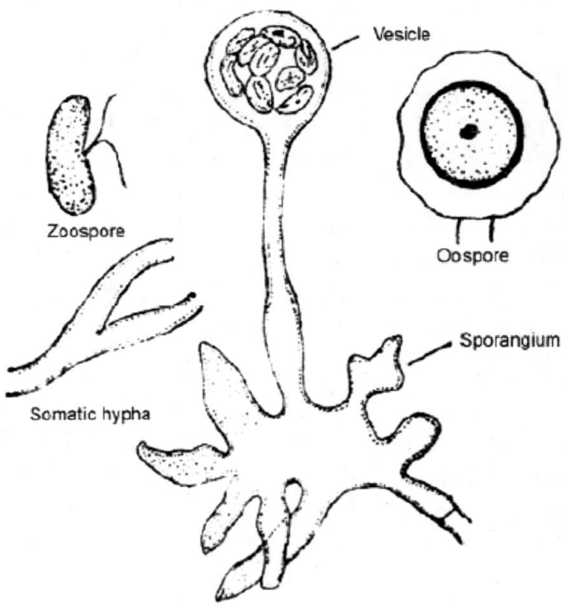

Fig. 1. Microscopic structures of *Pythium* spp.

The occurrence and distribution of *Pythium spp. were determined by collecting isolates of Pythium from common bean plants showing root or hypocotyl disease symptoms from different areas of Western Australia in 2012 (Li et al.*, 2014). Eight different *Pythium species (Pythium conidiophorum, P. diclinum, P. intermedium, P. irregulare, P. lutarium, P. mamillatum, P. pachycaule* and *P. perplexum*) were isolated and identified according to molecular sequences. *P. irregulare was the most widespread Pythium sp. All species, except P. perplexum*, were pathogenic to the hypocotyl and root of common bean. This constitutes the first report of *P. intermedium as a pathogen on common bean* world over.

Eleven species of the *Pythium viz., P. aphanidermatum, P. splendens, P. ultimum, P. attrantheridium, P. graminicola, P. oligandrum, P. dissotocum, P. irregurale, P. camurandrum, P. paroecandrum* and *P. acanthophoron* were reported to cause

Pythium root rot of common bean (Binagwa *et al.*, 2016 a). In another study, *Pythium aphanidermatum, P. myriotylum, P. ultimum, P. ultimum* var. *sporangiiferium*, and *P. ultimum* var. *ultimum* were reported as most virulent species causing seed rot and pre-emergence damping-off of dry bean (Rossman *et al.*, 2017).

HOST RANGE

The pathogen *Pythium* spp. are polyphagous and has been reported pathogenic on several crops (Maiti and Sahambi, 1979).

DISEASE CYCLE

The fungus perennates in soil in the form of oospores. In favourable weather conditions, the oospores germinate and form zoosporangia which in turn form vesicle and release biflagellate zoospores. The zoospores after swimming for some time in film of water become encysted and cause infection. The sporangia formed on the infected tissue serve as the source of secondary infection and the cycle continues.

EPIDEMIOLOGY

The fungus survives from one season to another in the form of oospores in soil. The optimum temperature for the disease is 25 °C and the progress of the disease was slow at 21 and 30° C.

MANAGEMENT

Cultural practices

Cultural practices like collection and destruction of the infected plant material, crop rotation, clean cultivation and avoid injury to the pods should be followed to keep the disease under check.

Host resistance

568 accessions of common bean categorized into 12 groups according to colour of seedcoat, were inoculated with *P. ultimum* under greenhouse conditions and it was found that 48 were symptomless (Lucas and Griffiths, 2004). Accessions with light seedcoats showed higher level of symptoms whereas cream-coloured beans exhibited higher level of resistance, Accessions PI 430207, PI 527803, PI 290996, PI 299021, PI 194574, PI 304110, PI 207129, PI 313601, PI 352748 and PI 352759 were resistant. In another study, 194 common bean varieties from 37 countries were screened for root disease severity caused by *Pythium irregular* (Li *et al.*, 2016). Varieties Quqa lecherita and PAC 18 showed the greatest resistance to *Pythium*. Navy Veracruz, Negro argel, and Phavul 77 had combined resistance to *P. irregulare* and tolerance to waterlogging. This is the first study to identify combined abiotic and biotic stress resistances in common bean.

Among 100 cultivars of common bean in Tanzania, two bean genotypes ADP-014 and ADP-080 showed promising trait of resistance, 38 genotypes showed moderate resistant trait and 57 were susceptible to Pythium root rot disease (Binagwa *et al.*, 2016 b). A genome-wide association analysis of resistance to Pythium root rot in the common bean of Andean gene pool using single nucleotide polymorphism (SNP) markers was conducted and about 260 genotypes of the Andean diversity panel (ADP) were evaluated under screen house conditions using *Pythium ultimum* in Uganda (Dramadri *et al.*, 2020). Sixteen significant signals for resistance to Pythium root rot were detected on chromosomes Pv01, Pv02, Pv04, Pv05 and Pv09 using 260K GBS-based and 6K Beadchip SNPs. Common significant signals were detected on Pv02 and Pv09 for PRR. Positional candidate genes associated with significant SNPs on Pv02 were *Phvul.002G119700*, 16.97 kb near marker S02_25507837 (25.50 Mb), encoding Subtilase family protein, and *Phvul.002G278400* near marker ss715645959 (44.79 Mb) encoding Defensin-like (DEFL) protein involved in plant defense responses. Based on the relatively high heritability estimates observed for Pythium root rot in this study, significant SNP markers associated with genomic regions for resistance to Pythium root rot could be validated for marker-assisted breeding in Andean beans.

Chemical control

Various systemic and no-systemic fungicides like metalaxyl + mancozeb, cymoxanil + mancozeb, mancozeb, copper oxychloride and propineb should be sprayed at 10-14 days interval to combat the ravages of this disease.

Biological control agents

Application of biocontrol agents *viz. Trichoderma harzianum* and *Pseudomonas fluorescens* as soil application and seed treatment were effective to reduce the incidence of damping-off in French bean cause by *Pythium ultimum* under field conditions (Dubey *et al.*, 2017).

REFERENCES

Balakrishnan, M.S. 1948. South Indian Phycomycetes II. Some little-known species of *Pythium* occurring in South India. *Proc. Indian Acad. Sci.* 28 B: 27-34.

Bhargava, K.S. and Gupta, S.C. 1957. Market diseases of fruits and vegetables in Kumaon I. Cottony leak of beans. *Indian Phytopath.* 10: 48-49.

Chew, P. S. and Hall, R. 1984. Effect of Pythium root rot on yield of white bean, *Phaseolus vulgaris* L. *Crop Protection* 3(4): 423-429.

Dramadri, I. O., Amongi, W., Kelly, J. D. and Mukankusi, C. M. 2020. Genome-wide association analysis of resistance to *Pythium ultimum* in common bean (*Phaseolus vulgaris*). *Plant Breeding* 139:1168-1180.

Dubey, S.C., Tripathi, Aradhika, Tak, R and Devi, I. 2017. Management of root rot and damping-off complex in French bean by biocontrol agents. *Indian Phytopathology* 70(3): 386-390.

Li, Y. P., You, M. P. and Barbetti, M. J. 2014. Species of *Pythium* associated with seedling root and hypocotyl disease on common bean (*Phaseolus vulgaris*) in Western Australia. *Plant Dis*. 98:1241-1247.

Li, Y.P., You, M.P., Norton, S. and Barbetti, M. J. 2016. Resistance to *Pythium irregulare* root and hypocotyl disease in diverse common bean (*Phaseolus vulgaris*) varieties from 37 countries and relationships to waterlogging tolerance and other plant and seed traits. *Eur J Plant Pathol* 146:147-176.

Lucas, B. and Griffiths, P. D. 2004. Evaluation of common bean accessions for resistance to *Pythium ultimum*. *Hort Science* 39(6): 1193-1195.

Maiti, S. and Sahambi, H.S. 1979. Aerial infection of *Pythium aphanidermatum* and *Corticium sasakii* on French bean in Meghalaya. *Indian J. Mycol. Pl. Pathol.* 9(1): 99-100.

Olson, J. D., Damicone, J. P. and Kahn, B.A. 2016. Identification and characterization of isolates of *Pythium* and *Phytophthora* spp. from snap beans with cottony leak. *Plant Dis.* 100: 1446-1453.

Binagwa, P. H., Bonsi, C. K., Msolla, S. N. and Ritte, I. I. 2016 a. Morphological and molecular identification of *Pythium* spp. isolated from common beans (*Phaseolus vulgaris*) infected with root rot disease. *African J. Plant Sci.* 10(1): 1-9.

Binagwa, P. H., Bonsi, C. K. and Msolla, S. N. 2016 b. Evaluation of common bean (*Phaseolus vulgaris*) genotypes for resistance to root rot disease caused by *Pythium aphanidermatum* and *Pythium splendens* under screen house conditions. *Journal of Natural Sciences Research* 6(6): 36-43.

Rossman, D. R., Rojas, A., Jacobs, J. L., Mukankusi, C., Kelly, J. D. and Chilvers, M. I. 2017. Pathogenicity and virulence of soilborne Oomycetes on *Phaseolus vulgaris*. *Plant Dis.* 101(11): 1851-1859.

19

COMMON BLIGHT

Common bacterial blight of bean or more readily referred to as common blight, was first recognized by Beach in 1892 in New York. In the late 1920's, Burkholder isolated a bacterium that was very similar to common bacterial blight pathogen from bean plants collected in Switzerland, but this bacterium produced brown pigment in culture media (Zaumeyer and Thomas, 1957). This bacterium induced symptoms very similar to common blight, but Burkholder noticed differences in its ability to cause disease in some bean varieties when compared to *Bacterium phaseoli* (Burkholder and Bullard, 1946). The disease induced by this brown - pigment producing bacterium was named fuscous blight (Zaumeyer and Thomas, 1957) and the bacterium was referred to as *Xanthomonas campestris* pv. *phaseoli* var. *fuscans* (Burkh.) Starr. The distribution of fuscus blight is considered to be somewhat more limited than that of common blight (Hayward and Waterston, 1964), but because of the similarity of symptoms there is little practical value in trying to separate the two diseases. Dye (1962) concluded that there was little justification for separating *X. phaseoli* from *X. phaseoli* var. *fuscans*, since pigment production is common among *Xanthomonas* species not pathogenic to bean and may not be a stable character (Basu, 1974). For these reasons, common and fuscous blights will be treated as essentially the same disease and referred to as common blight.

GEOGRAPHICAL DISTRIBUTION

The disease was first reported from New York, U.S.A. (Beach, 1892) and subsequently, reported from Australia (Magee, 1930), Bulgaria (Atanasoff and Kovachevski, 1929), Canada (Suit, 1934), China (Reinking, 1919), France (Delacroix, 1901), Hungary (Kern, 1929), Japan (Ideta, 1936), Madagaskar (Bouriquet, 1937), Norway (Jorstad, 1922), Phillipines (Reinking, 1919), Russia (Galachyan, 1936), Spain (Alfaro and Silvan, 1940), Turkey (Bremer, 1947), South Africa (Doidge, 1919), Uruguay (Bertelli, 1947), Yugoslavia (Tesic, 1946), Argentina, Brazil, Colombia, Paraguay, Venezuela, Guatemala, Nicaragua, Honduras, Mexico, the Dominican Republic, Uganda, Burundi, Zaire, Rawanda and Zambia (Allen, 1983; Saettler, 1989). In India, this disease was reported for the first time from Bombay by Uppal *et al.* (1946). The pathogen distributed in most regions where common bean is cultivated except in arid tropical areas. It is a major disease in African countries such as Malawi, Uganda, Kenya, Burundi, Egypt and Tanzania. It is also present in other South-Eastern and Southern Africa countries. In Ethiopia, it is ranked among the most important diseases of common bean (Leta *et al.*, 2017).

LOSSES

Common blight occurs in almost every bean growing areas including temperate, subtropical and tropical regions of the world and can cause significant yield losses if conditions are favourable for disease development. Warm and humid weather favours severe outbreaks of the diseases, thus the disease is more serious in the tropics (Allen, 1983; Saettler, 1989). Epidemics of common and fuscous blight occurred in Ontario, Canada in early 1960's and in 1962, when warm weather along with high humidity occurred (Sutton and Wallen, 1970) with more incidence of later (60% fields) than the former (8%), respectively, which was due to the introduction of bean variety Sanilac, which was particularly susceptible to fuscous blight. In Colombia, yield losses were estimated upto 22 and 45 per cent in natural and artificial infected fields, respectively (Yoshi *et al.*, 1976). It has also been estimated that common blight is responsible for an average 17 per cent yield losses in the Cauca Valley of Colombia (Saettler, 1989). In India, upto 84 per cent yield losses have been reported in French bean cvs. Contender and Arka Komal at different crop growth stages (Kishun, 1988).

SYMPTOMS

The disease affects stem, branches, leaves, pods and seed. The symptoms appear as characteristic water soaked spots on the underside of the leaves that enlarge and coalesce together to form large blighted areas. Infected areas often become flaccid and turn brown giving it a sunscald appearance with bright yellow margins (Plate 1). If infection occurs through the vascular tissues, reddish discolouration of veins and midribs is observed. In advanced stages, tissues become necrotic and defoliation may also take place. Lesions on stems appear as sunken water soaked areas, which gradually enlarge to form reddish streaks extending longitudinally in the stem. In early stages of pod formation, characteristic stem girdle or joint rot frequently develops. Infection on pods may occur at any place as small water soaked spots, which are surrounded by distinct zones of narrow reddish brown or brick red band of tissue. On seeds, discoloration appears especially near the hilum region. Plants arising from infected seeds exhibit characteristic wilting (Saettler, 1989).

Plate 1. Symptoms of Bacterial blight of bean

(*Courtesy: Dr. S. Kansal*)

THE PATHOGEN

The disease is caused by *Xanthomonas axanopodis* pv. *phaseoli* (Smith) Dye (XAP). The bacterium is a motile, non-capsulated, non-spore forming, gram-negative rod, measuring 1.9 x 0.8 µm in size with a single polar flagellum. The pathogen forms characteristic yellow coloured colonies due to a water insoluble eacotenoid and a mucoid growth on nutrient glucose agar medium. The characteristic yellow colour of the bacterial colonies is due to the production of brominated aryl polyenes, more frequently referred to as xanthomonadins, that have absorption peaks in between 443 and 446 nm (Starr, 1983). Some other useful properties of this bacterium include hydrolysis of starch, geletin and casein, lipolysis of Tween 80, and lack of pectinase activity. Acid is produced as a metabolic by product when cells grow on media containing arabinose, glucose, mannose, galactose, trehlose or cellubiose. It also causes proteolysis of milk. The pathogen is a strict aerobe, and metabolism of glucose is through oxidative pathway.

VARIABILITY

Pazos *et al*. (1986) failed to differentiate in Cuban isolates of *X. axanopodis* pv. *phaseoli* isolated from bean and by using bacteriophages, 3 lysotypes were distinguished among 35 isolates. Highly significant differences among fifteen isolates of this pathogen were observed among varieties and isolates in relation to resistance and pathogenecity (Rava and Romeiro, 1990) and isolate Xp CNP 28, characterized by high degree of pathogenecity, was suggested as a new standard isolate to replace Xp CNF 15.

Report regarding serological variability among *X. axanopodis* pv. *phaseoli* also exists in literature and 43 isolates from Sao Paulo and Perana States of Brazil were classified into 3 serological groups of which 22 did not produce melanin *in vitro* and reacted only with the homologous antiserum AS-Feij I (serogroup I), 17 reacted only with antiserum Feij 20 (serogroup II) and four did not react with the antisera tested (serogroup III) by using the agar double diffusion test (Maringoni *et al*., 1994). The strains of Xap were also grouped in 3 groups through profiles of total proteins (PAGE-SDS) and the analysis revealed in the first group, only the producers of a brown pigment; in the second group, producers and not producers of the brown pigment and in the third group, no producers of the brown pigment, suggested that there is a variability in the protein patterns among Xap strains (Maringoni and Bach, 1999).

On *Phaseolus acutifolius*, 30 isolates of Xap interacted differently and 8 physiologic races were defined suggesting an underlying gene- for-gene relationship (Opio *et al*., 1996). They further suggested that levels of resistance already available within *P. vulgaris* are adequate, so that breeder incorporating new complete resistance from *P. acutifolius* could risk destabilizing the host-bacterium relationship by introducing race-specific resistance, that could be transient in agriculture. Genotypes useful for differentiation of Xap physiological races have been identified (Zapata, 1997).

DNA probes were also used to differentiate *X. axanopodis* pv. *phaseoli* and *X. axanopodis* pv. *phaseoli* var. *fuscans* isolates originating from different geographical locations and RFLP analysis revealed DNA polymorphism within *X. axanopodis* pv. *phaseoli* isolates from the same geographical location (Otoya and Pastor-Corrales, 1992).

One hundred and forty three common bacterial blight isolates from 44 localities in four countries of Southern Africa, were inoculated onto eight *Phaseolus acutifolius* lines that differentiate between pathogenic races (Fourie and Herselman, 2011). This differential set was expanded to include resistant genotypes, XAN 159, GN 1 Nebr. sel 27, Wilk 2, Wilk 6, Vax 4, Vax 5 and Vax 6 and cv. Teebus as susceptible check. Genetic variation within nine selected *X. axanopodis* pv. *phaseoli* and *X. axonopodis pv. phaseoli var.* fuscans isolates and a non-pathogenic *Xanthomonas* isolate, was studied using RAPD and AFLP analysis. Genotypes XAN 159, Wilk 2, Wilk 6, Vax 4, Vax 5 and Vax 6 were resistant to all isolates, while GN #1 Nebr. sel 27 and cv. Teebus were susceptible. Isolates varied in aggressiveness on cv. Teebus; however, the pathogenic reaction on the set of differentials indicated that all, but one isolate, grouped in what has been reported as race 2. Thus, results based on reaction of the majority isolates, suggested the absence of different races. However, the distinct differential reaction recorded for a single isolate, may prove to represent another, as yet unrecorded, race of this pathogen. Both RAPD and AFLP analyses revealed high frequency of DNA polymorphism among isolates and could distinguish between *X. axanopodis* pv. *phaseoli*, *X. axonopodis pv. phaseoli var.* fuscans and a non-pathogenic isolate. Differences between *X. axanopodis* pv. *phaseoli* and *X. axonopodis pv. phaseoli var.* fuscans isolates demonstrate that these are two distinct groups of bacteria.

HOST RANGE

The host range of this pathogen includes both cultivated and weed host species viz., *Dolichos lablab, Emilia sonchifolia, Glycine max, Ipomoea crassicaulis, Lupinus polyphyllus, Phaseolus acutifolius, P. aureus, P. coccineus, Pisum sativum, Strophostyles helvola, Vigna mungo, V. radiata var. radiata,* (Saettler, 1989; Rodriguez *et al.*, 1991; Chand, 1992). The bacterium was also isolated from diseased leaves of peas, cowpeas, *Ipomoea purpure, Malva sysiaca* and *Vigna luteola* which may play significant role in the harbouring and spread of the pathogen in the neighbouring bean fields (Khlaif and Qadous, 1995)

DISEASE CYCLE

The bacterium overwinters in infected seed, crop debris and weed hosts (Gilbertson *et al.,* 1990; Irigoyen and Garbagnoli, 1998) of which contaminated seed has been observed to be the most efficient mode of survival and dispersal over long distances (Saettler, 1989). The pathogen survived for more than five months in the plant debris on the ground surface (Arnaud-Santana *et al.*, 1991) but the survival of the pathogen is influenced by tillage (Gilbertson *et al.*, 1990). Survival of the pathogen on the weed host *Parthenium hysterophorus* and subsequent transmission of the bacterium to *Phaseolus vulgaris* and vice versa has also been reported (Ovies and Larsinaga, 1988). Upon germination, the bacterium from infected seed infects the developing cotyledons and can then enter the vascular system of the plant. Alternatively the bacterium can be splashed to the plants from the plant debris. Once inside the plant, the bacteria invade and cause breakdown of the middle lamella (Zaumeyer and Thomas, 1957). Bacterial multiplication takes place, and masses of bacteria form that expand into what macroscopically appear as water-soaked spot. The bacteria also exude to the surface of the tissue and can be disseminated in the field by rain and wind to initiate new infections. Bacteria enter the plants via wounds caused by leaf eating insects or natural openings such as stomata or hydathodes (Saettler, 1989). The bacteria can infect stem via stomata or via the vascular system. Most vascular infections probably initiate from seedborne inoculum. In some of these cases, the bacteria can cause plugging of xylem elements, causing wilting. The bacteria can infect seed via the vascular system, moving through the pod pedicel into the funiculus and through the raphe into the seed coat, where they remain beneath the seed coat until seed germination.

EPIDEMIOLOGY

Common bacterial blight disease is a seed-borne, and the pathogen survives as long as the seed remains viable. Seed transmission is the primary means by which the pathogen is disseminated. Internally and externally infested seeds are important sources of primary inocula. Infection of the pedicels of flower buds and small pods resulted in transmission of the bacteria through the vascular tissues of the pod to the seeds, causing internal infection without any external symptoms on either pods or seeds (Aggour *et al.*, 1989). The minimum population of *X. axanopodis pv. phaseoli* required to initiate infection was 10^2 c.f.u./seed while seed infection levels of 0.2 per cent resulted in serious disease incidence (Opio *et al.*, 1993). The bacterium also enters plants through wounds and/or natural openings and infects stems via stomata, infected cotyledons or vascular system. The incidence and severity of the diseases increase with the increase in rainfall and temperature (Mayek-Perez, 1995). Warm and humid conditions favour disease development (Bradbury, 1984) and the disease can develop at temperatures ranging from 15-40° C and RH > 30 per cent with optimum 24°C and > 50 per cent RH (Qadous and Khlaif, 1997). Wind splashed rains help to spread the pathogen from infected to healthy plants. Infected seed from resistant genotypes had no visible symptoms and such seeds may play an important role in the epidemiology of this disease because they are difficult to detect and may occur at low frequency in seed lots (Mabagala, 1997). The disease was more severe under short photoperiod and high temperature than longer photoperiod and lower temperature (Santana *et al.*, 1993). Akhavan *et al.* (2009) showed that the epiphytic population size and disease severity were higher on spray inoculated plants irrigated with overhead sprinkler system during pods filling.

MANAGEMENT

The common blight of beans is particularly a difficult disease to manage due to its intricate association of the pathogen with the seed. An integrated approach is, therefore, useful to keep the disease under check.

Cultural practices

Use of pathogen free seed is must because it eliminates an important source of primary inoculum for common blight (Irigoyen and Garbagnoli, 1997), which is an important component of effective disease management strategy for this disease. The disease free seed can be produced in arid zones with low rainfall but its use cannot gurantee a clean crop. Because blight bacteria can colonize symptomless plants and the bacteria can systemically invade seed via vascular elements, certified pathogen-free seed may still be contaminated by blight bacteria (Weller and Saettler, 1980). The reduction of inoculum achieved by the use of the pathogen free seed greatly reduces the probability of common blight epidemics developing in the field.

Since very low levels of seed infestation may lead to development of epidemics in the field, it is desireable to have some efficient methods of seed testing for the presence of bacteria, particularly in programs that produce pathogen free seed. Some of the methods, which have been used, include pathogenecity test, bacteriophage typing and serology (Saettler, 1989) but their reliability is still doubtful. Use of nuclear approaches like recombinant DNA methods, possibly through the XCP DNA probe (Gilbertson *et al.*, 1989) or by PCR (polymerase chain reaction), a method in which minute amounts of a specific DNA sequence are amplified to readily detectable levels, may provide rapid and specific tools for detection of the bacteria in seed.

Deep ploughing, 2-3 years crop rotation and destruction of crop debris/weed hosts, avoidance of irrigation with sprinklers help to reduce the probability of epidemics (Panagopoulos and Biris, 1988; Saettler, 1989; Gilbertson *et al.*, 1990). Bacterial population in plant debris decline over time (Gilbertson *et al.*, 1990) and atleast in temperate regions, the bacteria probably cannot survive for more than one winter. In such situations, crop rotation will be an effective management strategy. But this may not be an effective approach in tropical region particularly in case of the farmers with small land holdings,. Moreover, the pathogen is having a vast host range; then again rotation will be less effective strategy.

Mixed intercropping of bean with maize was also found effective in reducing the incidence of this disease as compared to sole cropping and row intercropping, respectively (Fininsa, 1996). Delay in planting time to avoid frequent rains early in the season along with use of disease free seed has also been recommended to alleviate considerable yield loss and quality deterioration (Tu, 1999).

Host resistance

Since there is no satisfactory chemical control is available to combat the ravages of this disease, the use of resistant varieties/cvs. is an important control measure (Rodrigues *et al.*, 1999). The need for resistance to blight bacteria was recognized a long time ago, but little resistance has been found in *P. vulgaris* despite considerable efforts made by many workers. Surveys were conducted to collect bean germplasm from Africa, South and Central America, Europe and the United States for resistance to bacterial blight (Rands and Brotherton, 1925) and identified a number of number of lines from Eastern

Africa, Guatemala, Venezuela, Europe and the United States that apparently possessed reasonable levels of resistance to this disease, but developed slight levels of blight under field conditions. Many of African lines were not considered to have desirable horticultural traits and unlikely to be used for hybridization with American types (Leakey, 1973). Subsequent efforts to find good sources of resistance or immunity to common blight in *P. vulgaris* were also unsuccessful, however moderate level of resistance were identified in *P. vulgaris* by various workers (Coyne *et al.*, 1963; Coyne and Schuster, 1973). Various varieties/ lines viz., Fullgreen, PLS-63, 101, MS 9471, DL 2434 (Patel and Jindal, 1973); P 597 (*P. acutifolius*) Feijao 60 Diaz, PI 207262, Coll. 10 B, Mexico, 168, Mexico 240, Desconhecido Amarelo (Rava *et al.*, 1990), BR-40, Contista, Saginan, WW 55-0, Kuovskaya, Yushchoyasya (Ovchonnckova and Zaitsev, 1992); OAC 88-1 (Scott and Michaels, 1992); XR 235 (Wright and Wood, 1993); Great Northern Jules, GN Nebr1, Sel. 27, Pinto Nebr, EP1, Walley, Xr1, Xr2 (Velch *et al.*, 1994); BAC-6, XAN-159 (Mohamed *et al.*, 1995); IIHR 220 (Mohan *et al.*, 1997); T-589, T-769; T-352, T-647 (Vizgarra, 1998); Cut 54, XAN 280, 9356-26 (Rodriguez *et al.*, 1999); HR45 (Shi *et al.*, 2011) which possess resistance to this disease has been reported from different parts of the world. Five interspecific hybrids of *P. vulgaris* (Nim 1) x *P. coccineus* and cv. IPR96-4 were resistant to this disease (Kumar *et al.*, 2005, Khati and Hooda, 2006). Cultivars Bayo Madero, Negro 8025 and Pinto Villa were tolerant against common blight pathogen (Prudencio-Sains *et al.*, 2008). Plants of 17 bean genotypes were evaluated under greenhouse conditions for their reaction to *X. campestris* pv. *phaseoli* following leaf-spray inoculation with bacterial suspension (10^8 cfu/ml) and the cultivar Oreol was the most resistant (Todorovic *et al.*, 2008).

Sometimes varieties behave differently with the stage of infection as has been demonstrated in varieties BAT 1297 and BAT 93 (Murillo *et al.*, 1990). Similarly, determinate red mottled genotypes had almost twice as much disease as indeterminate genotypes, which may serve as useful sources of resistance to common bacterial blight (Beaver *et al.*, 1992). The varieties viz., A-417, A-420 without exhibiting foliage symptoms, bacteria can be transmitted to the seed of these varieties (Torres and Maringoni, 1997). Modified back cross method has been used successfully to develop high levels of resistance to *X. c.* pv. *phaseoli* by intercrossing genotypes with moderate levels of resistance (Zapata *et al.*, 1991). The number of genes controlling foliage and pod resistance is variable. Velich *et al.*, (1994) found that foliage resistance in 9-parent diallel cross was controlled by 4-5 gene blocks, while resistance in a Hungarian line and the Cv. Walley was conditioned by a single dominant gene. They further observed that pod resistance was associated with 2-3 gene blocks, in a diallel set of crosses.

Randomly amplified polymorphic DNA (RAPD) molecular markers have been effectively utilized in the identification of regions on the linkage map associated with resistance to common bacterial blight. The chromosomal region (a BCM interval in linkage group 5 was significantly associated with resistance to Xap strains DR-7 and EK-11 in leaves, pods and seeds. Moreover, regions in linkage group 1 were also significantly associated with resistance to Xap stains in more than one plant organ. The V locus of the flower colour gene (v^{lae}) was linked to a QTL (quantitative trait locus) with major effect on CBB resistance (Jung *et al.*, 1997).

According to Montoya *et al.* (1997) different mechanisms operate in controlling pod and leaf resistance and are independent of each other. Based on heritability estimates, Avila *et al.* (1998) have opined that the reaction of *Phaseolus vulgaris* to Xap and its variant fuscans is easy to manipulate genetically. Significant association between QTLs and genes controlling the length of second internode allows the use of this trait as a morphological marker for selection of resistant individuals in a plant population (Boscariol *et al.*, 1998). RFLP markers have yielded hidden favourable alleles, which can serve as an indirect selection tool to increase nodule number and resistance to the disease (Tsai *et al.*, 1998). Rodrigues *et al.*, (1998) observed additive genic effects for resistance of leaves to this disease. Tar'an *et al.* (1998) demonstrated that in a plant-breeding programme, marker distance resistance association remains stable and therefore, can be utilized for marker-assisted selection of CBB-resistant beans.

RAPD markers associated with genes for resistance to *Xanthomonas campestris* pv. *phaseoli* (Xcp) could be used to transfer the resistance genes from *Phaseolus acutifolius* to a susceptible *P. vulgaris* cultivar (Park *et al.*, 1998). Marker assisted selection has also been proposed for introgression of resistance genes into susceptible backgrounds by Yu *et al.* (1998). Four QTLs associated with resistance to *Xanthomonas* has been identified (Souza *et al.*, 2000).

The level of resistance to Xcp can also be enhanced by utilization of marker BC 437.1054 and Vgene, along with resistance genes from other germplasm to pyramid the different genes into a single line/ cultivar (Park *et al.*, 1999). Marker BC409.1250 has been found to be significantly associated with genes for CBB resistance (Jung *et al.*, 1999).. Significant general combining ability for both leaf and pod infection and specific combining ability for disease infection in pods in diallel crosses between three snap bean genotypes (Alesse, Hab 52 and Hab 198) and two dry bean genotypes (Bac-6 and A 794) have been observed (Rodrigues *et al.*, 1999).

Chemical control

Different chemicals have been used both as seed treatment or foliage application in attempts to control this disease. Although some success has been achieved but complete control is still intractable. Some chemical used by various workers both seed and/ or foliage application is mentioned as below:

Seed treatment: Hot water seed treatment at 50°C for ten minutes followed by diping in Streptocycline (10 μg/ml) solution has been found effective in reducing the seed infection (Jindal, 1991).

Foliar sprays: Several fungicides/antibiotics have been applied to foliage to control this disease but complete control is still lacking. The disease to some extent can be checked by using copper based fungicides (Saettler, 1989). However, heavy use of copper based fungicides to control diseases caused by Xanthomonads can result in selection of copper-resistant strains of the bacteria (Stall *et al.,* 1986) hence need based use of copper fungicides is warranted. Moderate control of the disease coupled with higher yields have been achieved by three sprays of copper oxychloride + zineb or copper oxychloride + mancozeb (Maringoni, 1990). Foliar spray of carbendazim, mancozeb and metalaxyl MZ significantly reduced disease incidence and increased yield (Rajappan *et al.,* 2000).

The chemical can form the effective part of integrated approach for this dreaded disease. Sometimes abruptly the environment become favourable for the development/ multiplication of bacteria, spot application of chemicals can keep the bacterial populations below damage threshold level.

Biological control agents

Bacterial isolates DFs093 offered 100% control, DFs041 and DFs1297 offered 90% and DFs490, DFs769, DFs831, DFs842 and DFs843 offered 80% control of common blight pathogen in the cotyledonary leaves when bean seeds immersed in suspension of each one of these isolates were agitated for 5 hours at 10°C and sowed in non-sterilized soil (Zanatta *et al*., 2007). *Paenibacillus polymyxa* strain HKA-15 showed strong antagonism against *Xanthomonas campestris* pv. *phaseoli* strains M-5 and CP-1-1. In agar diffusion assay, the antibacterial metabolite from *P. polymyxa* HKA-15 showed a clear zone of inhibition against M-5 and CP-1-1. Under phytotron conditions, the biocontrol activity of *P. polymyxa* HKA-15 against bacterial common blight pathogen *X. campestris* pv. *phaseoli* M-5 was assessed. At four days after inoculation of M-5, lowest mean disease rate (MDR) (1.13) and percent disease incidence (PDI) (28.25) were recorded in streptomycin sulphate at 100 ppm as positive chemical treatment.The application of crude metabolite from *P. polymyxa* HKA-15 at 100 ppm was at par with positive chemical control in suppression of bacterial common blight disease in French bean plants (Mageshwaran *et al*., 2012).

Biopesticides

Neem (*Azadirachta indica*) oil and neem oil + pungam (*Pongamia glabra* [*P. pinnata]*) oil based emulsifiable concentrate (EC) formulations were evaluated for their efficacy to control bacterial blight of French bean under field conditions. These formulations were able to manage the disease to some extent but were inferior to fungicides (Rajappan *et al*., 2000).

Induced resistance

Rhizobium leguminosarum bv. *phaseoli* was evaluated for its capacity to trigger resistance to common bacterial blight (CBB) of common bean caused by *Xanthomonas axonopodis* pv. *phaseoli* (Xap), under greenhouse and field conditions (Osdaghi *et al*., 2011). A common bean cultivar and three lines including two tolerant lines (Ks51103 and BF13607) and the susceptible cultivar Khomein and line Ks21479, were used. *R. leguminosarum* bv. *phaseoli*, was applied as a seed treatment and its effect on disease severity (DS) was compared with untreated (control) plants and with urea fertilizer treated plants in both the greenhouse and the field. *R. leguminosarum* bv. *phaseoli* in common bean roots tended to reduce CBB severity and to improve plant growth, particularly the 100-seed weight, in the field. The greatest decrease in CBB in the greenhouse occurred 15 days after inoculation (DAI) of Xap in the cultivar Khomein and line Ks21479, and 30 DAI in lines Ks51103 and BF13607. In the field, however, the greatest decrease occurred 35 DAI in the cultivar and all lines. Furthermore, in the field the greatest improvement in 100-seed weight occurred in the cultivar Khomein treated with *R. leguminosarum* bv. *phaseoli*.

The ability of yeasts *Rhodotorula glutinis* and *Sporidiobolus johnsonii* to reduce the severity of bean common bacterial blight was evaluated under field conditions with two and three sprays (Carvalho *et al.*, 2020). *R. glutinis* and *S. johnsonii* isolates reduced area under disease progress curve by 66.84 and 58.42%, respectively with three applications. For productivity, *R. glutinis* and *S. johnsonii* showed no difference between the number of applications. The results indicated that the *R. glutinis* and *S. johnsonii* reduced the severity of bean common bacterial blight and the most appropriate number of applications are two for both isolates.

REFERENCES

Aggour, A.R., Coyne, D.P., Vidaver, A.K. and Eskridge, K.M. 1989. Transmission of the common blight pathogen in bean seed. *J. Amer. Soc. Hort. Sci.* 114(6): 1002-1008.

Alfaro, A. and Silvan, A. 1940. La Tabaquera de las judias. *Bol. De Patol. Veg. Y Ent. Agr*. 9: 9-20.

Akhavan, A., Bahar, M., Saeidi, G. and Lak, M. (2009). Factors affecting epiphytic population of *Xanthomonas axonopodis* pv. *phaseoli* in epidemiology of bean common blight. *J. Sci. Technol. Agric. Nat. Res.*13(47): 265-277.

Allen, D.J. 1983. *The pathology of tropical food legumes: Disease resistance in crop imporovement*, Wiley, Chichester, England, 413pp.

Arnaud-Santana, E., Pena- Matos, E., Coyne, D.P. and Vidaver, A.K. 1991. Longevity of *Xanthomonas compestris* pv. *phaseoli* in naturally infested dry bean (*Phaseolus vulgaris*) debris. *Plant Dis*. 75(9): 952-953.

Atanasoff, D. and Kovachevski, J.K. 1929. New parasitic fungi of Bulgaria. *Soc. Bot. De Bulgarie Bul*. 3: 45-52.

Avila, J.R.de, Souza, R.M.de, Santos, J.B.dos, Souza, P.E.de and Castro, A.M.S. 1998. Reaction of common bean cultivars and lines to different isolates of *Xanthomonas campestris* pv. *phaseoli* and its variant *fuscans*. *Fitopatologia Brasileira* 23(1): 18-22.

Basu, P.K. 1974. Glucose inhibition of the characteristic melanoid pigment of *Xanthomonas phaseoli* var. *fuscans*. *Canadian J. Bot*. 52: 2203-2206.

Beach, S.A. 1892. Some bean diseases. *New York State Agric. Expt. Stn. Bull*. 48: 305-333.

Beaver, J.S., Steadman, J.R. and Coyne, D.P. 1992. Field reaction of landrace components of red mottled beans to common bacterial blight. *Hort. Sci.* 27(1): 50-51.

Bertelli, L.K. de. 1947. Cura de la semilla de arveja. *Uruguay Min. de Ganad. Y. Agri. Bol. Inform*. 4: 132.

Boscariol, R.L., Souza, A.A., Tsai, S.M. and Camargo, L.E.A. 1998. Mapping of genomic regions associated with resistance to two strains of *Xanthomonas axonopodis* pv. *phaseoli* in common bean. *Fitopatologia Brasileira* 23(2): 132-138.

Bouriquet, G. 1937. Madagascar: list of the parasites and diseases of cultivated plants. *Int. Inst. Agri.,Bul. Pl. Prot.* 11: 66-68.

Bradbury, J.F. 1984. *Xanthomonas* Dowson. 1939. In: *Bergeys Manual of Systematic Bacteriology* vol. I. (Eds. Kreig, N.R. and Holts, J.G.). Williams and Wilkin, USA.

Bremer, H., Ismen, H. and Karel, G. 1947. Beitrage zur kenntnis der parasitischen pilze der turkei. I. *Istanbul Univ. Faculte Sci. Rev., Ser. B*, 12: 307-334.

Burkholder, W.H. 1930. The bacterial diseases of the bean: a comparative study. *New York (Cornell) Agri. Expt. Sta. Mem.* 127: 1-88.

Burkholder, W.H. and Bullard, E.T. 1946. Varietal susceptibility of beans to *Xanthomonas phaseoli* var. *fuscans*. *Pl. Dis. Reptr*. 30: 446-448.

Carvalho, J.., Kuhn, O.J., Barabasz, R.F., Silva, R. M. F., Sustakowski, M. C., dos Reis, W., da Silva, R. H., Taís R. Kohler, Lorenzet, E., Heling, A. L., de Oliveira, V. H. D., Stangarlin, J. R. and Viecelli, C. A. 2020. Management of the common bacterial blight of the bean by *Rhodotorula glutinis* and *Sporidiobolus johnsonii*. *Journal of Agricultural Science* 12 (11):141-149.

Chand, R. 1992. Occurrence of *Xanthomonas compestris* pv. *phaseoli* (Smith) Dye on pea (*Pisum sativum* L.)in India. *Zeitschrift fur Pflanzenkrankheiten Pflanzenschutz* 99(4): 426-429.

Coyne, D.P. and Schuster, M.L. 1973. Phaseolus germplasm tolerant to common blight bacterium (*Xanthomonas phaseoli*). *Pl. Dis. Reptr*. 57: 111-114.

Coyne, D.P., Schuster, M.L. and Al-yasiri, S. 1963. Reaction studies of bean species and varieties to common blight and bacterial wilt. *Pl. Dis. Reptr*. 47: 534-537.

Delacroix, G. 1901. La Graisse Maladie Bacterienne des haricots. *Inst. Nat. Agron. Paris, Ann. (1897-1900)* 16: 151-160.

Doidge, E.M. 1919. The bacterial blight of beans: *Bacterium phaseoli* Erw. Sm. *Soc. African J. Sci*. 15: 503-505.

Dye, D.W. 1962. The inadequacy of the usual determinative tests for the identification of *Xanthomonas* spp. *New Zealand J. Sci*. 5: 393-416.

Fininsa, C. 1996. Effect of intercropping bean with maize on common bacterial blight and rust diseases. *Int. J. Pest Mgt*. 42(1): 51-54.

Fourie, D. and Herselman, L. 2011. Pathogenic and genetic variation in *Xanthomonas axonopodis pv. phaseoli* and its fuscans variant in Southern Africa. *African Crop Science Journal* 19(4): 393-407.

Galachyan, R. 1936. Bacteriosis of beans, their noxiousness, distribution, and ways of infecting. *Lenin Acad. Agri. Sci. Inst. Pl. Prot., Sum. Sci. Res. Work Inst. Pl. Prot.* 1935: 513-515.

Gilbertson, R.L., Rand, R.E. and Hagedorn, D.J. 1990. Survival of *Xanthomonas compestris* pv. *phaseoli* and pectolytic strains of *Xanthomonas compestris* in bean debris. *Plant Dis*. 74(4): 322-327.

Gilbertson, R.L., Maxwell, D.P., Hagedorn, D.J. and Leong, S.A. 1989. Development and application of a plasmid DNA probe for detection of bacteria causing common bacterial blight of bean. *Phytopathology* 79: 518-

Gilbertson, R.L., Rand, R.E. and Hagedorn, D.J. 1990. Survival of *Xanthomonas campestris* pv. *phaseoli* and pectolytic strains of *Xanthomonas campestris* in bean debris. *Plant Dis*. 74: 322-

Hayward, A.C. and Waterston, J.M. 1965. C.M.I. description of pathogenic fungi and bacteria No. 48: *Xanthomonas phaseoli*; No. 49: *Xanthomonas phaseoli* var. *fuscans*.

Ideta, A. 1936. Handbuch der pflanzenkrankheiten. Shakwabo, 1909-1911, 926pp. Tokyo, Japan.

Irigoyen, E.D. and Garbagnoli, C. 1997. Common bacteriosis in bean (*Xanthomonas compestris* pv. *phaseoli* (E.F. Smith) Dowson): detection, infection and transmission through seeds. *Fitopatologia* 32: 166-172.

Irigoyen, E.D. and Garbagnoli, C. 1998. Diagnosis identification and transmission of bacteria carried by bean seeds (*Phaseolus vulgaris* L.) in Argentina. *Fitopatologia* 33(3): 146-152.

Jindal, K.K. 1991. Physical and chemical agents for the eradication of *Xanthomonas compestris* pv. *phaseoli* from bean seeds. *Pl. Dis. Res*. 6(1): 68-71.

Jorstad, I. 1922. Beretning om planteykdommer I land-og havebruket 1920-21. 1. Landbruksvekster og Gronnsaker. *Norway Rep. Min. Agri*., 79pp.

Jung Geun Hwa, Scrotch, P. W., Coyne, D.P., Nienhuis, J., Arnaud-Santana, E., Ariyarathen, H.M., Kaeppler, S.M. and Bassett, M.J. 1997. Molecular marker based genetic analysis of tepary bean derived common bacterial blight resistance in different developmental stages of common bean. *J. Amer. Soc. Hort. Sci.* 122(3): 329-337.

Jung, G., Skroch, P.W., Neinhuis, J., Coyne, D.P., Arnaud-Santana, E., Ariyarathen, H.M. and Marita, J.M. 1999. Confirmation of QTL associated with common bacterial blight resitance in four different genetic backgrounds in common bean. *Crop Science* 39(5): 1448-1455.

Kern, M.H. 1929. Hungary: important or new plant diseases observed during 1926-28. *Int. Inst. Agri. Int. Bu. Pl. Prot*. 3(6): 82-83.

Khati, Pinkey and Hooda, K.S. 2006. Multiple disease resistance in rajmash genotypes. *Indian Phytopath*. 59(4): 521-522.

Khlaif, H.M. and Qadous, N.S. 1995. Common blight of beans in the Jordon Valley: sources of inoculum. *Dirasat. Series B, Pure and App. Sci*. 22(3): 717-728.

Kishan, R. 1988. Loss in yield of French beans affected with common blight. Pp, 29-31. In: *Advances in Research on Plant Pathogenic Bacteria*. (Gnanamanickan, S.S. and Mahadevan, A. Eds). Today and Tomorrow's Printers and Publishers, New Delhi.

Kumar, A., Kaushal, R. P. and Sharma, R. 2005. Transfer of resistance to floury leaf spot and bacterial blight in french bean from alien germplasm. *Indian Phytopathology* 58(2): 136-142.

Leakey, C.L.A. 1973. A note on *Xanthomonas* blight of beans (*Phaseolus vulgaris* (L.) Savi) and prospects for its control by breeding for tolerance. *Euphytica* 22: 132-140.

Leta A, Lamessa F and Ayana G. 2017. Occurrence and Importance of *Xanthomonas axonopodis* pv. *phaseoli* in common bean (*Phaseolus vulgaris* L) seed produced under different seed production system in central rift valley of Ethiopia. *J Plant Pathol Microbiol* 8: 406.

Mabagala, R.B. 1997. The effect of populations of *Xanthomonas compestris* pv. *phaseoli* in bean reproductive tissues on seed infection of resistant and susceptible genotypes. *European J. Pl. Pathol.* 103(2): 175-181.

Magee, C.J. 1930. Bacterial blight of beans. *Agri. Gaz. N.S. Wales* 41: 529-531.

Mageshwaran V., Mondal K.K., Kumar U. and Annapurna K. (2012). Role of antibiosis on suppression of bacterial common blight disease in French bean by *Paenibacillus polymyxa* strain HKA-15. *African J. Biotech.* 11(60): 12389-12395.

Maringoni, A.C. 1990. Chemical control of the common bacterial blight of bean and its effect in the transmission of *Xanthomonas compestris* pv. *phaseoli* (Smith) Dye by seeds. *Pesquisa Agropecuaria Brasileira* 25(8): 1151-1156.

Maringoni, A.C. and Bach, E.E. 1999. Characterization of *Xanthomonas axonopodis* pv. *phaseoli* strains through electrophoretic analysis of total proteins. *Summa Phytopathologica* 25(2): 173-177.

Maringoni, A.C., Kimati, H. and Kurozawa, C. 1994. Serological variability among isolates of *Xanthomonas compestris* pv. *phaseoli. Summa Phytopathologica* 20(3/4): 164-167.

Mayek-Perez, N., Pedroza-Flores, J.A., Villneal-Garcia, L.A. and Valdes-Lozano, C.G.S. 1995. Genetic and environmental factors related to temporal dynamics and diseases effects on dry beans (*Phaseolus vulgaris* L.) at Marin, Nuero, Mexico. *Revista Mexicana de Fitopatologia(RPP 98/ 415)*

Mohan, N., Aghora, T.S. and Somkumar, R.G. 1997. Identification of French bean lines resistant to bacterial blight. *J. Maharashtra Agri. Univ*. 21(2): 309-10.

Mohamed, M.F., Coyne, D.P. and Read, P.E. 1995. A radiation-induced mutant with resistance to common bacterial blight disease in common beans. *Hort. Sci.* 30(3): 577-578.

Montoya, C.A., Beaver, J.S. and Milkas, P.N. 1997. Evaluation of bean (*Phaseolus vulgaris* L.) for its reaction to bacterial common blight (*Xanthomonas campestris* pv. *phaseoli*). *Fitopatologia Columbiana* 21(1): 18-24.

Murillo, R.J.E., Guzmann, V.P. and Munoz, F.J.E. 1990. Reaction in leaves and pods of six genotypes of beans to *Xanthomonas compestris* pv. *phaseoli*. *Acta Agronomica, Universidad Nacional de Columbia* 40(1-2): 64-73.

Opio, A.F., Teri, J.M. and Allen, D.J. 1993. Studies on seed transmission of *Xanthomonas compestris* pv. *phaseoli* in common beans in Uganda. *African Crop Sci J.* 1: 59-67.

Opio, A.F., Allen, D.J. and Teri, J.M. 1996. Pathogenic variation in *Xanthomonas compestris* pv. *phaseoli*, the causal agent of common bacterial blight in *Phaseolus* beans. *Plant Pathol.* 45(6): 1126-1133.

Osdaghi, E., Shams-Bakhsh, M., Alizadeh, A., Lak, M., and Maleki, H. 2011. Induction of resistance in common bean by Rhizobium leguminosarum bv. phaseoli and decrease of common bacterial blight. *Phytopathologia Mediterranea,* 50(1): 45-54.

Otoya, M.M. and Pastor-Corrales, M.A. 1992. DNA probes for studies of genetic diversity of pathogens causing common bacterial blight of beans. *Fitopatologia Columbiana* 16(1/2): 149-157.

Ovichinnikova, A.M. and Zaitsev, V.N. 1992. Resistance of French bean to diseases in the south of the Non-chermozem zone. *Selektsiya I Semenovodstvo (Moskva)* 4-5: 17-18.

Ovies, J. and Larrinaga, L. 1988. Transmission of *Xanthomonas compestris* pv. *phaseoli* by a wild host. *Ciencia y Tecnica en la Agricultura Proteccion de Plantas* 11(2): 23-30.

Panagopoulos, C.G. and Biris, D.A. 1988. Susceptibility of various bean cultivars to bacterial blights. In: *Progress on pest management in field vegetables. Proc. CEC/ IOBC Experts group meeting, Rennes, 20-22 Nov., 1985* (Eds. Cavalloro, R. and Pelerents, C.), Rotterdam, Netherlands: A.A. Balkema: 275-276.

Park, S.O., Coyne, D.P., Dursun, A. and Jung, G. 1998. Identifying randomly amplified polymorphic DNA (RAPD) markers linked to major genes for common bacterial blight resistance in tepary beans. *J. Amer. Soc. Hort. Sci.* 123(2): 278-282.

Park, S.O., Coyne, D.P., Mutlu, N., Jung, G. and Steadman, J.R. 1999. Confirmation of molecular markers and flower colour associated with QTL for resitance to common bacterial blight in common beans. *J. Amer. Soc Hort. Sci.* 124(5): 519-526.

Patel, P.N. and Jindal, J.K. 1972. Occurrence of common and fuscous blight diseases of bean (*Phaseolus vulgaris*) in India. *Indian Phytopath.* 25: 387-391.

Patel, P.N. and Jindal, J.K. 1973. Studies on resistance in crops to bacterial diseases in India. VI. Reaction of moth bean (*Phaseolus acutifolius*) and hycinth bean (*Dolichos lablab*) lines to bacterial blight and leaf spot diseases caused by *Xanthomonas phaseoli. Indian Phytopath.* 26: 607-608.

Pazos, V., Lagomasino, I. And Martinez, R. 1986. Sensitivity of *Xanthomonas compestris* pv. *phaseoli* isolates to bacteriophages. *Revista de Proteccion Vegetal* 1(1): 21-24.

Prudencio Sains, J.M., Navarrete Maya, R., Navarrete Maya, J. and Acosta Gallegos, J.A. (2008). Dynamics of the bean common and halo blights in the valley of Mexico. *Agricultura Tecnica en Mexico*34(2): 201-212.

Qadous, N. and Khlaif, H. 1997. Common blight of beans in Jordon valley: disease development and response of different host cultivars to the disease. *Dirasat. Agri. Sci.* 24(2): 260-267.

Rajappan,K.; Ramaraj,B. and Natarajan,S. 2000. Field evaluation of fungicides and neem formulations against bacterial blight of Frenchbean. *Indian Phytopathology*: 53(3): 303-304

Rands, R.D. and Brotherton, W. 1925. Bean varietal tests for disease resistance. *J. Agric. Res.* 31: 110.

Rava, C.A. and Romeiro, R.S. 1990. Variability among isolates of *Xanthomonas compestris* pv. *phaseoli* in relation to their pathogenecity on *Phaseolus vulgaris* cultivars. *Summa Phytopathologica* 16(3/4): 225-232.

Rava, C., Sartorato, A. and Romeiro, R.S. 1990. Evaluation of bean cultivars for resistance to *Xanthomonas compestris* pv. *phaseoli* in greenhouse and field conditions. *Summa Phytopathologica.* 16(2): 83-91.

Reinking, O.A. 1919. Diseases of economic plants in Southern China. *Philippine Agri.* 8: 109-134.

Reinking, O.A. 1919. Philippine plant diseases. *Phytopathology* 9: 114-140.

Rodriguez, O., Faure, B., Benitez, R., Carballo, R.M. and Capote, I. 1999. Advances in the study of common bean resistance to bacterial diseases in Cuba. *Agronomia Mesoamericana* 10(1): 55-58.

Rodrigues, R., Leal, N.R., Pereira, M.G. and Lam-sanchez, A. 1999. Combining ability of *Phaseolus vulgaris* L. for resistance to common bacterial blight. *Genetics and molecular Biology* 22(4): 571-575.

Rodriguez, R. Leal, N.R. and Lam-Sanchez, A. 1998. Diallel analysis in bean cultivars for common bacterial blight resistance. *Horticultura Brasileira*16(1): 61-64.

Rodriguez, A.M., Ovies, J. and Cruz, R. 1991. Two weed hosts of *Xanthomonas compestris* pv. *phaseoli*. *Proteccion de Plantas* 1(3/4): 43-48.

Saettler, A.W. 1989. Common bacterial blight. In bean production problems in the tropics, 2nd ed. (Eds. Schwatz, H.F. and Pasto-corrales, M.A.), CIAT, Cali, Colombia, 726pp.

Santana, E.A., Coyne, D.P., Beaver, J.S. and Zaiter, H.Z. 1993. Effect of photoperiod and temperature on common bacterial blight disease of common beans (*Phaseolus vulgaris* L.). *Euphytica* 66(3): 211-216.

Scott, M.E. and Michaels, T.E. 1992. *Xanthomonas* resistance of *Phaseolus* interspecific cross selections confirmed by field performance. *Hort. Sci.* 27(4): 348-50.

Souza, A.A., boscariol, R.L., Moon, D.H., Camargo, L.E.A. and Tsai, S.M. 2000. Effects of *Phaseolus vulgaris* QTL in controlling host-bacteria interactions under two levels of nitrogen fertilization. *Genetics and Molecular Biology* 23(1): 155-161.

Stall, R.E., Loschke, D.C. and Jones, J.B. 1986. Copper resistance and avirulence loci are linked to a self-transmissible plasmid in *Xanthomonas compestris*. *Phytopathology* 76: 240-243.

Starr, M.P. 1983. The genus *Xanthomonas*, *In the Prokaryotes: A handbook on habitats, isolation, and identification of bacteria,* (Ed. Starr, M.P.), Springer-Verlag, New York, 742pp.

Suit, R.F. 1934. Preliminary report on investigation of bacterial blight of beans. *Quebec Soc. Prot. Pl. Ann. Rep. (1932-34)*: 25-26, 75-79.

Sutton, M.D. and Wallen, V.R. 1970. Epidemiological and ecological relations of *Xanthomonas phaseoli* var. *fuscans* in Southwestern Ontario, 1961-1968. *Canadian J. Bot.* 48: 1329-1334.

Taran, B., Michaels, T.E. and Pauls, K.P. 1998. Stability of the association of molecular markers with common bean bacterial blight resistance in common bean (*Phaseolus vulgaris* L.). *Plant Breeding* 177(6): 553-558.

Tesic, Z. P. 1946. Les bacterioses de notre haricot. Yugoslavia. *Ann. Des Trav. Agri. Sci. et Tech., Belgrade, N.S.*, 1: 18-61.

Todorovic, B., Milijasevic, S., Rekanovic, E., Potocnik, I. and Stepanovic, M. (2008). Susceptibility of bean genotypes to *Xanthomonas campestris* pv. *phaseoli* in greenhouse conditions. *Pesticidi i Fitomedicina*23(3): 167-173.

Torres, J.P. and Maringoni, A.C. 1997. Reaction of dry bean to *Xanthomonas compestris* pv. *phaseoli* and seed transmission of the bacterium. *Fitopatologia Brasileira* 22(4): 546-549.

Tsai, S.M., Nodari, R.O., Moon, D.H., Camargo, L.E., Venkovdsky, R. and Gepts, P. 1998. QTL mapping for nodule number and common bacterial blight resistance in *Phaseolus vulgaris* L. *Plant and Soil* 204(1): 135-145.

Tu, J.C. 1999. Relationship between disease severity and yield loss of navy bean caused by bacterial blight (*Xanthomonas campestris* pv. *phaseoli*). In: 51st International Symposium on crop protection, Gent, Belgium, 4th May, 1999, Part II. *Mededelingen-Faculteit Landbouwkundige en Toegepaste Biologische Wetenschappen, Universiteit Gent* 64(3b): 637-642.

Uppal, B.N., Patel, M.K. and Nigam, B.G. 1946. Bacterial blight of French bean in Bombay province. *Proc. Natl. Inst. Sci.* 12: 351-359.

Velich, I., Szarka, J. and Neda, P. 1994. Biotic and abiotic stresses in the bean: I. Genetic background of complex resistance to bacterial diseases. *Kerteszeti Tudomany* 26(1): 49-54.

Vizgarra, O.N. 1998. Carioca beans: beans of untra diti oval colour for Argentina. *Avance Agroindustrial* 19(75): 24-26.

Weller, D.M. and Saettler, A.W. 1980. Evaluation of seedborne *Xanthomonas phaseoli* and *Xanthomonas phaseoli* var. *fuscans* as primary inocula in bean blights. *Phytopathology* 70: 148-

Wright, D.G. and Wood, B.A. 1993. Testing three cultivars of French bean (*Phaseolus vulgaris*) from Florida for resistance to common blight caused by *Xanthomonas compestris* pv. *phaseoli.Australian Pl. Pathol.* 22: 58-61.

Yoshii, K., Galvez, G.E. and Alvarez, G. 1976. Estimation of yield losses in beans caused by common blight. *Proc. Amer. Phytopath*. Soc. 3: 298-299.

Yu, Z.H., Stall, R.E. and Vallejos, C.E. 1998. Detection of genes for resistance to common bacterial blight of beans. *Crop Sci.* 38(5): 1290-1296.

Zanatta, Z.G.C.N., Moura, A.B., Maia, L.C. and Santos, A. 2007. Bioassay for selection of biocontroller bacteria against bean common blight (*Xanthomonas axonopodis* pv. *phaseoli*). *Brazilian J. Microbiol.* 38(3): 511-515.

Zapota, M. 1997. Identification of *Xanthomonas campestris* pv. *phaseoli* races in *Phaseolus vulgaris* leaves. *Agronomia Mesoamericana* 8(1): 44-52.

Zapota, M., Wilkinson, R., Freytag, G.F., Velez, H., Ortiz, F.H. and Lopez-Rosa, J.H.1991. Incorporating resistance to *Xanthomonas compestris* pv. *phaseoli* into bean using the latent period as a resistance marker. *J. Agri. Univ. Puerto Rico* 75(4): 345-352.

Zaumeyer, W.J. and Thomas, H.R. 1957. A monographic study of bean diseases and methods for their control. *U.S.D.A. Tech. Bull. No.* 868 : 255 pp.

20

HALO BLIGHT

Halo blight has been recognized as a serious disease of beans known to occur world over, causing considerable crop losses. Due to the appearance of greasy water-soaked lesions on affected organs, the disease is also known as 'greasy spot disease'.

GEOGRAPHICAL DISTRIBUTION

Halo blight has been reported during the year 1926 from New York, USA (Burkholder, 1926). Subsequently it was reported from the Netherlands (Wieringa, 1930), Switzerland (Burkholder, 1930), New Zealand (Reid, 1931), France (Labrousse, 1931), Tasmania (Dowson, 1932), Australia ((Adam, 1936), Denmark (Gram, 1933), Poland (Zaleski, 1933), Brazil (Muller, 1934), Germany (Boning, 1934), Russia (Galachyan, 1936), South Africa (Wager, 1936), Belgium (Marchal, 1937), England (White, 1937), French Morocco (Anonymous, 1937), Spain (Sardina, 1939), Canada (Conners and Savile, 1943), Uruguay (Bertelli, 1947), Kenya (Nattrass, 1950), Mauritius (Orian, 1951), Chile (Dubin and Ciampi, 1974) and Italy (Bazzi and Calzolari, 1988).

LOSSES

Halo blight is slightly less important as compared to common blight because it is more prevalent in those areas where temperature is low. It is a major constraint of dry bean production in South Africa and Poland especially in the moderate to cooler areas of the country (Marcinkowska, 1994; Fourie, 1998). The disease is one of the most common diseases of bean in Italy (Randhi and Parisi, 1997). Yield losses in the range of 23-43 per cent in Michigan fields have been reported by Saettler and Potter (1970).

SYMPTOMS

Small, isolated, water-soaked spots appear on the under surface of the leaves which gradually increase in size and coalesce. Later, these spots are surrounded by a characteristic greenish yellow halo measuring 2-3 cm in diameter, thereby giving the disease its name. Systemic spread of the pathogen leads to plant chlorosis with leaf yellowing and malformation without much exterior infection (Zaumeyer, 1932). Later, the pathogen has been isolated from such leaves (Hildebrand and Schroth, 1971). Systemic chlorosis is more prominent and uniform at 20° C (Coyne and Schuster, 1974). During infection process, the pathogen produces a non-specific toxin "phaseolotoxin' (Mitchell and Bieleski, 1977) which in turn produces systemic chlorosis and typical halo in the plant. Infections are initiated on the lateral veins which progress to the main vein often accompanied by a marked water- soaking and reddish discolouration of the interveinal regions. On young stems, sunken water- soaked spots appear which gradually enlarge to form reddish streaks, extending longitudinally. Stem epidermis split open and light cream to silver coloured bacterial exudations appear in advanced stages of infection. Brown to red zonate lesions surround water- soaked spots on affected pods. The bacterium soon invades the tissue and progress to the developing seed often leading to shriveling, discoloration and even rotting of seed.

THE PATHOGEN

Halo blight disease is caused by *Pseudomonas savastanoi* pv. *phaseolicola* (Burkholder) Garden *et al.* (Young *et al.*, 1996). It is a strictly aerobic, Gram -tive, rod shaped, motile bacterium occurring singly or in pairs. The organism exhibits a smooth to rough colony variation in culture. The presence of D- gluconate, L (+) arabinose, sucrose, succinate, DL-ß-OH butyrate, transaconitase, L-serine, L-alanine and P-hydroxybenzoate favour growth of the bacterium in culture while it does not utilize glutarate, mesotartarate, DL-glycerate, isoascorbate, erythritol, sorbitol and meso-inositol. White to cream coloured colonies with a bluish hue and accompanied by a green fluorescent pigment appear on agar medium. Sensitivity of bacterial strains to bacteriophages 11P, 12P, and 48P and serological tests also been suggested for the confirmation of identity of *P.s.* pv. *phaseolicola* (Taylor, 1970) which were later successfully used in the identification of this pathogen (Mabagala and Saettler, 1992; Fourie, 1998).

The genome sequencing of *P. syringae* pv. *phaseolicola* strain 1448A was done in 2005 (Joardar *et al.*, 2005). The genome size of *P. syringae* pv. *phaseolicola* is 6.5-Mbp. Genome consisted of one circular chromosome and two plasmids which are predicted to code for 5353 open reading frames (ORFs) (Joardar *et al.*, 2005).

VARIABILITY

Variation among the populations of this pathogen is well known (Mabagala and Saettler, 1992). Jenson and Livingstone (1944) were the first to demonstrate host specificity among 13 strains of this pathogen. Later, the occurrence of races 1 and 2 based on the reaction as resistant and susceptible, respectively on bean cultivar Red Mexican U.I.-3 had been reported in the United States (Patel and Walker, 1965). The occurrence of these two races has also been documented from Tanzania, Colombia, New Zealand and Kenya (Hale and Taylor, 1973; Buruchara and Pastor-Corrales, 1981; Kinyua and Mukunya, 1981; Silbernagel and Mills, 1987; Taylor *et al.*, 1987). The existence of a third race in Nabraska was first suggested by Schuster *et al.* (1979) and Coyne *et al.* (1979), who recovered strains of halo blight bacteria that were virulent to the usually resistant Great Northen bean cultivars such as Great Northen U.I.-59 and California Pink. The existence of race 3 of *P. savastanoi* pv. *phaseolicola* has been confirmed in South Africa by Taylor *et al.* (1987). Race 3 causes a hypersensitive reaction in cv. Tendergreen, which is susceptible to other two races i.e. race 1 and race 2. Mabgala and Saettler (1992) also recorded the subsistence of three races i.e. race 1, race 2 and race 3 in northern Tanzania. Besides these, some strains of race 2, designated as 2P, can grow at 34° C and produce a brown, diffusible pigment on agar media, whereas the non-pigment producing strains of race 2 and of races 1 failed to grow at 34°C temperature. Taylor *et al.* (1996) extended the range of differentials to eight cultivars/ lines, and accordingly identified nine races of this pathogen. Recently, in South Africa, seven races (1, 2, 4, 6, 7, 8 and 9) were recorded among 1128 isolates, with race 8 the most prevalent (Fourie, 1998). Races 1, 2, 6 and 8 were widely distributed through the whole production area, while races 4, 7 and 9 were restricted to 1 or 2 localities.

HOST RANGE

The bacterial pathogen has a vast host range which includs *Neonotonia wightii, Parthenium hysterophorus, Phaseolus acutifolious, P. angularis, P. coccineus, P. lunatus, P. polyanthus, P. polystachyus, P. radiatus, Pueuaria hirsuta, P. thunbergia* and *Glycine max.* (USDA, 1970; Mabagala and Saettler, 1992). Natural infections have been recorded on several other legume species, including all members of the tribe Phaseoleae with the exception of *Desmodium* spp. and *Pisum sativum* (Arnold *et al.*, 2011).

PHASEOLOTOXIN

P. syringae pv. *phaseolicola* secretes a nonspecific phytotoxin named phaseolotoxin. Due to the secretion of phaseolotoxin, characteristic chlorotic halo which surrounds water-soaked lesions are formed on infected plant parts. The toxin phaseolotoxin inhibits the enzyme ornithine carbamoyl transferase which plays a critical role in the urea cycle. The production of phaseolotoxin is dependent on temperature. Optimum phaseolotoxin production occurs at 18-20°C whereas at 30 °C, no detectable amounts of the toxin are found. Transcription profiling showed that the optimum expression of phaseolotoxin not only requires a low temperature, but also specific plant components present in leaf and apoplast (Hernández-Morales *et al.*, 2009). The genes involved in phaseolotoxin biosynthesis are organized into the *argK tox* gene cluster, which includes five transcriptional units (Aguilera *et al.*, 2007). This gene cluster is flanked by insertion sequences and transposases and is located in a highly conserved 38-kb DNA fragment with the characteristics of a GI (Aguilera *et al.*, 2007). The gene cluster consisted of 23 genes. *argK* encodes a phaseolotoxin-resistant ornithyl carbamoyl transferase (Mosqueda *et al.*, 1990). The *argK tox* gene cluster was conserved in both *P. syringae* pv. *phaseolicola* and *P. syringae* pv. *actinidiae* and hence it is assumed that it has been inherited by horizontal transfer (Sawada *et al.*, 2002). The significance of phaseolotoxin in bean- *P. syringae* pv. *phaseolicola* pathosystem was studied in a collection of 138 isolates of *P. syringae* pv. *phaseolicola* in Spain and it was found that 95 isolates did not produce phaseolotoxin determined by an *E. coli* inhibition bioassay (Rico *et al.*, 2003). 94 isolates didn't amplify the phaseolotoxin gene cluster in PCR with specific primers. This was the first report of the widespread occurrence of Tox^- isolates in the field, and has been confirmed by San José *et al.* (2010). However, earlier it was believed that Tox^- strains were of little or no epidemiological significance.

DISEASE CYCLE

The bacterium is an important seed-borne pathogen of this crop (Calzolari, 1997). Seed borne inoculum is carried on the surface of the seed or beneath the testa and in between the cotyledons. Pathogen also survives in bean debris and volunteer bean plants (Mabgala and Saettler, 1992). Within the same genotype, *P. savastanoi* pv. *phaseolicola* tended to survive longer in the stems than in leaf debris. The bacterium enters directly through wounds or through stomata during periods of high relative humidity and free moisture and start multiplying inside the host tissue. Symptoms of blight appear on the leaf surface.

EPIDEMIOLOGY

The bacterium is seed borne and also overwinters in infected plant debris. In the periods of high humidity and free moisture, the bacterium enters the host through stomata or wounds. Symptoms appear in 6-10 days at 24-28º C but may be delayed for 2- 3 days more at higher temperature (Zaumeyer and Thomas, 1957). Under high soil moisture conditions and low pH, survival of this bacterium did not exceed 2-3 months. The bacterium apparently cannot survive in colder regions in soil or plant debris.

MANAGEMENT

Cultural practices

Cultural practices include use of disease-free seed, a 2-3 year crop rotation, ploughing in plant debris immediately after harvest and avoiding irrigation with sprinklers (Calzolari, 1997).

Host resistance

Breeding for halo blight resistance is generally considered the best method of disease control (Mabgala and Saettler, 1992) and has been used successfully in several areas of the world (Baggett and Frazier, 1967; Coyne *et al.*, 1967; Hagedorn *et al.*, 1974; Schwartz, 1989). The development and deployment of resistant cultivars requires knowledge of the distribution races of the pathogen in a particular area. Resistance against this disease has been reported in *Phaseolus vulgaris* hybrid/ cvs. Ph-64-23-144, Ph-65-9-9, Ph-64-9-61 and Ph-64-9-56 (Radkov *et al.*, 1978), *P. radiatus*, *P. calcaratus*, *Vigna angularis*, PI 150414, cv. Edmund (Kiessling and Skiebe, 1987), V4141had higher tolerance to race1 and 2 of the pathogen. (Hanan and Khan, 1999). Accession CF 800375 has also been reported to be resistant to this disease (Rava *et al.*, 2003). Common bean 12 accessions had resistance to halo blight from the Iberian Peninsula (Monteagudo *et al.*, 2006). Cv. 'Chase' has been found resistant to halo blight pathogen (Singh *et al.*, 2007).

Inheritance of resistance studies revealed that independent genes govern resistance for the leaf, pod and plant systemic chlorotic reactions (Baggett and Frazier, 1967,

Coyne and Schuster, 1974). Pod susceptibility may occur frequently in plants, which possess leaf resistance. Existence of some linkage between different genes controlling the leaf and plant systemic chlorotic reactions has also been detected (Hill *et al.*, 1972). However, resistance controlled by a single recessive gene has also been reported (Velich *et al.*, 1988). Resistance to race 3, which causes hypersensitive reaction in cv. Tendergreen, susceptible to race 1 and 2, is governed by a single dominant gene, which is also present in several cultivars of U.S. origin, including Seafarer and Tendercrop (Harper *et al.*, 1987). Resistance to the halo blight is governed by two phenomena i.e. resistance to growth of the bacterial cells *in vivo* and resistance to toxin production (Russell, 1977) which is essential for pathogenecity (Gnanamanickam and Patil, 1976). Three resistance genes were detected in 3 Tanzanian landrace mixtures like race-specific R_4 was found at a high frequency (43-63% of the mixtures) while other gene R_1 and a race non-specific resistance was present only at a low frequency (Taylor and Teverson, 1996).

In general, mature plants are more resistant to infection (Omer and Wood, 1969). The pathogen is known to multiply in the xylem vessels of resistant and susceptible plants (Omer and Wood, 1969), and field resistance may occur when there is a reduced rate of bacterial multiplication in vascular tissues and a necrotic response of parenchymatous tissues or meristems to the bacterial toxins (Hubbeling, 1973). Physiological examination of resistant and susceptible genotypes showed that the concentration of carbohydrates was lowest in primary leaves and highest in the youngest upper leaves while glucose: sucrose quantitative ratio decreased with ageing of the leaves until flowering (Sardi *et al.*, 1999). The glucose concentration decreased considerably in the susceptible leaves after infection with *P. savastanoi* pv. *phaseolina*. It has been further shown that glucose plays an important role in the formation of bacterial extracellular polysaccharide (EPS). Because there is a positive correlation between age dependent bacterial resistance and low sugar (especially glucose) content in older leaves of the originally susceptible plants, suggesting that in old leaves, there is not enough glucose for production of the EPS coat. Lacking EPS coat, the bacterial and plant cell walls come in direct contact, which permits the induction of hypersensitive response characteristic of resistance (Sardi *et al.*, 1999). The decrease of glucose levels in susceptible variety Fonix was more significant as compared to Cherokee (resistant) while leaf samples exhibiting systemic chlorosis, showed the increase in carbohydrate levels (Vegvari *et al.*, 2000).

Chromosomal regions related to race 5 halo-blight resistance were localized on a genetic map of RAPD and AFLP molecular markers and constructed by the analysis of a 'Jules' · 'Canela' F2 progeny (Yaish *et al.*, 2006). 'Jules' showed quantitative resistance to halo-blight and 'Canela' was a very appreciated but susceptible Spanish bean landrace. Two QTL for resistance to halo-blight were mapped in two linkage groups. There were four large groups, with 14-22 molecular markers each, five with 4–8 markers each, and three with 2 or 3 markers each.

Chemical control

Seed treatment

Hot air treatments (70°C for 2 h for moistened seed or 50°C for 3 h for dried seed) reduced 72-74 per cent seed infection. Hot water soaks at >48°C reduced the number of bacteria present in the seeds but tended to decrease germination and prolonged soaking rendered the seed unfit for sowing. Antibiotics, kasugamycin and streptomycin eradicated seed infection when used as a hot water soak (60 min) and at relatively higher concentration while a successive tap water rinse and a further 30 min soaking in sodium hypochlorite (0.5%) improved the effectiveness of the streptomycin used at low dosage and reduced its phytotoxic effects (Tamietti and Garibaldi, 1984).

Foliage treatment

Sprays of Cupric hydroxide significantly reduced or limited *P. savastanoi* pv. *phaseolicola* on bean foliage. Copper oxychloride (0.2-0.25%) sprays starting from seedling stage to flowering is also found effective in reducing pod infection. Application (3-4 times) of copper products after seedling emergence was also recommended (Calzalari, 1997). Though, streptomycin is effective in reducing foliage infection, but continuous use can induce resistant mutants.

Biopesticides

The antibacterial activity of essential oils of *Tagetes minuta* against *Pseudomonas savastanoi* pv. *phaseolicola* was studied and mean inhibition zone diameters of 41.83 and 44.83 mm was observed after 24 and 48 hours, respectively (Gakuubi *et al.,* 2006). The minimum inhibitory concentrations and minimum bactericidal concentrations of the essential oils on the test bacteria were in the ranges of 24-48 mg/mL and 95-190 mg/mL, respectively.

REFERENCES

Adam, D.B. 1936. Halo blight in French beans. A report on measures for its control. *Victoria Dept. Agri. J.* 34 : 34-45.

Aguilera, S., López-López, K., Nieto, Y., Garcidueñas-Piña, R., Hernández-Guzmán, G., Hernández-Flores, J.L., Murillo, J. and Álvarez-Morales, A. 2007. Functional characterization of the gene cluster from *Pseudomonas syringae* pv. *phaseolicola* NPS3121 involved in synthesis of phaseolotoxin. *J. Bacteriol.* 189: 2834-2843.

Anonymous, 1937. La graisse du haricot. *Defense des Veg. Mem.* 50, 6 pp.

Arnold, D. L., Lovell, H. C., Jackson, R. W. and Mansfield, J. W. 2011. *Pseudomonas syringae* pv. *phaseolicola*: from 'has bean' to supermodel. *Molecular Plant Pathology* 12: 617-627.

Bazzi, C. and Calzolari, A. 1998. Bacterial diseases of bean: two years of observations in the Po valley. *Informatore Fitopatologico* 38(9): 49-51.

Beebe, S.E. and Pastor-Corralles, M.A. 1991. Breeding for disease resistance. In: *Common Beans* (eds. Schoonhoven, A.V. and Voysest, O.), CAB International, Wallingford. Pp. 561-610

Bertelli, L.K. de. 1947. Cura de la semilla de arveja. *Uruguay Min. de Ganad. Y Agri. Bol. Inform*. 4: 132.

Boning, K. 1934. The grease spot disease of beans. *Prakt. Bl. F. Pflanzenban u. Pflanzenschutz* 11: 265-269.

Burkholder, W.H. 1926. A new bacterial disease of the bean. *Phytopathology* 16: 915-927.

Burkholder, W.H. 1930. The bacterial diseases of the bean: A comparative study. *N.Y. (Cornell) Agri. Expt. Sta. Mem.* 127: 1-88.

Buruchara, R.A. and Pastor-Corrales, M.A. 1981 Variation, virulence of *Pseudomonas syrige* pv. *phaseolicola* on bean in Colombia. Pp. 341-351 in: Proc. 5th Int. Conf. Plant Pathog. Bact..

Calzolari, A. 1997. Halo and common spot of beans. *Informatore Agrario* 53(25): 66-67.

Conners, I.L. and Savile, D.B.O. 1943. Bean. *Canada Dept. Agri., Canad. Pl. Dis. Sur*. (1947) 22: 37-39.

Coyne, D.P. and Schuster, M.L. 1974. Breeding and genetic studies of tolerance to several bean (*Phaseolus vulgaris* L.) bacterial pathogens. *Euphytica* 23: 651-656.

Coyne, D.P., Schuster, M.L. and Erwin, D.C. 1979. Reaction of *Phaseolus vulgaris* germplasm to a new virulent strain of *Pseudomonas phaseolicola. Ann. Rep. Bean Improv. Coop*. 22: 20-21.

Coyne, D.P., Schuster, M.L. and Fast, R. 1967. Sources of tolerance and reaction of bean to races and strains of halo blight bacteria. *Pl. Dis. Reptr*. 51: 20-24.

Dowson, W.J. 1932. Notes on some bacterial plant diseases in Tasmania. *J. Pomol. And Hort. Sci.* 10: 301-305.

Dubin, H.J. and Ciampi, L.R. 1974. *Pseudomonas phaseolicola* en Chile. *Fitopatologia* 9: 91-92.

Fourie, D. 1998. Characterization of halo blight races on dry beans in South Africa. *Plant Dis*. 82(3): 307-310.

Gakuubi, M. M., Wagacha, J., M., dossaji, S. F. and Wanzala, W. 2006. Chemical composition and antibacterial activity of essential oils of *Tagetes minuta* (Asteraceae) against selected plant pathogenic bacteria. *International Journal of Microbiology* 2016: Article ID 7352509, 9 pages.

Galachyan, R. 1936. Bacteriosis of beans, their noxiousness, Distribution, and ways of infecting. *Lenin Acad. Agri. Sci. Inst. Pl. Prot., Sum. Sci. Res. Work Inst. Plant Prot.* 1935: 513-515.

Gnanamanickam, S.S. and Patil, S.S. 1977. Phaseotoxin suppresses bacterially induced hypersensitive reaction and phytoalexin synthesis in bean *cultivars. Physiol. Plant Path.* 10: 169-179.

Gram, E. 1933. Plantesggdomme I Denmark 1932. Oversigt, Samlet ved Statens Palntenpatologiske forsg. *Tidsskr. For Planteav* 39: 453-506.

Hagedorn, D.J., Walker, J.C., and Rand, R.E. 1974. Wis. HBR 40 and Wis. HBR 72 bean germplasm. *Hort. Sci.* 9: 402.

Hale, C.N. and Taylor, J.D. 1973. Races of *Pseudomonas phaseolicola* causing halo blight of beans in New Zealand. *N.Z.J. Agric. Res*. 167: 147-149.

Hanan, F. and Khan, I. 1999. Resistance of common bean (*Phaseolus vulgaris*) cultivars to two races of halo blight (*Pseudomonas phaseolicola* Burkh.). *Sarhad J. Agri.*

Harper, S., Zewdie, N., Brown, I.R. and Mansfield, J.W. 1987. Histological, physiological and genetical studies of the response of leaves and pods of *Phaseolus vulgaris* to three races of *Pseudomonas syringae* pv. *coronafaciens*. *Physiol. Mol. Plant Pathol.* 31: 153-172.

Hernández-Morales, A., De la Torre-Zavala, S., Ibarra-Laclette, E., Hernandez-Flores, J.L., Jofre-Garfias, A.E., Martinez-Antonio, A. and Alvarez-Morales, A. 2009. Transcriptional profile of *Pseudomonas syringae* pv. *phaseolicola* NPS3121 in response to tissue extracts from a susceptible *Phaseolus vulgaris* L. cultivar. *BMC Microbiol.* 9: 257-272.

Hildebrand, D.C. and Schroth, M.N. 1971. Isolation of *Pseudomonas phaseolicola* from bean leaves exhibiting systemic symptoms. *Phytopathology* 61: 580-581.

Hill, K. Coyne, D.P. and Schuster, M.L. 1972. Leaf, pod and systemic chlorosis reactions in *Phaseolus vulgaris* to halo blight controlled by different genes. *J. Amer. Soc. Hort. Sci*. 97: 494-498.

Hubbeling, N. 1973. Field resistance in Phaseolus beans to *Pseudomonas phaseolicola. Mededelingen Fakulteit Landbouwwetenschappen, Gent.* 38: 1351-1363.

Jensen, J.H. and Livingstone, J.E. 1944. Variation in symptoms produced by isolates of *Phytomonas medicaginis* var. *phaseolicola. Phytopathology* 34: 471-480.

Joardar, V., Lindeberg, M., Jackson, R.W., Selengut, J., Dodson, R., Brinkac, L.M., Daugherty, S.C., DeBoy, R., Durkin, A.S., Gigilio, M.G., Madupu, R., Nelson, W.C., Rosovitz, M.J., Sullivan, S., Crabtree, J., Creasy, T., Davidsen, T., Haft, D.H., Zatar, N., Zhou, L., Halpin, R., Holley, T., Khouri, H., Feldblyum, T., White, O., Fraser, C.M., Chatterjee, A.K., Cartinhour, S., Schneider, D.J., Mansfield, J., Collmer, A. and Buell, C. 2005. Whole-genome sequence analysis of *Pseudomonas syringae* pv. *phaseolicola* 1448A reveals divergence among pathovars in genes involved in virulence and transposition. *J. Bacteriol.* 187: 6488-6498.

Kiessling, D. and Skiebe, K. 1987. Studies on the causes of variability in incomplete resistance of bean, *Phaseolus vulgaris* L. to *Pseudomonas syringae* pv. *phaseolicola* (Burkholder) Young *et al., Arch. Zichtungs-forschung* 17: 191-200.

Kinyua, G.K. and Mukunya, D. 1981. Variability in isolates of *Pseudomonas phaseolicola* (Burk.) Young, Die and Wilkie in Kenya and genetic studies on resistance in dry food beans. Pp. 352-357351 in: *Proceedings of 5th International Conference on Plant Pathogenic Bacteria.*

Labrousse, F. 1931. Observations sur quelques maladies ses plantes maraicheres. *Rev. de Path. Veg. et d'Ent. Agri. de France* 18: 286-289.

Mabgala, R.B. and Saettler, A.W. 1992. Races and survival of *Pseudomonas syringae* pv. *phaseolicola* in Northern Tanzania. *Plant Dis.* 76(7): 678-682.

Marchal, E. 1937. Observations et recherches effectuees a la station de Phytopathologie de l'etat pendant l'annee 1936. *Inst. Agron. De I'Etat Bull.* 6: 73-80.

Marcinkowska, J. 1994. Diseases of bean recorded in Poland. *Biuletyn Instytutu Hodowli I Aklimatyzagi Roslin* 190: 169-175.

Mitchell, R.E. and Bieleski, R.L. 1977. Involvement of phaseolotoxin in halo blight of beans: transport and conversion to functional toxin. *Plant Physiol.* 60: 723-729. Baggett, J.R. and Hagedorn, D.J. 1967. Sources of resistance to halo blight in *Phaseolus vulgaris. Pl. Dis Reptr.* 51: 661-665.

Monteagndo, A.B., Rodino, A.P., Lema, M., Fuente, M de la, Santalla, M., Ron, A. M. de and Singh, S.P. 2006. Resistance to infection by fungal, bacterial and viral pathogens in a common bean core collection from the Iberian Peninsula. *Hort. Sci.* 41: 319-322.

Mosqueda, G., Van den Broeck, G., Saucedo, O., Bailey, A.M., Álvarez-Morales, A. and Herrera-Estrella, L. 1990. Isolation and characterization of the gene from *Pseudomonas syringae* pv. *phaseolicola* encoding the phaseolotoxin-insensitive ornithine carbamoyltransferase. *Mol. Gen. Genet.* 222: 461-466.

Muller, A.S. 1934. Brazil: preliminary list of diseases of plants in the State of Minas Geraes. *Inst. Agri. Int. Bul. Pl. Prot.* 8: 193-198.

Nattrass, R.M. 1950. Ann. Rep. Of the Senior Plant Pathologist, 1948. *Kenya Colony Dept. Agri. Ann. Rep. 1948*: 95-98.

Omer, M.E.H. and Wood, R.K.S. 1969. Growth of *Pseudomonas phaseolicola* in susceptible and in resistant bean plants. *Ann. Appl. Biol.* 63: 103-116.

Orian, G. 1951. Division of Plant Pathology. *Mauritius Department of Agriculture-Annual Report* 1950: 80-85.

Patel, P.N. and Walker, J.C. 1965. Resistance in *Phaseolus vulgaris* to halo blight. *Phytopathology* 55: 889-894.

Radkov P., Nedelchev NK and Dekov D. 1978. Study of the field resistance of some bean hybrids and cultivars to bacteriosis under non-irrigated and irrigated conditions. *Rastenievdni-Nauki 15*(9-10): 149-153.

Randhi, P. and Parisi, B. 1997. Viral and bacterial diseases of French beans. *Informatore Agrario* 53(32): 56-57.

Rava, C.A., Costa, J.G.C.-da, Fonseca, J.R., Salgado, A.L. and da-Costa, J.G.C. 2003. Identification of sources of resistance to anthracnose, common bacterial blight and common wilt in common bean. *Revista Ceres* 50: 797-802.

Reid, W.D. 1931. A bacterial wilt disease of beans, occurrence in Marlborough and measures for control. *N.Z. J. Agri*. 49: 164-169.

Rico, A., Lopez, R., Asenio, C., Aizpun, M.T., Asensio, S., Manzanera, M.C. and Murillo, J. 2003. Nontoxigenic strains of *Pseudomonas syringae* pv. *phaseolicola* are a main cause of halo blight of beans in Spain and escape current detection methods. *Phytopathology* 93: 1553-1559.

Russell, P.E. 1977. Observations on the in vivo growth and symptom production of *Pseudomonas phaseolicola on Phaseolus vulgaris. J. Appl. Bacteriol*. 43: 167-170.

Saettler, A.W. and Potter, H.S. 1970. Chemical control of halo bacterial blight in field beans. Pp. 1-8, *Michigan Agri. Expt. Sta. Res. Rep. 98*.

San José, M., Rodicio, M.R., Argudín, M.A., Mendoza, M.C. and González, A.J. 2010. Regional variations in the population structure of *Pseudomonas syringae* pathovar *phaseolicola* from Spain are revealed by typing with *Pme*I pulsed-field gel electrophoresis, plasmid profiling and virulence gene complement. *Microbiology* 156: 1795-1804.

Sardi, E., Velich, I., Hevesi, M. and Klement, Z. 1999. Ontogenesis and biotic stress-dependent variability of carbohydrate content in snap bean (*Phaseolus vulgaris*). *Zeitschrift fur Naturforschung. Sec. C, Biosciences* 54(9/ 10): 782-787.

Sardina, J.R. 1939. La grassa de las judias (debida a *Bacterium Medicaginis* var. *phseolicola*) en espana. *Bol. De Patol. Veg. Y Ent. Agri*. 8: 231-264.

Sawada, H., Kanaya, S., Tsuda, M., Suzuki, F., Azegami, K. and Saitou, N. 2002. A phylogenomic study of the OCTase genes in *Pseudomonas syringae* pathovars: the horizontal transfer of the *argK–tox* cluster and the evolutionary history of OCTase genes on their genomes. *J. Mol. Evol.* 54 437-457.

Schuster, M.L., Coyne, D.P. and Smith, C. 1979. New strains of halo blight bacterium in Nebraska. *Ann. Rep. Bean Improv. Coop*. 22: 19-20.

Schwartz, H.F. 1989. Halo blight. Pp. 285-301. in Bean Production Problems in the Tropics. 2nd ed. (Eds. Schwatz, H.F. and Pastor-Corralles, M.A.), CIAT, Cali, Colombia.

Silbernagel, M.J. and Mills, M.J. 1987. Halo blight of *Phaseolus vulgaris*. Pp. 81-83 in: *Bean Research*. Vol. 3. (Eds. Salema, M.P. and Minjas, A.N.). Benedictine Publications, Nadanda, Peramiho.

Singh, S.P., Teran, H., Lema, M., Webster, D.M., Strausbaugh, C.A., Miklas, P.N., Schwartz, H.F. and Brick, M.A. 2007. Seventy-five years of breeding dry bean of the Western USA. *Crop Sci.* 47(3): 981-989.

Tamietti G and Garibaldi A. 1984. Eradication of *Pseudomonas syringae* pv. *phaseolicola* in bean seeds in Italy. In *Phytopathology and plant bacterial diseases of quarentine significance*, pp.69-76.

Taylor, J.D. 1970. Bacteriophage and serological methods for identification of *Pseudomonas phaseolicola* (Burkh.) Dowson. *Ann. App. Biol*. 66: 387-395.

Taylor, J.D. and Teverson, D.M. 1996. Interactions between pseudomonads and Phaseolus beans and exploitation of genotype mixtures in subsistence agriculture. In: *Brighton Crop Protection Conference: Pests and Diseases*-1996 Vol. 2.

Taylor, J.D., Teverson, D.M. and Davis, J.H.C. 1987. Diseases of *Phaseolus* beans-Biology, resistance and control. Pp. 63-64 in: *Annu. Rep. Inst. Hort. Res.*, Wellsbourne, Warwick, England.

USDA. 1970. Index of plant diseases in the United States. Plant pests of importance to North American agriculture In *USDA Handbook* 165 p.

Vegvari, A., Sardi, E., Csoke, B., Stefanovitis-Benyai, E., Szarka, J. and Velich, I. 2000. Changing of carbohydrates by inoculation of *Pseudomonas savastanoi* pv. *phaseolicola* on bean lines with different resistance. *Int. J. Hort. Sci.* 6(1): 82-85.

Velich, I., Szarka, J. and Horvath, I. 1988. Inheritance of resistance to bacterial diseases in bean. *Zoldsegtermesztesi Kutato Intezet Bulletinje* 21: 17-21.

Wager, V.A. 1936. Two common diseases of beans. *Farming in South Africa* 11: 279-281.

Weiringa, K.T. 1930. De Vetvlekkenziekte een voor Nederland nieuwe ziekte bij bruine boonen (*Phaseolus vulgaris*). *Tijdschr. Over Plantenziekten* 36: 84-87.

White, H.L. 1937. Halo blight of the glasshouse runner bean. In Plant Diseases. *Chestnut Expt. and Res. Sta. Ann. Rep. (1936)* 22: 55-57.

Yaish, M. W. F., Sosa, D., Vences, J. F. and Vaquero, F. 2006. Genetic mapping of quantitative resistance to race 5 of *Pseudomonas syringae* pv. *phaseolicola* in common bean. *Euphytica* 152:397-404.

Young MW, Saddler GS, Takikawa Y, DeBoer SH, Vauterin L, Gardan L, Gvozdyak RI and Stead DE. 1996. Names of plant pathogenic bacteria 1864-1995. *Rev. Pl. Pathol.* 75(9):721-763.

Zaleski, K. 1933. Relative varietal susceptibility of beans grown in Poland to halo-blight disease. *Roez. Nauk, Rolincz. Lesnych* 30: 39-116.

Zaumeyer, W.J. 1932. Comparative pathological histology of three bacterial diseases of bean. *J. Agri. Res*. 44: 605-632.

Zaumeyer, W.J. and Thomas, H.R. 1957. A monographic study of bean diseases and methods for their control. *U.S.D.A. Tech. Bull. No*. 868: 255 pp.

21

BACTERIAL WILT

Bean wilt is yet another major disease occurring world over and causing serious yields reductions.

GEOGRAPHICAL DISTRIBUTION

Florence Hedges first reported the disease in 1922 (Hedges, 1922). Since then, it has been reported from various countries like France (Hedges, 1926), Germany, Belgium (Stapp, 1927), Bulgaria (Kovachevsky, 1929), New Zealand (Reid, 1931), Australia (Adam and Pugsley, 1934), South Africa (Wager, 1936), U.S.A. (Zaumeyer and Thomas, 1957), North Dakota State of U.S.A. (Venette *et al.*, 1995), Sau Paulo, Brazil (Maringoni and Rosa, 1997), Spain (González *et al.*, 2005) and Iran (Osdaghi *et al.*, 2015).

LOSSES

This disease has been serious in many bean-growing areas of the world. Sometimes the disease occurs in association with common and halo blight pathogens, then almost total loss of the crop occurs. Bacterial wilt has not been considered an important pathogen of beans for nearly 20 years in North Dakota- Minnesota where it has threatened the certified bean seed production (Venette *et al.*, 1995).

SYMPTOMS

The distinguishing symptoms appear early in the seedling stage. The diseased leaves appear flaccid and hang limply along the stem and branches of the plant, thereby giving a dull greenish or brownish discolouration (Plate 1). During the warmest part of the day, leaves become flaccid but regain their turgor during the cooler hours. Sudden death of young seedlings occurs at temperatures above 30 °C due to augmented growth of bacterial pathogen in the vascular tissues (Huang *et al.,* 2009). Plugging of vascular bundles by bacteria results in browning of foliage, which later wilt and die. Early infection of the plant leads to complete death of the plant whereas late infection of plants with the bacterium results in stunted growth of the plant with severe yellowing and wilting without fertile pod production.

Plate. 1. Symptoms of bacterial wilt of dry beans caused by *Curtobacterium flaccumfaciens* pv. *flaccumfaciens* on common bean. Symptoms first appear as interveinal chlorotic areas (a,b), leading to leaf wilting (c) and flabby or flaccid areas on the leaves (d,e). Subsequently, tissue necrosis on leaves surrounded by a chlorotic margin are observed (f–h). Severe infections on common bean lead to overall wilting and defoliation as well as plant mortality (i)

(Source: Osdaghi *et al.*, 2020)

THE PATHOGEN

The actinobacterium responsible for this disease is *Curtobacterium flaccumfaciens* pv. *flaccumfaciens* (Hedges) Collins & Jones. It is a gram-positive, non-spore-forming coryneform bacteria which belongs to family *Microbacteriaceae* and order *Actinomycetales.* The term *Curtobacterium* is derived from a Latin word "curtus" or "shortened" which depicts the short and rod-shaped bacterial cells, whereas the word "flaccumfaciens" is a two-part term (flaccum + faciens) derived from the Latin word "flaccus" which means "flabby or flaccid" and "faciens" means "making". Hence, the name of the bacterial pathogen explains the morphology of bacterial cell and associated symptoms as "wilt-inducing, short, rod-shaped bacterium". The bacterium, a strict aerobe is rod shaped and gram-positive possessing one or two polar or subpolar flagella. Its optimum temperature for growth in culture is about 37°C and produces smooth wet slimy yellow or orange-coloured colonies on agar medium.

C. flaccumfaciens pv. *flaccumfaciens* is not a well-studied and no information regarding the virulence repertoires and pathogenicity determinates is available in literature (Osdaghi *et al.,* 2020). Transformation vectors and technologies for genetic manipulation have not been developed yet in for the bacterium.

HOST RANGE

The bacterium *C. flaccumfaciens* pv. *flaccumfaciens* has a broad host range including *Phaseolus lunatus* f. sp. *macrocarpus*, *P. vulgaris, P. coccineus, P. aureus*, P. *mungo*, *Glycine max*, *Vigna sinensis*, *V. sesquipedalis*, *Lablab niger* and *Dolichos lablab*.

DISEASE CYCLE

The pathogen overwinters in plant disease debris and the bacterium is seed borne also. Infection of the plants takes place either through infected seed or wounds or physical openings on roots and aerial plant parts (Zaumeyer and Thomas, 1957; Rikard and Walker, 1965) or root injury caused by nematode feeding (Schuster, 1959) or by cultural/ field operations (Fig. 1). Seed infection occurs when the bactreium enters the ovule through the vascular elements of the funiculus and raphe or through the micropyle. Young plants are more susceptible and systemic development occurs rapidly once the bacteria reach the vascular system in the stem or petiole (Rikard and Walker, 1965). The bacterium multiplies in the vascular bundles and systemically colonizes the host plants. The bacterium can survive saprophytically on common bean seeds at room temperature for up to 24 years (Burkholder, 1945; Osdaghi *et al.*, 2020). The bacterium forms a thick layer under the seed coat or superficially on seed-surface.

EPIDEMIOLOGY

The pathogen overwinters in infected seed as well as crop debris. The bacterium does not overwinter well in soil but can survive between growing seasons in plant debris or on weeds. Dry weather coupled with warm temperatures especially above 32°C favour rapid spread of the disease. Irrigation water/rain and hailstorm help to spread the pathogen to newer areas. Entry of the pathogen into the host tissues is accompanied through stomata or wounds.

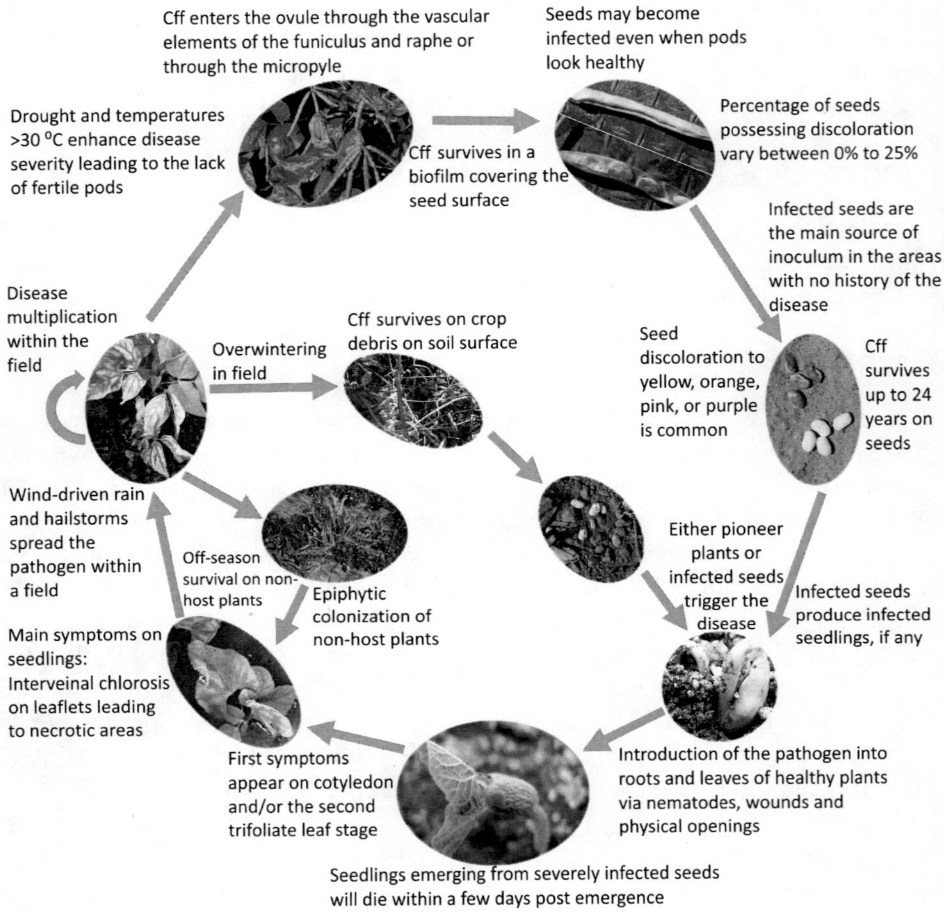

Fig. 1. Disease cycle of bean bacterial wilt

(Source: Osdaghi *et al.*, 2020)

MANAGEMENT

Cultural practices

Cultural practices like planting of pathogen free seed, as well as crop rotations have been successfully used to check infections. Sometimes bacteria survive and multiply in resistant plants and could be transmitted through infected seed of resistant cultivars and in field help in spreading to the neighbouring fields where susceptible plants are grown (Schuster *et al.*, 1964) thus indicating the need for clean seed even cultivar is presumed to be resistant to the bacterial infection. Of late a rapid method Biolog identification system (BIS) with Microstation™ V 3.5 software has been developed which is rapid, accurate and posses high standards (Zhao *et al.*, 1997). This detection software is used for rapid identification of bean wilt bacteria in the seed for plant quarantine regulations. Such methods can be used to reduce the chances of infection under field conditions from the seed that is apparently healthy or from resistant varieties, which carry incipient infections.

Host resistance

The use of host resistance has been found as the most effective method to manage this disease. Bean accessions viz., PI 136677, 136725, 165078, 177510, 204600 from *Phaseolus vulgaris*; PI 21304, 214332 from *P. acutifolius* and PI 247686 from *P. calcaratus* have been reported to possess resistance (Coyne *et al.*, 1963). Resistant cultivars viz., Great Northern Star and Emmerson (Coyne and Schuster, 1971, 1976) have been used successfully. Urrea and Harveson (2014) tested 427 accessions from the North American Phaseolus National Plant Germplasm System's bean collection for resistance to the pathogen. They reported that only one accession, PI 325691 (a wild common bean accession collected in Mexico), exhibited disease resistance. However, the seed was too small and not commercially acceptable, thus it still needed to be crossed with the others having good horticultural characters to obtain resistant, high yielding cultivars. In Brazil under greenhouse conditions, resistance to bacterial wilt was observed in dry bean cultivars IAC Diplomata, IAC Alvorada, IAC Imperador, IAC Carioca Tybatã, IAC Carioca Pyatã, IPR Corujinha, and IPR Tangará, and in the lines P5-4-4-1 and C4-5-4-1-2 (Maringoni *et al.,* 2015).

Chemical control

Seed treatment with antibiotics is effective to reduce the surface infection of bacteria on seeds. Seed treatments with slurry of Streptomycin Agri-Strep 500 (50,000 ppm) were found to be effective to reduce *C. flaccumfaciens* pv. *flaccumfaciens* infection on seeds of common bean (Schwartz, 2007). Copper-based chemicals (e.g., copper oxychloride, copper hydroxide, and copper sulphate) were not effective for the management of bacterial wilt on common bean and hence there is need for alternative, sustainable disease-management strategies (Martins *et al.,* 2013; Harveson, 2019). The integration of crop sanitation and cultural practices with chemical compounds is more effective for managing bacterial wilt (Lamichhane *et al*., 2018).

Harveson (2019) evaluated the performance of newly available antimicrobial chemicals along with copper hydroxide and copper sulphate for managing the four major bacterial diseases of dry beans i.e., common blight, halo blight, brown spot, and wilt under field conditions for seven years during 2010–2016 at 13 sites. Treatments with SaniDate (hydrogen peroxide 18.5% a.i. and peroxyacetic acid 12% a.i.) and ecoAgra 300 (plant-based fatty acids, a.i.) resulted in the production of higher seed yields compared to the controls. However, there was no reduction in disease incidence after the treatment with SaniDate and ecoAgra 300. These products do not appear to be curative in action but act as protectants by decreasing or prohibiting disease progress. The treatments performed better under high disease pressure compared to low levels of disease.

Biological control agents

Although there are many reports on the efficiency of biological control agents for the management of bacterial wilt of beans in greenhouse conditions, but none of biological control agents has been commercialized on a large scale. Seed treatment with combination of *Bacillus cereus* (strains DFs093 and DFs769) and *Pseudomonas fluorescens* (strain DFs831) were evaluated to reduce the severity of bacterial wilt, Fusarium wilt, charcoal rot, and angular leaf spot of dry beans and significant reduction in the severity of all four diseases was obtained using *B. cereus* strains DFs093 and DFs769 and *P. fluorescens* strain DFs831 (Corrêa *et al.*, 2014) Application of of *B. cereus* resulted in reduction in bacterial wilt incidence which ranged from 42 to 76 per cent (Martins *et al.*, 2013). It has been observed that although the colonization of plants with the pathogen and antagonist is influenced by environmental temperature (20 vs. 30 °C) but similar suppression of bacterial wilt using the *B. cereus* strain ALB629 was found at two temperatures i.e. 20°C and 30°C (Martins *et al.*, 2014).

Soaking seeds of common bean in a suspension of *Pantoea agglomerans* at 3×10^8 cfu/ml resulted in intense colonization of the complete bean seedling from root to apical stem in 7 days after inoculation and leads to reduction of bacterial wilt under greenhouse conditions (Hsieh *et al.*, 2005). Application of *P. agglomerans* as a soil drench 24 hr after planting was effective in reducing bacterial wilt but was not as successful as seed treatment. Furthermore, seed treatment using *Rhizobium leguminosarum* was found to be effective in reducing the incidence of bacterial wilt and common bacterial blight (Huang *et al.*, 2007; Osdaghi *et al.*, 2011; Pastor-Bueis *et al.*, 2019).

REFERENCES

Adam, D.B. and Pugsley, A.T. 1934. Bacterial plant diseases in Victoria. *Victoria Dept. Agri. J.* 32: 304-311.

Burkholder, W.H. 1945. The longevity of the pathogen causing the wilt of the common bean. *Phytopathology 34:* 430-432.

Corrêa, B.O., Schafer, J.T. and Moura, A.B. 2014. Spectrum of biocontrol bacteria to control leaf, root and vascular diseases of dry bean. *Biological Control* 72: 71–75.

Coyne, D.P. and Schuster, M.L. 1971. 'Emmerson' the new large seeded Great Northern dry bean variety tolerant to bacterial wilt disease. *Nebraska Agr. Expt. Sta. Bull. No* 516. 11p.

Coyne, D.P. and Schuster, M.L. 1976. Great Northern Ster dry bean tolerant to bacterial diseases. *Hort. Sci.* 11:621.

Coyne, D.P., Schuster, M.L. and Al-Yasiri, S. 1963. Reaction studies of bean species and varieties to common blight and bacterial wilt. *Plant Disease Reporter.* 47: 534-537.

González, A.J., Tello, J.C. and Rodicio, M.R. 2005. Bacterial wilt of beans (*Phaseolus vulgaris*) caused by *Curtobacterium flaccumfaciens* in southeastern Spain. *Plant Disease* 89: 1361.

Harveson, R.M. 2019. Managing dry bean bacterial diseases in Nebraska with new copper-alternative chemicals. *Plant Health Progress* 20**:** 14-19.

Hedges, F. 1922. A bacterial wilt of the bean caused by *Bacterium flaccumfaciens* Nov. sp. *Science 55*: 433-434.

Hedges, F. 1926. Bacterial wilt of beans (*Bacterium flaccumfaciens* Hedges) including comparisons with *Bacterium phaseoli. Phytopathology* 16: 1-22.

Hsieh, T.F., Huang, H.C., and Erickson, R.S. 2005. Biological control of bacterial wilt of bean using a bacterial endophyte, *Pantoea agglomerans*. *Journal of Phytopathology* 153: 608-614.

Huang, H.C., Erickson, R.S., Balasubramanian, P.M., Hsieh, T.F. and Conner, R.L. 2009. Resurgence of bacterial wilt of common bean in North America. *Canadian Journal of Plant Pathology*, 31: 290-300.

Huang, H.C., Erickson, R.S. and Hsieh, T. 2007. Control of bacterial wilt of bean (*Curtobacterium flaccumfaciens* pv. *flaccumfaciens*) by seed treatment with *Rhizobium leguminosarum*. *Crop Protection* 26:1055-1061.

Kovachevsky, I. C. 1929. Etiology of the bacterial diseases of beans in Bulgaria. *Sofia. Univ. Faculte Agron. Ann.* 7: 429-444.

Lamichhane, J.R., Osdaghi, E., Behlau, F., Köhl, J., Jones, J.B. and Aubertot, J.N. 2018. Thirteen decades of anti-microbial copper compounds applied in agriculture. *A Review. Agronomy for Sustainable Development* 38: 28.

Maringoni, A.C. and Rosa, E.F. 1997. Occurrence of *Curtobacterium flaccumfaciens* pv. *flaccumfaciens* on bean in the State of Sao Paulo, Brazil. *Summa Phytopathologica* 23(2): 160-162.

Maringoni, A.C., Ishizuka, M.S., Silva, A.P., Soman, J.M., Moura, M.F. and Silva Júnior, T.A.F. 2015. Reaction and colonization of common bean genotypes by *Curtobacterium flaccumfasciens* pv. *flaccumfasciens*. *Crop Breeding and Applied Biotechnology* 15: 87-93.

Martins, S.J., Medeiros, F.H.V., Souza, R.M. and Vilela, L.A.F. 2014. Is *Curtobacterium* wilt biocontrol temperature dependent Acta Scientiarum. *Agronomy Maringá* 36: 409-415.

Martins, S.J., Medeiros, F.H.V., Souza, R.M., Resende, M.L.V. and Junior, P.M.R. 2013. Biological control of bacterial wilt of common bean by plant growth-promoting rhizobacteria. *Biological Control* 66: 65-71.

Osdaghi, E., Young, A. J. and Harveson, R. M. 2020. Bacterial wilt of dry beans caused by *Curtobacterium flaccumfaciens* pv. *flaccumfaciens*: A new threat from an old enemy. *Molecular Plant Pathology* 21(5): 605-621.

Osdaghi, E., Taghavi, S.M., Fazliarab, A., Elahifard, E. and Lamichhane, J.R. 2015. Characterization, geographic distribution and host range of *Curtobacterium flaccumfaciens*: An emerging bacterial pathogen in Iran. *Crop Protection* 78: 185-192.

Osdaghi, E., Alizadeh, A. , Shams-bakhsh, M. , Lak, M.R. and Hatami Maleki, H. 2011. Induction of resistance in common bean by *Rhizobium leguminosarum* bv. *phaseoli* and decrease of common bacterial blight. *Phytopathologia Mediterranea* 50: 45-54.

Pastor-Bueis R., Sánchez-Cañizares C., James Euan K., González-Andrés F. 2019. Formulation of a highly effective inoculant for common bean based on an autochthonous elite strain of *Rhizobium leguminosarum* bv. *phaseoli* and genomic-based insights into its agronomic performance. *Frontiers in Microbiology* 10: 2724

Reid, W.D. 1931. A bacterial wilt disease of beans. Occurrence in Marlborough and measures for control. *New Zealand J. Agri*. 43: 408-415.

Rikard, S.F. and Walker, J.C. 1965. Mode of inoculation and host nutrition in relation to bacterial wilt of bean. *Phytopathology* 55: 174-178.

Schuster, M.L. 1959. Relation of root-knot nematodes and irrigation water to the incidence and dissemination of bacterial wilt of bean. *Plant Disease Reporter* 43: 27-32.

Schuster, M.L., Coyne, D.P. and Singh, K. 1964. Population trends and movement of *Corynebacterium flaccumfaciens* var. *aurantiacum* in tolerant and susceptible beans *Plant Disease Reporter* 48: 823-827.

Schwartz, H.F. 2007. *Dry Bean Bacterial Wilt. High Plains IPM Guide, a Cooperative Effort of the University of Wyoming, University of Nebraska, Colorado State University and Montana State University*. Nebraska, USA: University of Nebraska.

Stapp, C. 1927. Die bakterielle welkekrankheit der bohnen. *Nachrichtenbl. F. den deut. Pflanzenschutzdienst* 7: 88-90.

Urrea, C.A. and Harveson, R.M. 2014. Identification of sources of bacterial wilt resistance in common beans (*Phaseolus vulgaris* L.). *Plant Disease* 98: 973-976.

Venette, J.R., Lamppa, R.S. and Gross, P.L. 1995. First report of bacterial wilt caused by *Curtobacterium flaccumfaciens* subsp. *flaccumfaciens* in North Dakota. *Plant Dis*ease 79(9): 966.

Wager, V.A. 1936. Bacterial wilt and blight of French beans. *Union South Africa Dept. Agri. & Forestry, Plant Indus. Ser. 14, Sci. Bull*. 149: 19pp.

Zaumeyer, W.J. and Thomas, H.R. 1957. A monographic study of bean diseases and methods for their control. *U.S.D.A. Tech. Bull*. 868: 255 pp.

Zhao, Y.F., Wei, Y., Gao, C.S., Zhao, L.Z., Huang, G.M. and Huang, Q.L. 1997. Using the Biolog identification system for rapid identification of bean wilt bacterium (*Curtobacterium flaccumfaciens* pv. *flaccumfaciens*). *Acta Phytopathologica Sinica* 27(2): 139-144.

22

BEAN COMMON MOSAIC

Bean Common Mosaic is one of the most commonly prevalent viral diseases of French bean, causing considerable losses in yield and quality (Harter, 1927; Zaumeyer, 1930; Horsfall, 1930; Harrison and Burkholder, 1936; Zaumeyer, 1946; Zaumeyer and Thomas, 1947).

GEOGRAPHICAL DISTRIBUTION

Ever since its first record from Russia (Iwanowski, 1894) the disease has been reported from various countries viz., U.S.A. (Clinton, 1909; Stewart and Reddick, 1917), Canada (Stone and Howitt, 1920); Bermuda (Ogilvie, 1924); Agrentina, Belgian Congo, Chile, Columbia, Czechoslovakia, Ecuador, France, Guatemala, Honduras, Italy, Japan, Java, The Netherlands, Peru and Uruguay (Rands and Brotherton, 1925); China (Porter, 1926) England (Grainger, 1930); New Zealand (Chamberlain, 1939) Australia (Cass Smith, 1945); Brazil (Costa and Forster, 1941) Germany (Quantz, 1962); Iran (Kaiser *et al.*, 1968); Venezuella (Ordosgoitty, 1973).

Yaraguntaiah and Nariani first reported the occurrence of this disease in India in 1963 from New Delhi and subsequently from Shimla, Himachal Pradesh (Nagaich and Vashisth, 1963); Bangalore, Karnataka (Muniyappa, 1976), Pantnagar, Uttarkhand (Singh, 1976) and Pune, Maharashtra (Capoor *et al.*, 1985).

LOSSES

Yield losses amounting to 26 to 84% had been reported to be caused by the disease (Sastry *et al.*, 1981; Sharma and Dhar, 1994).

SYMPTOMS

Typical symptoms consist of well-defined dark green areas of foliar tissue against a lighter green background of affected leaf lamina (Plate 1) (Moralles, 1989). During early stages, affected leaves appear crinkled, chlorotic, stiff (Harrison, 1935) and accompanied by downward curling (Pierce, 1934). In advance stages of infection much reduced leaves with sinuous or filiform leaflets are produced (Fajardo, 1930) whereby plants become stunted (Dickson, 1922) and appear dwarfed and bushy. Seedlings from infected seed show characteristic mottling, blistering, downward cupping, and chlorosis on the fruit and succeeding compound leaves (Fajardo, 1930; Moralles, 1989). Such plants usually fail to set pods or develop undersized pods (Nelson, 1932).

The pods on severely infected plants are usually undersized and contain fewer ovules than those produced on normal plants. They are occasionally covered with small, dark green spots and often mature very late. The seed may be aborted, smaller or malformed. The symptoms are temperature dependent, and infection of this disease causes delayed seed maturity (Zaumeyer and Meiners, 1975). The roots of affected plants turn dark and during periods of severe infection the tap root system turns almost black due to vascular necrosis giving rise to what are typically called Black root' symptoms (Jenkins, 1940). Plants affected by black root, initially exhibit slight wilting of the leaflets near the top of the plants, which subsequently lose their bright green colour and appear greyish, green. A permanent general wilting of all the leaves follows often resulting in death of the plants.

Plate 1. Symptoms of Bean common mosaic

THE VIRUS

At least 20 viruses belonging to 12 taxonomic groups (Table 1) are shown to infect French bean under natural conditions. Dhanju *et al.* (1995d) reported that the viruses associated with mosaics of French bean in Himachal Pradesh consisted mostly strains of *Bean Common Mosaic* and *Bean Yellow Mosaic Virus* of which the former is most predominant.

The *Bean Common Mosaic Virus* (BCMV) is a ubiquitous pathogen of worldwide occurrence. Taxonomically the virus belongs to the potyvirus group. Earlier two serotypes of the virus namely A and B had been described to cause mosaic diseases. However, those once grouped as serotype A are now considered to be isolates of a separate potyvirus species and several strains (Drijfhout *et al.*, 1978) once considered to be distinct have now been shown to be strains of this virus (McKern *et al.*, 1992). BCMV consists of filamentous flexuous particles measuring 847-886 X 12-15 nm in size. The physico-chemical properties relating to this virus has been summarized in Table 2.

Table 1. Virus affecting French beans

Sr. No.	Taxonomic Group	Name of Virus
1.	*Alfalfa Mosaic Virus*	Alfalfa Mosaic
2.	*Bromovirus*	Cowpea chlorotic mottle
3.	*Comovirus*	Bean pod mottle Bean rugose mosaic Quail pea mosaic
4.	*Cucumovirus*	Cucumber Mosaic Peanut stunt
5.	*Geminivirus*	Abultion Mosaic Bean golden mosaic Euphorbia mosaic Mungbean yellow mosaic Tobacco yellow dwarf
6.	*Luteovirus*	Pea leaf roll
7.	*Potyvirus*	Bean common mosaic Bean yellow mosaic
8.	*Sobemovirus*	Southern bean mosaic
9.	*Tobomovirus*	Tobacco mosaic
10.	*Tobacco necrosis virus*	Tobacco Necrosis
11.	*Tomato spotted wilt virus*	Tomato spotted wilt
12.	*Carmovirus*	Bean mild mosaic

(Source: Matthews, 1982)

Table 2. Physicochemical properties of *Bean Common Mosaic Virus*

Genome	ssRNA, Monopartite
Transmission	Sap, Seed, Insects
Properties of particles in sap:	
TIP	60°C
LIV	1-4 days
DEP	10^{-3} - 10^{-4}
Physical Properties:	
Sedimentation Coefficient	154-158 S
Density	1.31-1.32 g cm^{-3} in CsCl

TIP = Thermal inactivation point; LIV =Longevity *in vitro*; DEP = Dilution end point

(Source Brunt *et al.*, 1996)

TRANSMISSION

Transmission of the virus through sap and seed was first reported by Reddick and Stewart (1918). The efficient transmissibility of the virus through bean seed is responsible for the worldwide distribution of the disease (Schmidt, 1992). Nelson (1932) reported the role of an aphid vector (*Macrosiphum solanifolii* Ashm.) for transmission of the virus in a non- persistent manner. However, subsequent workers conclusively established the seed/sap transmissibility of the virus (Archibald, 1921; Pierce and Hungerford, 1929; Fajardo, 1930; Nelson, 1932; Harrison, 1935; Skotland and Burke, 1960; Yaraguntaiah and Nariani, 1963; Calot *et al.* 1971; Hagita *et al.*, 1975; Capoor *et al.*, 1986; Morales and Castano, 1987). Aphid vectors viz., (*Aphis craccivora, A. euonymi; A fabae* sub. sp. *solenella; A. gossypii; A. medicaginis; A. rumicis; A. spiraecola; Acyrthosiphum pisum; Brevicoryne brassicae; Hyalopterus atriplicis; Lipaphis erysimi; Macrosiphum ambrosiae; M. pisi; M. solanifoli; Myzus persicae* and *Rhophalosiphum psuedobrassicae* have also been reported to transmit BCMV (Fajardo, 1930; Zaumeyer and Kearns, 1936; Capoor *et al.*, 1986).

HOST RANGE

The virus is known to infect various cultivated species as well as weed hosts *viz.*; *Cajanus cajan; Cassia tora; Chenopodium amaranticolor; C. quinoa; Cicer arietinum; Crotolaria spectabilis; C. striata; Cyamopsis tetragonoloba; Glycine max; Gomphrena globosa; Lens esculenta; Lupinus albus; L. angustifolius; Macroptilium lathyroides; Melilotus alba; Nicotiana benthamiana; N. clevelandii; Phaseolus lathyroides; P. lunatus; P. vulgaris; Rhynchosia minima; Sesbania exaltata; Tetragonia expansa; Trifolium incarnatum; T. subterraneum; Vicia faba; V. sativa; V. villosa; Vigna angularis; V. radiata* and *V. unguiculata* (Zaumeyer and Goth, 1964; Drijfhout and Bos, 1977).

EPIDEMIOLOGY

Infected seed, plants of susceptible French bean cultivars and weed hosts serve as natural sources of primary inoculum. (Robertson, 1962; Robertson and Klostermeyer, 1961, 1962). Aphid vectors are responsible for secondary field spread of BCMV (Anonymous, 1975). Gupta and Chowfla (1988) observed that nymphs and apterous aphids are more efficient as compared to winged adults. Air temperature and relative humidity significantly affected vector populations (Thomas and Vevai, 1940; Berg, 1984).

MANAGEMENT

No single method has proved effective to manage the disease. However, an integrated approach involving alteration of planting dates, manipulation of plant population density, use of insecticides to control insect vectors and breeding for disease resistance have been found to be effective (Jeyarajan and Doraisamy, 1986).

Cultural practices

(i) Use of virus free seed is essential by virtue of efficient transmissibility of the virus through seed (Schmidt, 1992). The production of BCMV free seed is taken up in isolated regions away from other bean plantings and aphid vectors or under screenhouse / greenhouse conditions (Morales, 1989, Sharma *et al*., 2006).

(ii) Alteration of planting dates coinciding with periods of reduced vector activity has also been reported to be effective (Pierce and Hungerford, 1929).

(iii) Rouging of infected bean plants from the field or weed hosts from in and around the field help to eradicate natural reservoirs of the virus.

(iv) Planting of non host barrier crops especially maize and closer spacing help in minimizing virus incidence by interfering with flight of the vector and thereby inactivating the virus (Van Rheenenet *et al.*, 1981; Dhanju *et al.*, 1995b, Sharma *et al*., 2006).

Host resistance

Phaseolus vulgaris cvs./lines viz., Wisconsin Refugee, Idaho Refugee (Walker, 1956); Idagreen, Improved New Stringless, Logan (Wade, 1943); Pure gold, Rival, Tenderlong 15, Top crop (Zaumeyer, 1950); Wade (Hoffman and Kanapaux, 1952); Hyscore, Kingreen, Processor Seminole (Wolf and Hills, 1954); 225 (del Barco), 355 (Troncon), No. 70 (de la Virgen), S-27 (San Fiacre) (Puerto and Alongo, 1958); Coche, Cubagua, Margarita (Trujillo and Saettler, 1972); Bean 1 49-3, Bean 2 119-3, Bean Br II 42-1, Contender, HPK 69, Iran 431, Kentucky Wonder, Master, Pusa Parvati, Turkey 326, Turkey 434 (Gupta and Chowfla, 1990); ESAL 547, ESAL 550, Jalo (Pellicano *et al.*, 1993); Topcrop, Masterpiece (Devanathan *et al.*, 1994); Premier (Sharma and Dhar, 1994); PDR 14 (Bhagawati and Bhagabati, 1994); EC94469, EC 94929 (Dhanju *et al.*, 1995a); IC 41654, IC 41661, P414, EC 19505, EC 27518, EC 24939, EC 43636, EC 271562, EC 271523, EC 271491, EC 241107 (Dhanju *et al.*, 1995c), EC156011, EC156059, HUR91 and HUR120 (Vishwa and Gurha, 1998), EC 57080, Hans, KRC 18, KRC 21, KRC 22, Kentucky Wonder, KU 25, KU 34, KU 54, KU 73, P 2, P 19, P 26, P 29, Pant Anupama and Premier (Sharma *et al*., 2006) have also been reported to possess resistance to BCMV. Drijfhout (1978) reported resistance to BCMV in differential cultivars viz., Dubbele Witte, Imuna, RG-B, Michelite, Pinto 114, Great Northern 31, Widusa, Jubila, Topcrop and Amanda to be conditioned by non-specific recessive genes, strain specific recessive genes and a dominant necrosis gene.

The genetic control towards *Bean common mosaic virus* (BCMV) and *Bean common mosaic necrosis virus* (BCMNV) is assured by one dominant *I* gene and several recessive (*bc-u, bc-1, bc-1*^{2}*, bc-2, bc-2*^{2} and *bc-3*) genes (Kelly *et al.,* 1995; Strausbaugh *et al.,* 1999). Pasev *et* al (2014) evaluated resistance gene combinations in advanced F8 breeding lines, derived from two crosses (A-8-40-7-2-1 × IVT 7214) and (Zaria × RH 26D) against BCMV and BCMNV and found that a valuable gene combination (*Ibc-3)* was established in seven breeding lines with immune reaction to BCMNV.

Chemical control

Soil application of granular insecticides (Thimet @ 1.5 kg a.i./ha) at sowing followed by 2-3 foliar sprays of either monocrotophos (0.05%) or dimethoate (0.05%) (Singh *et al.*, 1981) has been seen effective to manage aphid population resulting in reduced spread of the virus. Foliar spraying of Metasystox at 0.1% (4 sprays at 15-day intervals) was effective against the disease (Sharma *et al.*, 2006). Presently, Nimitz (25 kg/ha) is available as granular insecticide which can be used at the time of sowing and imidacloprid can be used as foliar spray for the management of aphid vectors.

Biopesticides

Plant extracts from *Plectranthus tenuiflorus*, *Azadirachta indica*, *Clerodendrum inerme*, *Schinus terebinthifolius* and *Mirabilis jalapa* were evaluated as antiviral materials against *Bean common mosaic virus* (BCMV) in bean plants (Elsharkawy and El-Sawy, 2015). The results showed that all the tested plant extracts were effective in reducing the number of local lesions formed by BCMV. The mixing of *P. tenuiflorus* extracts with BCMV inoculum achieved the highest reduction in BCMV infection (92%), while the mixing of *S. terebinthifolius* extracts with BCMV inoculum recorded the lowest reduction in BCMV infection (68%). Seed treatment with 1% (w/v) of *P. tenuiflorus* showed the highest reduction in disease incidence (93% and 85%) under greenhouse and field conditions, respectively. Spray treatments of *P. tenuiflorus* and *M. jalapa* reduced disease incidence to 12% and 17% under greenhouse conditions, whereas under field conditions they reduced disease incidence to 17% and 23%, respectively.

REFERENCES

Anonymous, 1975. Bean production program. *Ann.Rpt.Centro International de Agriiculture*, Cali Columbia.

Archibold, E. S. 1921. Report of the acting Dominion Botanist, *Canada Dept. Agri. Expt. Farms Rept.* 1919-1920: 62.

Berg, G. M. 1984. Nonpersistent viruses: epidemiology and control. *Adv. Virus Res.* 31: 387-429.

Bhagawati, R. and Bhagabati, K. N. 1994. Incidence of bean common mosaic virus of French bean and its effect on grain yield. *Indian J. Virol.* 10: 141-143.

Brunt, A. A., Crabtree, K., Dallwitz, M. J., Gibbs, A. J. and Watson, L. 1996. In *Viruses of Plants, Descriptions and lists from the vide Database*, CAB International, University Press, Cambridge, 1484p.

Calot, L., Leon, C. and Teyssandier, E. E. 1971. Common mosaic virus of bean (*Phaseolus vulgaris*) in farms of the outskirts of Buenos Aires. *Revista de la Facultad de Agronomie Y Veterinaria de Buenos Aires* 19: 25-31.

Capoor, S.P., Rao, D. G. and Sawant, D. M. 1985. A note on a virus disease of French bean. *Indian Phytopath.* 38: 152-154.

Capoor, S.P., Rao, D. G. and Sawant, D. M. 1986. Seed transmission of French bean mosaic virus. *Indian Phytopath.* 39: 343-345.

Cass Smith, W. P. 1945. The bean mosaic menance *West Australian J. Sci. Technol.* 22: 20-25.

Chamberlain, E. E. 1939. Bean Mosaic (*Phaseolus* virus 1 of Smith, 1937) *New Zealand J. Sci. Technol.* 20: 381-388.

Clinton, G. P. 1909. Notes on fungus diseases, etc. for 1908 *Conn. State Agr. Sta. Bien. Rpt.* 1907-1908: 849-871.

Devanathan, M., Muniyappa, V. and Peter, G. B. 1994. Screening of French bean (*Phaseolus vulgaris* L.) varieties for bean common mosaic virus (BCMV) resistance. *South Indian Hort.* 42: 53-54.

Dhanju, K. S., Chowfla, S. C. and Garg, I. D. 1995d. Characterization of viruses causing French bean mosaic in Himachal Pradesh. *Int. J. Trop. Pl. Dis.* 13: 33-41.

Dhanju, K. S., Chowfla, S.C. and Handa, A. K. 1995a. Screening of French bean germplasm under field and glasshouse conditions against bean common mosaic virus. *Legume Res.* 18:53-55.

Dhanju, K. S., Chowfla, S. C. and Handa, A. K. 1995b. Effect of barrier crops and spacings on the incidence of mosaic disease and yield of French bean. *Legume Res.* 18: 113-116.

Dhanju, K. S., Chowfla, S. C. and Handa, A. K. 1995c. Relative response of French bean germplasm against common mosaic under glass house conditions. *Legume Res.* 18: 135-136.

Dickson, B. T. 1922. Studies concerning Mosaic diseases *McGill Univ. Tech. Bull.* 2, 125pp.

Drijfhout, E. and Bos, L. 1977. The identification of two new strains of bean common mosaic virus. *Netherlands J. Pl. Pathol.* 83: 13-25.

Drijfhout, E. 1978. Genetic interaction between *Phaseolus vulgaris* and bean common mosaic virus with implications for strain identification and breeding for resistance. Centre for Agricultural Publishing and Documentation, Wegenniingen, Netherlands. 98p.

Drijfhout, E., Silbernagel, M.J. and Burke, D.W. 1978. Differentiation of strains of bean common mosaic virus. *Neterlands J. Pl. Pathol.* 84: 13-26.

Elsharkawy, M. M. and El-Sawy, M. M. 2015. Control of Bean common mosaic virus by plant extracts in bean plants. *Int. J. Pest Mgt.* 61 (1): 54-59.

Fajardo, T. G. 1930. Studies on the mosaic of the bean (*Phaseolus vulgaris* L.) *Phytopathology* 20: 469-494.

Grainger, J. 1930. The appearance of bean mosaic in England. *Leeds Phil. Lit. Soc. Proc.* 2: 1-32.

Gupta, Y. and Chowfla, S. C. 1988. Susceptibility of French bean to Bean Common Mosaic virus in relation to plantage, population age and age of *Myzus persicae. Indian J. Mycol. Pl. Pathol.* 18: 93-94.

Gupta, Y. and Chowfla, S. C. 1990. Screening of French bean germplasm for resistance to bean common mosaic virus. *Indian Phytopath.* 43: 434-436.

Hagita, T., Senboku, T., Kajinig, M., Shikata, E. and Murayama, M. 1975. Studies on legume virus diseases in Hokkaido II. Seed transmission of bean common mosaic virus. *Memoirs of the Faculty of Agriculture, Hokkaido Univ.*, Japan.

Harrison, A. L. 1935. Mosaic of the Refugee bean. *New York State Agr. Expt. Sta. Bul.* 656-119.

Harrison, A. L. and Burkholder, W. H. 1936. Canning bean diseases in New York in 1936. *Pl. Dis. Reptr.* 16: 164-166.

Harter, L. L. 1927. Bean Diseases in the West. *Pl. Dis. Reptr.* 11: 148-149.

Hoffman, J. C. and Kanapaux, M. S. 1952. Wade, a new all purpose snap bean. *Seed World* 70(2): 38-39, 46p.

Horsfall, J. G. 1930. Diseases of Canning Crops. *Pl. Dis. Reptr. Supl.* 76: 83-85.

Iwanowski, D. 1894. Ueber die Mosaikkrankheit der tabaksflanze. *Acad. Imp. Sci. St. Petersbourg Bul.* 35(3): 67-70.

Jenkins, W.A. 1940. A new virus disease of snap beans. *J. Agr. Res.* 60: 279-288.

Jeyarajan, R. and Doraisamy, S. 1986. Virus diseases of pulses and their management. In *Pest* and *Disease Management Centre for Pl. Protect Studies*, TNAU, Coimbatore, 109-114.

Kaiser, W.J., Danesh, D and Okhovat, M. 1968. Diseases of pulse crops (edible legumes) in Iran. *Pl. Dis. Reptr.* 52: 687-691.

Kelly JD, Afanador L, Haley SD. 1995. Pyramiding genes for resistance to Bean common mosaic virus. *Euphytica* 82:207-212.

Mathews, R. E. F. 1982. Classification and nomenclature of virus. Report of the International Committee on Toxonomy of viruses. *Intervirology* 17: 1-199.

McKern, N. M., Mink, G. I., Barnet, O. W., Mishra, A., Whittaker, L. A., Silbernagel, M. J., Ward, C.W. and Shukla, D. D. 1992. Isolates of bean common mosaic virus comprising two district potyviruses. *Phytopathology* 82(5): 923-929.

Moralles, F.J. 1989. Bean common Mosaic-Screening for Disease Resistance. *Centro Internacional de Agricultura Tropical (CIAT) Publ.* No.33. Cali, Colombia. 28p.

Moralles, F. J. and Castano, M. 1987. Seed transmission characteristics of selected bean common mosaic virus strains in differential bean cultivars. *Plant Dis.* 71: 51-53.

Muniyappa, V. 1976. Studies on a mosaic disease of French bean (*Phaseolus vulgaris*) *Madras Agric. J.* 63: 69-70.

Nagaich, B. B. and Vashisth, K. S. 1963. A virus causing a typical mosaic disease of bean. *Indian J. Microbiol.* 3: 113-116.

Nelson, R. 1932. Investigations in the mosaic of beans (*Phaseolus vulgaris* L.) *Mich. Agr. Expt. Sta. Tech. Bul.* 118, 71p.

Ogilvie, L. 1924. Preliminary Report of the Plant Pathologist for the period Sep' 27 to Dec' 31, 1923. *Bermuda Dept. Agric. Rpt.* 1923, 28-34.

Ordosgoitty, A. 1973. Identification of bean (*Phaseolus vulgaris*) common mosaic virus in Venezuela. *Agronomictropical Venezuela* 22: 29-43.

Pasev, G., Kostova, D. and Sofkova, S. 2014. Identification of genes for resistance to *Bean Common Mosaic Virus* and *Bean Common Mosaic Necrosis Virus* in Snap Bean (*Phaseolus vulgaris* L.) breeding lines using conventional and molecular methods. *J Phytopathol:* 162: 19-25.

Pellicano, I. J., Ramalho, M. A. P., Figueira, A. and Dos, R. 1993. Evaluation of bean (*Phaseolus vulgaris*) cultivars for resistance to bean common mosaic virus. *Giencia Pratica* 13: 112-115.

Pierce, W. H. 1934. Viruses of the bean. *Phytopathology* 24: 87-115.

Pierce, W. H. and Hungerford, C. W. 1929. Symptomotology, transmission, infection and control of bean mosaic in Idaho. *Idaho Agr. Expt. Sta. Res. Bul.* 7, 37pp.

Porter, R. H. 1926. A preliminary report of surveys for plant diseases in East China. *Pl. Dis. Reptr. Supl.* 20: 279-280.

Puerto, J. and Alonso, A. 1958. Tests for resistance to different diseases in various French bean varieties. *Bol. Inst. Invest. agron. Madr.* 18: 37-48.

Quantz, L. 1962. Studies on bean common mosaic virus and soyabean mosaic *Phytopath. Z.* 43: 79-101.

Rands, R. D. and Brotherton, W. 1925. Bean varietal tests for disease resistance *J. Agric. Res.* 31: 101-154.

Reddick, D. and Stewart, V. B. 1918. Varieties of beans susceptible to mosaic. *Phytopathology* 8: 530-534.

Robertson, R. S. 1962. The role of seed transmission in epidemiology of bean mosaics in central Washington. *Pl. Dis. Reptr.* 56: 71-72.

Robertson, R. S. and Klostermeyer, E. C. 1961. Aphid transmission of bean viruses in field beans in Central Washington. *J. Econ. Entomol.* 54: 414-416.

Robertson, R. S. and Klostermeyer, E. C. 1962. The role of alternate plant hosts in the transmission of bean mosaic in central Washington. *J. Econ. Entomol.* 55: 460-462.

Sastry, K. S. M., Singh, S. J. and Rajendran, R. 1981. Screening of different bean cultivars against bean common mosaic virus. *Pulse Crop. Newsletter* 1: 30.

Schmidt, H. E. 1992. Bean Mosaics. In: *Plant Diseases* of *International Importance Vol II. Diseases of Vegetables and Oilseed Crops*. H.S. Chaube, J. Kumar, A.N. Mukhopadhyay and U.S. Singh (Ed.) Prentice Hall Inc. 40-73p

Sharma, O. P. and Dhar, J. 1994. Studies on mosaic disease of French bean (*Phaseolus vulgaris* L.) in Himachal Pradesh. In: *Virology in the Tropics*. (Eds. N. Rishi; K. L. Ahuja and B. P. Singh), Malhotra Publishing House. New Delhi. pp. 319-325.

Sharma, P. K., Sharma, O. P. and Sharma, P. 2006. Management of bean mosaic disease through host resistance and physicochemical means. *Environment and Ecology* 24S (Special 2): 312-314.

Singh, R. N. 1976. A new strain of bean common mosaic virus on bean in India. *Indian J. Mycol. Pl. Pathol* 6: 156-159.

Singh, S. J., Sastry, K. S. and Sastry, K. S. M. 1981. Field tests with insecticides and mineral oil for the protection of French beans from yellow mosaic virus disease. *Gartenbauwissenchaft* 46: 88-91.

Skotland, C. B. and Burke, D. W. 1960. A virus disease of wide host range seed borne in *Phaseolus vulgaris. Phytopathology* 54: 655.

Stewart, V.B. and Reddick, D. 1917. Bean Mosaic *Phytopathology* 7: 61 (Abstr.)

Stone, R.E. and Howitt, J. E. 1920. Plant Diseases and fungi comparatively new or rare in Ontario. *Phytopathology* 10: 317-318 (Abstr.).

Strausbaugh CA, Myers JR, Forster RL, McClean PE. 1999. Bc-1 and Bc-u - two loci controlling *Bean common mosaic virus* resistance in common bean are linked. *J Am Soc Hortic Sci* 124:644-648.

Thomas, I. and Vevai, E. J. 1940. Aphid migration. An analysis of the results of five reasons trapping in North Wales. *Ann. appl. Biol.* 27: 393-405.

Trujillo, G. E. and Saettler, A. W. 1972. Reaction of Venezuelan varieties of bean (*Phaseolus vulgaris* L.) to different plant viruses. *Agronomie Tropical* 23: 379-391.

Van Rheenen, H. A., Hasselbach, O. E. and Muigai, S. G. S. 1981. The effect of growing beans together with maize on the incidence of bean diseases and pests. *Netherlands J. Pl. Path.* 87: 193-199.

Vishwa, Dhar and Gurha, S. N. 1998. Sources of resistance to bean common mosaic virus in French bean. *Indian J. Pulses Res.* 11(1): 138-139.

Wade, B.L. 1943. Logan, a new, hardy snap bean. *Seed World* 53(5): 12-13, 40-41.

Walker, J.C. 1936. Discussions on disease free and disease resistant beans. *Canning Agr.* 17: 171-172, 178.

Wolf, E.A. and Hills, W.A. 1954. A new disease resistant, green row podded bush bean. *Fla. Agr. Expt. Sta. Cir.* 5-73, 6pp.

Yaraguntaiah, R. C. and Nariani, T. K. 1963. Bean Mosaic virus in India. *Indian J. Microbiol.* 3: 147-150.X

Zaumeyer, W.J. 1930. Bean diseases in Western United States in 1929. *Pl. Dis. Reptr.* 14: 38-43.

Zaumeyer, W.J. 1946. Bean diseases in some of the intermountain states in 1945. *Pl. Dis. Reptr.* 30: 97-105.

Zaumeyer, W.J. 1950. Topcrop, a new processing snap bean. *Canner* 110(2): 12-15.

Zaumeyer, W. J. and Goth, R. W. 1964. A new severe symptom inducing strain of common bean mosaic virus. *Phytopathology* 54: 1378-1385.

Zaumeyer, W. J. and Kearns, C. W. 1936. The relation of aphids to the transmission of bean mosaic. *Phytopathology* 26: 614-629.

Zaumeyer, W. J. and Meiners, J. P. 1975. Disease resistance in beans. *Ann. Rev. Phytopath.* 13: 313-334.

Zaumeyer, W.J. and Thomas, H. R. 1947. Bean diseases in some of the mountain states in 1947. *Pl. Dis. Reptr.* 31: 432-442.

23

BEAN YELLOW MOSAIC

This disease also causes significant losses, if the favourable environment exists for its spread.

GEOGRAPHICAL DISTRIBUTION

The occurrence of *Bean Yellow Mosaic Virus* (BYMV) was first reported in 1934 from USA (Bos, 1970). Subsequently it was reported from East Africa, Germany, Canada, Holland, Italy, Switzerland and India (Sharma *et al.*, 2015).

LOSSES

The losses caused by BYMV are frequently very large. Since the symptoms of this virus and its strains and Common bean mosaic often overlap, it is almost impossible to assign any amount of loss caused by an individual virus.

SYMPTOMS

Mosaic symptoms consist of contrasting dark and yellowish green areas, often accompanied by bright yellow spots on leaves, which intensify with the age of the plants (Plate 1). Some isolates cause only a mild and diffuse chlorotic mottle and limited plant stunting, while others cause a prominently coarse mosaic, rugosity, malformation and severe stunting. Some varieties may show necrotic spots, veinal and apical necrosis, wilting and premature death. Late infection generally causes less prominent foliar symptoms and if the infection coincides with pod formation, the pods eventually will exhibit a light green mottle and slight malformation. BYMV infected plants are usually stunted and bushy due to reduction of internode length and proliferation of lateral branches. Plant maturity is delayed, and quality and quantity of the seed is affected. Certain strains of BYMV induce systemic necrosis that appear as a purplish discolouration at the base of the lower leaves, accompanied by veinal, stem and petiole necrosis. The necrosis at the terminal growing point consequently results into death of the plant. Some strains incite local lesions and others produce malformed pods bearing reddish brown necrotic spots. However, type and severity of symptoms vary according to strain of virus and cultivar of French bean involved.

Various metabolites of host tissue are also altered due to viral infection. Pigment contents, carbohydrates and free amino acids were decreased with time in common bean plants inoculated with BYMV whereas soluble proteins remained constant (Hamida, 2006).

THE VIRUS

Bean Yellow Mosaic Virus (BYMV) belongs to the potyvirus group. The virus is characterized by elongate, flexuous, filamentous particles about 750-810 x 11-15 nm (Plate 2). The virus has TIP of 65° C, LIV of 2 to 4 days and DEP of 10^{-3} to 10^{-5} (Brunt *et al.*, 1996).

Plate 1. Symptoms of bean yellow mosaic

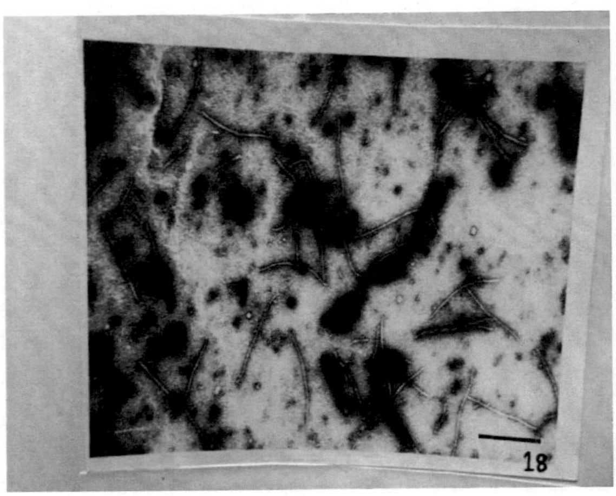

Plate 2. Flexuous, filamentous particles of *Bean Yellow Mosaic Virus*

TRANSMISSION

BYMV is readily transmitted by through sap inoculation and many species of aphids in non-persistent manner but not through the seeds of French bean though it has been reported in broad-bean and other leguminous hosts (Bos, 1970). BYMV has been reported to be experimentally transmitted by 47 aphid species in 25 different genera (Edwardson and Christie, 1991) but *Acyrthosiphum pisum, Macrosiphum euphorbiae, Myzus persicae, Aphis craccivora* and *A. fabae* are the important vectors of this virus under field conditions.

HOST RANGE

BYMV has a wide host range and can infect 233 species in 71 genera of 21 families (Edwardson and Christie, 1991). Besides French bean, its host range includes mung bean, lima bean, pigeon pea, chickpea, lentil, peas, broadbean, soybean, fenugreek, *Cucurbita pepo*, *Gladiolus* spp. and *Robinia pseudoacacia* (Raizada *et al*., 1991).

EPIDEMIOLOGY

BYMV is not a seed borne in French bean though its seed borne nature has been reported in other leguminous hosts. The virus has a wide host range and aphid vectors transmit it. Field spread by aphid vectors from infected French bean and other collateral hosts is primarily responsible for the epidemics of this disease.

MANAGEMENT

Cultural practices

Destruction of collateral host of this virus from in and around French bean fields, plantation of barrier crops like maize and crop rotation reduces the infection of this disease.

Host resistance

This is the most reliable method for the control of this disease. Sources of resistance to a BYMV strain including pod malformation have been identified in various Great Northern lines like GNUI No. 31, 59, 123 and 1140. Numerous dwarf bean accessions were screened for resistance to three strains of *Bean Common Mosaic Virus* (BCMV) (Walkey and Innes, 1979). It was found that many lines were resistant to a strain of BCMV isolated at Wellesbourne (NVRS) and the Dutch NL4 strain, but only a few lines showed resistance or tolerance to the Dutch NL3 strain. Further, hybridization experiments showed that resistance to the NVRS strain of BCMV in the dry bean cultivars Seafarer and Aurora was controlled by a dominant gene and that in Seafarer this was probably the dominant *I* gene originally found in Corbett Refugee.

French bean Recombinant Inbred Lines (RIL's) population comprising of 44 genotypes (F5), which were developed by crossing different lines, along with two checks viz. Ring bean and Arka Suvidha were evaluated against BCMV under field conditions (Patil *et al.*, 2017). Out of 46 genotypes, 4 genotypes (Line 3-2, Line 6-1, Line 7-1 and Ring bean) were found to be highly resistant with no CBMV infection. Ten genotypes were resistant, 17 genotypes were moderately resistant, 10 genotypes were susceptible, and 5 genotypes were found to be highly susceptible to BCMV infection.

Chemical control

Spray application of systemic insecticide like quinolphos 25%EC (0.025%) or fipronil 5%SC (0.0075%) or imidacloprid 17.8%SL (0.01%) also help in keeping the vector population in check in French bean fields or other host crops and reduces the spread of virus. Sprays of mineral oils or neem formulations (2-3ml/lt) also help in checking vector population.

Integrated disease management

Manjunath *et al.* (2020) reported that intercropping with two rows of border crops of maize 30 days before sowing of pole bean, mulching with black silver mulch, seed treatment with thiamethoxam 25 WG - 5g/kg seeds during sowing, soil application of *Pseudomonas fluorescens* along with neem cake (1kg/100kg), installation of yellow sticky trap @ 10 no./acre and spraying of seaweed extract (1.5ml/L) 20 days after sowing, spraying of thiamethoxam 25% WG (0.5 g/L) and imidacloprid 17.8 SL (0.5ml/L) 30 and 45 days after sowing respectively recorded less disease incidence of Bean yellow mosaic and higher yield compared to recommended practice.

REFERENCES

Bos, L. 1970. Bean yellow mosaic virus. In: *Description of Plant Viruses*. No. 73. CMI/ AAB, Kew, Surrey, England.

Brunt, A.A.; Crabtree, K; Dallwitz, M.J; Gibbs, A.J. and Watson, L. (1996). In *Viruses of Plants, Descriptions and lists from the VIDEDatabase*, CAB International, University Press, Cambridge, 1484p.

Edwardson, J.R. and Christie, R.G. 1991. *The potyvirus Group*. Vol. I-IV, Fl. Agric. Exp. Stn. Monograph 16.

Hemida, S.K. 2008. Effect of *Bean Yellow Mosaic Virus* on physiological parameters of *Vicia faba* and *Phaseolus vulgaris*. *International Journal of Agriculture and Biology* 7(2): 154-157.

Manjunath. B., B. S. Rajendra Prasad, S. Pavithra, R. Manjunath, A. P. Mallikarjuna Gowda, Savita S. Manganavar, B. Gayathri and Chithra. Y. D. 2020. Assessment on management of *Yellow Mosaic Virus* in pole beans through integrated approach. *International Journal of Current Microbiology and Applied Science* 9(5): 172-179.

Patil, R. K., Satish, D. and Jagadeesha, R. C. 2017. Screening of recombinant inbreed lines of French bean (*Phaseolus vulgaris* L.) for *Bean Common Mosaic Virus* (BCMV) resistance. *Plant Archives* 17(1): 479-482.

Raizada, R.K. Mohd Aslam QMR Haq and Singh, B.P. 1991. *Vicia birsuta*: a natural host of bean yellow mosaic virus. *International Conference on Virology in Tropics*. Dec. 5, NBRI, Lucknow, pp.37.

Sharma, P. N., Sharma, V., Sharma, A., Rajput, K., and Sharma, S. K. 2015. Identification and molecular characterization of *Bean Yellow Mosaic Virus* infecting French bean in Himachal Pradesh. *Virus Disease* 26(4): 315-318.

Walkey, D., and Innes, N. 1979. Resistance to bean common mosaic virus in dwarf beans (*Phaseolus vulgaris* L.). *The Journal of Agricultural Science* 92(1): 101-108.

24

BEAN GOLDEN MOSAIC

It is an important disease of this crop and under favourable weather conditions cause considerable yield losses.

GEOGRAPHICAL DISTRIBUTION

The occurrence of the disease was first reported from southern Brazil during 1961 where it now ranks among the most serious diseases of the French bean (Costa, 1976). Subsequently it was reported from tropical and subtropical America and many other countries (Tripathi, 2004).

LOSSES

Considerable losses have been recorded in various French bean growing areas of the world. Costa and Cupertino (1976) noticed reduction of 85 and 48% if plants are inoculated 15 and 30 days after sowing, respectively.

SYMPTOMS

The disease is generally more severe on older plants than younger ones until and unless the population of viruliferous white flies is very high. Mostly the symptoms become more conspicuous on 3-4 trifoliate leaf stage. The first symptoms appear as downward rolling of young leaves which later show the golden mosaic symptoms. These may be predominantly on the veins or involve large areas of the leaf parenchyma with striking yellow- green mosaic pattern (Gilbertson *et al.*, 1991). Pods also show golden mosaic spots and may be malformed. Plant stunting, pod malformation, discoloration and reduction of seed size are also common features of the disease (Costa and Cupertino, 1976; Goodman and Bird, 1982).

THE VIRUS

Bean Golden Mosaic Virus (BGMV) is characterized by isomeric particles in pairs (geminate), 18-20 nm in dia. dimmers, 30 nm in length, angular in profile and contains colvalently closed circular single stranded DNA- genome (Goodman, 1981). The virus has TIP of 50 to 55 °C, LIV of 72 days and DEP of 10^{-2} (Brunt *et al.*, 1996). The virus has been successfully detected by polymerase chain reaction and serology (McLaughlin *et al.*, 1994; Iizuka and Charchar, 1994).

Nardo and Costa (1986) identified 5 strains of BGMV resulting probably from natural segregation. Based on symptoms FR-1, FR-2 and FR-pi caused milder symptoms than CN-1, while symptoms induced by MB-1 were moderate and by FO-1 were more severe. No mechanical transmission was observed.

HOST RANGE

Besides French bean (*Phaseolus vulgaris*) its host range includes *P. lunatus, P. acutifolius, P. polystachios, P. longependunculatus, P. aborigeneus, P. coccineus, Desmodium occuleatum, Macroptilus lathyroides, Terramnus urcinatus, Vigna radiata, V. unguiculata* and *C. muconoides* (Bird *et al.*, 1972).

TRANSMISSION

The virus is transmitted by whitefly (*Bemisia tabaci*) and inoculation with sap from young, infected leaves but not through the seeds of French bean (Pierre, 1975). Both male and female whiteflies are equally efficient vectors (Bird *et al.*, 1973). The virus has been transmitted by inoculation with sap to French bean from the same host and *Calopogenium mucunoides* and to lima bean and French bean from Lima beans (Meiners *et al.*, 1975; Goodman *et al.*, 1977). Since this virus is not seed borne, lima bean and other susceptible legumes including voluntary and cultivated beans and weeds act as reservoirs. Bean golden mosaic, like other whitefly-transmitted diseases, is more severe under conditions of warm temperatures and low rainfall.

MANAGEMENT

Cultural Practices

Collect and destroy all reservoir plants from in and around the field since this virus is not seed borne and overwinters in the reservoir plants like susceptible legumes including volunteer and cultivated beans and weeds. Adjustment of time of sowing avoiding peak vector activity period also provides some protection (Vicente *et al.*, 1988).

Host resistance

P. vulgaris accessions possessing partial resistance or tolerance to BGMV include Porrillo Sintetico, Porrillo 70, Turrialba 1, and ICA-Pijao. Resistance to BGMV was reported in black-seeded common bean cultivars in South America. BGYMV resistance genes discovered are *bgm* or *bgm-1* in bean genotypes of Mexican 'Durango' race or the *bgm-2* of Andean origin (Morales, 2003). A single dominant gene, *Bdm*, conferred BDMV resistance in crosses between Othello and Topcrop bean cultivars (Seo *et al.*, 2004).

Phaseolus coccineus L. germplasm accession G35172 was found, by scientists at International Center for Tropical Agriculture, U.S. Department of Agriculture, to be resistant to BGYMV. Populations derived from an interspecific cross between *P. vulgaris* and *P. coccineus* were evaluated to study the inheritance of resistance to BGYMV (Osorno *et al.,* 2007). Segregation ratios of F2 plants and other populations suggest that BGYMV resistance from *P. coccineus* is controlled by two genes. A recessive gene, with the proposed symbol bgm-3, confers resistance to leaf chlorosis and a dominant gene, with the proposed name Bgp-2, prevents pod deformation in the presence of BGYMV. Results from allelism tests with previously reported BGYMV resistance genes (bgm, bgm-2, and Bgp) and the absence of the SR-2 sequence-characterized amplified region marker for bgm support the hypothesis that bgm-3 and Bgp-2 are different genes for BGYMV resistance.

Chemical Control

Application of insecticides and mineral oils has also been reported effective in reducing the vector population (Meiners *et al*., 1975).

REFERENCES

Bird, J. Perez, J.E., Alconero, R., Vakili, N.C. and Melendez, P.L. 1972. A whitefly transmitted golden yellow mosaic virus of *Phaseolus lunatus* in Puerto Rico. *J. Agric. Univ. Puerto Rico* 56-64-74.

Bird, J., Sanchez, J., and Vakili, N.G. 1973. Golden yellow mosaic of beans (*Phaseolus vulgaris*) in Puerto Rico. *Phytopathology* 63: 1435.

Brunt, A.A.; Crabtree, K; Dallwitz, M.J; Gibbs, A.J. and Watson, L. 1996. In *Viruses of Plants, Descriptions and lists from the VIDE Database*, CAB International, University Press, Cambridge, 1484p.

Costa, A.S. 1976. Whitefly transmitted plant diseases. *Annual Review Phytopathology* 14: 429-449.

Costa, C.L. and Cupertino, F.P. 1976. Avaliacao das perdas na producao do feijoeiro causadas pelovirus do mosaico dourato. *Fitopatologia. Brassiliera* 1: 18-25.

Goodman, R.M. 1981. Geminiviruses. *Journal of General Virology* 54: 9-21.

Goodman, R.M. and Bird, J. 1982. Bean golden mosaic virus. In: Description of Plant Viruses. No. 102. CMI/ AAB. Kew Surrey, England.

Goodman, R.M., Bird, J. and Thongmeearkon, P. 1977. Unusual virus like particles associated with golden yellow mosaic of beans. *Phytopathology* 67: 37-42.

Gilbertson, R.L., Rojas, M.R., Russell, D.R. and Maxwell, D.P.1991. Use of the asymmetric polymerase chain reaction and DNA sequencing to determine genetic variability of bean golden mosaic geminivirus in the Dominican Republic. *Journal of General Virology* 72(11): 2843-2848.

Iizuka, N. and Caharchar, M.J. D'A. 1994. Identification of some viral diseases of leguminous crops in the Cerrados. In: Relatorio tecnico do projecto nipo-brasileiro de cooperacao em pesquisa agricola nos cerrados 1987/1992. *Centro de pesquisa Agropecuaria dos Cerrados* 237-247.

McLaughlin, W., Rojas, M.R., Nakhala, M. K., Hidayat, S.H. and Maxwell, D.P. 1994. Partial molecular characterization of bean golden mosaic virus isolates from Jamaica and Central America. *Plant Disease* 78: 1220.

Meiners, J.P., Lawson, R.H., Smith, F.F. and Diaz, A.J. 1975. Mechanical transmission of whitefly (*Bemisia tabaci*) borne disease agents of beans in El salvador. Pp. 61-69. In: *Tropical Diseases of Legumes* (Eds. Bird, J. and Maramorosch, K.). Academic Press, New York.

Morales F.J. 2003. Common Bean. In: Loebenstein G., Thottappilly G. (eds) Virus and Virus-like Diseases of Major Crops in Developing Countries. Springer, Dordrecht. https://doi.org/10.1007/978-94-007-0791-7_17.

Osorno, J., M., Munoz, C. G., Beaver, J. S., Ferwerda, F. H., Bassett, M.J., Miklas, P. N., Olezyk, T. and Bussey, B. 2007. Two Genes from *Phaseolus coccineus* confer resistance to *Bean Golden Yellow Mosaic Virus* in common bean. *Journal of American Society Horticulture Science* 132(4):530-533.

Pierre, R.E. 1975. Observations on the golden mosaic of bean (*Phaseolus vulgaris* L.) in Jamaika. Pp. 35-59. In: *Tropical Diseases of Legumes* (Eds. Bird, J. and Maramorosch, K.). Academic Press, New York.

Seo, Y. S., Gepts, P. and Gilbertson, R. L. 2004. Genetics of resistance to the geminivirus, *Bean dwarf mosaic virus*, and the role of the hypersensitive response in common bean. *Theoretical and Applied Genetics* 108:786-793.

Tripathi, D.P. 2004. *Introductory Plant Virology*. Kalyani Publishers, Ludhiana, 555 pp.

Vicente, M., Kanthack, R.D., Noronha, A.B. and Stradioto, M.F.S. 1988. Incidence of golden mosaic on bean grown in two planting seasons in the presi dente region. *Fitopatol. Bras*. 13: 373-376.

25

ROOT KNOT

Among pests of French bean, nematodes play important role by causing significant loss in crop production. They cause root dysfunction by reducing the rooting volume, nutrient and water use efficiency. Several plant parasitic nematodes are found to be associated with leguminous crops *viz.* root-knot nematodes (*Meloidogyne* spp.), lesion nematodes (*Pratylenchus* spp.), sheath nematodes (*Hemicycliophora* spp.), stubby root nematodes (*Trichodorus* spp.), *Tylenchus* spp., *Criconemella* spp., and *Aphelenchus* spp. Most devastating *Meloidogyne* spp. are of utmost importance because of their prevalence and wide distribution in the sub-tropical and tropical regions of the world, majority having subsistence agricultural systems.

GEOGRAPHICAL DISTRIBUTION

Root-knot nematodes prefer sandy to sandy loam soils and are generally more damaging in tropical regions than in temperate climates (De Waele and Elsen, 2007). Across the world, root knot is a major problem in bean production (Bridge and Starr, 2007). *Meloidogyne incognita, M. javanica* and *M. arenaria* are the most common root-knot species associated with *P. vulgaris* causing extensive losses in the America, Africa and Asia (Sikora and Greco, 1990; Di Vito *et al.*, 2005, Sikora *et al.,* 2005).

LOSSES

Yield losses of up to 45-60 per cent were noticed due to root-knot nematode in French bean (Mullin et al., 1991). An experiment was laid out in wet temperate high hill zone of Kinnaur, Himachal Pradesh consecutively for four years to work out avoidable yield losses induced by the infestation of root-knot nematode on Rajmash variety Jawala (Sharma and Dogra, 2014). Treated (phorate 10 G @ 0.3 g ai/m^2) and untreated beds (6 m^2 size) in the replication of six were maintained in the nematode infested field. The results revealed that with 18.7-28.0 per cent reduction in root gall index and 40.0-83.3 per cent decrease in the J2 (second stage juveniles) population of the nematode, 15.4-23.0 per cent yield increase in treated beds in comparison to control was recorded.

SYMPTOMS

Spectacular symptoms are produced by root-knot nematode on plant roots. Due to nematode feeding, large galls or 'knots' can form all over the root system of infected plants. Extreme infections on crop results in reduced yield. With the increase of nematode population in a particular field, the number of galls per plant will also raise. Huge numbers of nematodes penetrating roots in close proximity will upshot larger galls. Galls produced by *Meloidogyne hapla* are less than half the size of those formed by *M. incognita* on the same plant hosts. Heavily infected plants often wilt rapidly as galled roots have restricted ability to take up and transfer water and nutrients to the rest of the plant. Severely infected plants may wilt even in the presence of sufficient soil moisture, especially during the afternoon hours. Nutrient deficiency symptoms may also be exhibited by plants because of their reduced ability to absorb and transport nutrients from the soil. Additional fertilization will not generally result in remediation of root-knot nematode-induced chlorosis. Reduction in plant size is frequently observed on host crops grown in root-knot nematode-infested fields with reduced crop yields.

Fig.1. Root knot infection of bean plant

THE PATHOGEN

Root-knot nematodes belong to the kingdom Animalia, phylum Nematoda, class Nemata, subclass Sercenentea, order Tylenchida, suborder Tylenchina, family Meloidogynidae, and genus *Meloidogyne* (Karssen *et al.*, 2006). Presently, out of more than 80 nominal species of *Meloidogyne* (Karssen, 2002), four species *viz. M. incognita, M. javanica, M. arenaria* and *M. hapla* are of economic importance in bean production across the globe (Chitwood, 1956; Bridge and Starr, 2007). Besides these species, *M. paranaensis* and *M. enterolobii* may also be found associated with the crop (Machado, 2011). Recently, novel root-knot species were noticed in bean growing regions in Brazil, *M. inornata* (Machado *et al.*, 2013) and *M. luci* (Machado *et al.*, 2016).

Populations of root-knot nematode consist of males and females, which are readily distinguished morphologically. Length of the worm like males varies from 1.20 - 1.50 mm with 30 - 60μm of diameter (body width). Pear shaped mature females are about 0.40 - 1.30 mm long with 0.27 - 0.75 mm of diameter. Microscopic second-stage juveniles are vermiform in shape while third and fourth stage juveniles are sausage shaped (Siddiqi and Alam, 1987).

HOST RANGE

Meloidogyne spp. are one of the most important plant-parasitic nematodes in the world, showing wide distribution and a huge host range (Antes *et al.*, 2019). *Meloidogyne incognita* is probably the most widely distributed and economically important species of plant parasitic nematode in tropical and subtropical regions (Kofoid and White, 1919; Chitwood 1949). Two-thirds of the root-knot nematode samples obtained from a number of tropical countries were of *M. incognita* (Sasser, 1979). *M. incognita* is an endoparasite of more than 3000 plant species (Cheng *et al.*, 2015). About 232 plant genera in India alone have been found as hosts to *M. incognita* (Krishnappa, 1985) and worldwide the species is a parasite of a wide range of crop plants.

NEMATODE BIOLOGY AND LIFE CYCLE

Nematodes are multi-cellular, bilaterally symmetrical, non-segmented and worm-like, organisms with well-developed reproductive and digestive systems, primitive excretory and nervous systems and no respiratory and circulatory systems. These bisexual organisms undergo four molting stages from egg to adult. Almost in all agro-ecosystems, nematodes are found interacting directly and indirectly with plants and other micro-fauna. They regulate decomposition and release of nutrients to the plants.

Females of *Meloidogyne* spp. produce a few eggs or no eggs at all, when the host's temperature and surroundings are adverse. Under less congenial environment, like extreme temperature, a single female produces 300 - 500 eggs whereas under optimum temperatures of 27^{0}C it can produce more than 2800 eggs. Eggs are laid in sac-like gelatinous matrices by females (Evans and Perry, 2009). A minimum of 25 days are required for new generation to arise but under less favorable conditions, this life cycle may be extended to 30 or 40 days or development may stop entirely (Agrios, 2005). A thick outer covering is laid over each egg to protect it during the dormant period. Inside the egg, the first larval stage develops and undergoes the first moult to become a second-stage juvenile (J2). The later emerges from the egg into the soil where it moves until it finds a susceptible root by chemotaxis. Only the second-stage juvenile (J2) are infective (Agrios, 2005). The third-stage juvenile (J3) lacks stylet. Fourth-stage juvenile (J4) can be distinguished either as male or female. After the final molt, it turns into a free-living male nematode or a parasitic adult female. Plant penetration by second-stage juvenile (J2) of root knot nematode occurs between 10^{0}C and 35^{0}C, with 27^{0}C being the optimum temperature depending on the species. The root knot nematode's second stage juvenile (J2) usually enters the plant through roots and modifies some root cells into feeding cells. This phenomenon is necessary for the reproduction and development of the parasite. After penetration of the roots of susceptible plants by the root knot nematode juveniles, the cambial root cells are transformed into specialized feeding cells named 'Giant cells'. These metabolically active cells are induced and maintained in susceptible hosts by the feeding activities of the nematode (Hussey and Jansen, 2002). They act as permanent feeding sites for the parasite throughout its life cycle (Hussey *et al.*, 1994).

MANAGEMENT

The management of plant parasitic nematodes is a complicated task which primarily depends upon the chemical nematicides. Significant reduction of nematode population has been achieved through this method. Although soil nematicides are effective and fast-acting, they are currently being reappraised with respect to the environmental hazards and human health.

Cultural practices

Application of various organic amendments *viz.,* crop residues and animal manures, heat treatment, soil solarization and crop rotation with non-hosts helps to manage root knot nematode (Oka *et al.*, 2007).

Nematicidal effects of decomposing plant amendments are known. Various crops like Holly leaves (*Ilex aquifolium*), crotalaria (*Crotalaria spectabilis* Roth.), castor bean (*Ricinus communis* L.), velvet bean (*Mucuna deeringiana* L.) and rapeseed (*Brassica napus*) have been shown to curb root knot nematode (*M. arenaria* and *M. chitwoodi*), with varying degrees of suppression (Mojtahedi *et al.*, 1991).

Host resistance

Screening and evaluation of 302 cultivars of French bean for resistance to *M. incognita* in replicated trials in the field and glasshouse under Indian conditions was carried out and results showed 31 cultivars were resistant in both field and glasshouse conditions while 64 cultivars were moderately resistant (Singh *et al.*, 2009) .

Chemical control

Nematicides are chemically synthesized products that kill nematodes or adversely affect nematodes. These products can have nematicidal or nemastatic effects against nematodes. Nematicidal compounds are highly toxic and kill exposed nematodes, whereas nemastatic compounds do not kill nematodes but impede nematode movement toward host plant roots or delay nematode egg hatching. Nematicides are grouped into fumigant and non-fumigant nematicides based on their volatility in soil. Pre-plant soil fumigation with metham potassium, metham sodium or 1,3-dichloropropene (1,3-D 97), granular or micro-incapsulated liquid fenamiphos were effective treatments against *Meloidogyne incognita* on tobacco (Lamberti *et al.*, 2000). Emamectin benzoate soil application showed good efficacy against *M. incognita* and resulted in increased yield in tomato (Cheng *et al.*, 2015).

Recently a new non-fumigant nematicide named fluensulfone, of fluoroalkenyl chemical group, was evaluated against field populations of root-knot nematodes in commercial cucumber and tomato greenhouses and substantially reduction in the nematode multiplication either in tomato or cucumber cultivation at all doses above 24 ml 100 m^{-1} was found (Giannakou and Panopoulou, 2019). Application of fluensulfone in both crops resulted in significant increase in fruits.

Biological control agents

Different doses of potassium fertilizer (K_2O) along with the biofertilizers, *Trichoderma harzianum* and *Pochonia chlamydosporia* was evaluated for the management of the root-knot disease of French bean (Sharf *et al.*, 2014). Additionally, the effect of treatments on the growth and physiological parameters of bean plants were also studied. It was observed that the application of double dose of potassium and both fungal biocontrol agents improved growth and biochemical parameters *viz,* chlorophyll, protein, nitrate reductase, nitrogen and phosphorus contents. Further, reduction in galls per root system was found compared to other treatments and control.

Biopesticides

Oil-cake of pongamia (*Pongamia glabra*) and neem (*Azadiracta indica*) each at the rate of 2.5 Mt/ha was very successful in reducing root-knot of okra and tomato (Singh and Sitaramaiah, 1994). Nematicidal properties of *Crotalaria virgulata* subsp. *grantiana* leaf extracts were reported (Jourand *et al.*, 2004) and hence could be used as green manure and also natural alternative to synthetic chemicals in integrated pest management strategies. Nazli *et al.* (2008) reported that leaf extracts of quick stick plant (*Gliricidia sepium*) has some insecticidal, nematicidal and antibacterial activity resulting in 60% mortality of second stage juvenile of *M. incognita* (Pakeerathan *et al.*, 2009). In several parts of Western Kenya, plants of the genus *Desmodium* spp. have been widely accepted as green manures (Desaeger and Rao, 2000). Compounds present in Sun hemp (*Crotalaria jucea*) and Mexican marigold (*Tagetes minuta*) have been found to be lethal to root knot nematode (Kimenju *et al.*, 2008)

The efficacy of neem (*A. indica A. Juss*), Tithonia (*Tithonia diversifolia*) and Tephrosia (*Tephrosia purpurea*) leaf extracts was tested for the management of root-knot (Njenga and Mbaka, 2019). Controlled lath house and field trials were conducted at different concentrations of 25 ml/L, 50 ml/L and 100 ml/L of the treatments. Vydate (Oxamyl 10%) a synthetic nematicide served as a standard positive control while treatments with no extracts application and no nematode application served as negative controls. The extracts evaluated reduced root galling with their efficacy being similar to that of Vydate (Oxamyl 10%) which was used as a positive control. Neem extract treatments showed the lowest mean galling index, followed by Tithonia. In the untreated control, root-knot nematode galling indices were highest at 10 both in trial I and trial II. Treatment trials from Neem and Tithonia at a concentration of 100 ml/L showed highest yield of French beans compared to lowest yield in untreated plants.

Integrated disease management

Different integrated disease management (IDM) approaches including chemical (fumigant, non-fumigant, seed dressing, and seed dip), biological (the egg-parasitic fungus, *Paecilomyces lilacinus* and the mycorrhizal fungus *Glomus* sp.), physical (soil solarization), and cultural (chicken litter and urea) were compared to manage root-knot nematode of French bean under greenhouse (Al-Hazmi *et al.*, 2017). Plant response (plant growth and root galling) and nematode reproduction (production of eggs and the reproduction factor Rf) were considered for evaluation of nine different treatments. It was found that use of chicken litter as an organic fertilizer severely affected the growth and survival of the plants and hence was not evaluated further. Other eight treatments were found to be effective with varying level. The eight treatments decreased (38.9–99.8%) root galling, increased plant growth and suppressed nematode reproduction. Gall index (GI), egg mass index (EMI), and nematode reproduction factor (RF) were calculated for each treatment and treatments were grouped in three categories: (A) the relatively high effective group (GI = 1.0–1.4, Rf = 0.07–0.01), which included the fumigant dazomet, the non-fumigant fenamiphos, soil solarization, and seed dip with fenamiphos; (B) the relatively moderate effective group (GI = 3.4–4.0, Rf = 0.24–0.60), which included seed dressing with fenamiphos and urea; and (C) the relatively less effective group (GI = 5.0, Rf = 32.2–37.2), which included biocontrol agents *viz.*, *P. lilacinus* and *Glomus* sp.

REFERENCES

Agrios, G. N. 2005. *Plant Pathology*. 5th Ed. San Diego CA. Elsevier Academic Press. 15-70 pp.

Al-Hazmi, A., S., Dawabah, A. A. M., Al-Nadhari, S. N. and Al-Yahya, F. A. 2017. Comparative efficacy of different approaches to managing *Meloidogyne incognita* on green bean. *Saudi Journal of Biological Sciences* 24(1): 149-154.

Antes, V. A., Comerlato, A. P., Schuelter, A. R., Carneiro, R. M. D. G. and Furlanetto, C. 2019. Native-plant hosts of *Meloidogyne* spp. from Western Paraná, Brazil. *Brazilian Archives of Biology and Technology* 55(2): 213-219.

Bridge, J., and Starr, J. L. 2007. *Plant Nematodes of Agricultural Importance*. London, UK: Manson Publishing Ltd.

Cheng X, Liu X, Wang H, Ji X, Wang K, Wei M and Qiao K. 2015. Effect of emamectin benzoate on root-knot nematodes and tomato yield. PLoS ONE 10(10): e0141235.

Chitwood, B. G. 1956. Root-knot nematode. A revision of genus *Meloidogyne* Geoldi 1887. Proc. Helminthol. Soc. Was. 16:90-104.

De Waele, D., and Elsen, A. 2007. Challenges in tropical plant nematology. *Annual Review of Phytopathology* 45: 457-485.

Desaeger, J., and Rao, M. R. 2000. Parasitic nematode populations in natural fallows and improved cover crops and their effects on subsistence crops in Kenya. *Field Crop Research* 65:41-56.

Di Vito, M., Parisi, B. and Catalano, F. 2005. Effect of population densities of *Meloidogyne incognita* on common bean. *Nematologia Mediterranea* 32: 81-85.

Evans, A. A. F and Perry, R. N. 2009. Survival Mechanisms .in: "*Root-knot nematodes*" edited by Perry, R. N. Moens, M. and Starr, J. CABI 2009, 201-222 pp.

Giannakou, I. O. and Panopoulou, S. Moral, M. T. (Reviewing editor). 2019. The use of fluensulfone for the control of root-knot nematodes in greenhouse cultivated crops: Efficacy and phytotoxicity effects, *Cogent Food and Agriculture* 5:1, DOI: 10.1080/23311932.2019.1643819

Hussey, R. S, and Jansen, G. J. W. 2002. Root-knot nematodes: *Meloidogyne* species in Starr, J. L, Cook, R, and Bridge, J (eds) *Plant resistance to parasitic nematodes*. CAB International. pp 43-47.

Hussey, R. S., Davis, E. L. and Ray, C. 1994. Meloidogyne stylet secretions in Lamberti. F, De Gorgi. C., and D. M. Bird (eds) *Advances in molecular plant nematology*. Plenum Press, New York. pp 233-249.

Jourand, P., Giraud, E., Be'na, G., Sy, A., Willems, A., Gillis, M., Dreyfus, B. and de Lajudie, P. 2004. *Methylobacterium nodulans* sp. nov., for a group of aerobic, facultatively methylotrophic, legume root-nodule-forming and nitrogen-fixing bacteria. *Int. J. Syst. Evo. Microbiol.* 54: 2269-2273.

Karssen, G., Perry, R. N. and Moens, M. 2006. *Root knot nematodes – plant nematology* CAB international, Wallingford, UK.59-90 pp.

Kimenju, J. W., Kagundu, A. M., Nderitu, J. H., Mambala, F., Mutua, G. K. and Kariuki, G. M. 2008. Incorporating green manure into bean cropping systems. *Asian Journal of Plant Science*: 1682-3974.

Krishnappa, K. 1985. Nematology in developing countries: India - IMP Region VIII. In: Sasser, J.N., Carter, C.C. eds. *An Advanced Treatise on Meloidogyne. Vol.I. Biology and Control. Raleigh, North Carolina State Graphics*, 379- 398.

Lamberti, F., D'Addabbo, T., Sasanelli, N., Carella, A. and Greco, P. 2000. Chemical control of root-knot nematodes. *Acta Hortic*. 532: 183-188.

Machado, A.C.Z., Dorigo, O.F. and Mattei, D. 2013. First report of the root knot nematode *Meloidogyne inornata* parasitizing common bean in Paraná State, Brazil. *Plant Disease* 97: 431.

Machado, A.C.Z., Dorigo, O.F., Carneiro, R.M.D.G. and Araújo Filho, J.V. 2016. *Meloidogyne luci*, a new infecting nematode species on common bean fields at Paraná State, Brazil. *Helminthologia* 53: 207-210.

Mojtahedi, H., Santo, G. S., Hang, A. N. and Wilson, J. H. 1991. Suppression of root-knot nematode populations with selected rapeseed cultivars as green manure. *J. Nematology* 23: 170-174.

Mullin, B. A., Abawi, G. S., Pastor-Corrales, M. A., and Kornegay, J. L. 1991. Root-knot nematodes associated with beans in Colombia and Peru and related yield loss. *Plant Disease* 75:1208-1211.

Nazli, R., Akhter, M. Ambreen, S. Solangi, A. H and Sultana, N. 2008. Insecticidal, nematicidal and antibacterial activities of Gliricidia sepium. *Pak. J. Bot.* 40(6): 2625- 2629.

Njenga, J. K. and Mbaka, J. N. 2019. Efficacy of Neem, Tithonia and Tephrosia leaf extracts in management of root-knot nematodes in French beans (*Phaseolus vulgaris* L.). *Journal of Agricultural Studies* 7(3): 240-251.

Oka, Y., Shapira, N. and Fine, P. 2007. Control of root-knot nematodes in organic farming systems by organic amendments and soil solarization. *Crop Prot*. 26: 1556-1565.

Pakeerathan, K., Mikunthan, G. and Tharshani, N. 2009. Effect of different animal manures on *Meloidogyne incognita* (kofoid and white) on tomato. *World J. Agric. Sci.* 5 (4):432- 435.

Sasser, J. N. 1979. Economic importance of *Meloidogyne* in tropical countries. In: Lamberti, F. and Taylor, C.E. (eds) *Root-Knot Nematodes (Meloidogyne species). Systematics, Biology and Control*. London, UK: Academic Press, 359-374.

Sharf, R., Hisamuddin, A. and Akhtar, A. 2014. Management of root-knot disease in *Phaseolus vulgaris* using potassium fertilizer and biocontrol agents. *J Plant Pathol Microb* 5: 242.

Sharma, G. C. and Dogra, B. S. 2014. Assessment of avoidable yield losses in Rajmash, *Phaseolus vulgaris* due to root-knot nematode, *Meloidogyne incognita. Agric International* 1:24:26.

Siddiqui, M. A. and Alam, M. M. 1987. Control of plant-parasitic nematodes by intercropping with *Tagetes minuta. Nematol. Medit*. 15, 205-211.

Sikora, R.A. and Greco, N. 1990. Nematode parasites of food legumes. Pp. 181-235. In: *Plant Parasitic Nematodes in Subtropical and Tropical Agriculture* (Luc M., Sikora R.A. and Bridge J. eds). CAB International, Wallingford, U.K.

Sikora, R.A., Greco, N. and Silva, J.F.V. 2005. Nematode parasites of food legumes. In: Luc, M., Sikora, R.A., Bridge, J. (Eds) *Plant parasitic nematodes in subtropical and tropical agriculture*. CAB Publishing, Wallingford, USA, pp. 259- 318.

Singh, D. B., Reddy, P. P., Rao, V. R. and Rajendran, R. 2009. Cultivars of French bean resistant to root-knot nematode, *Meloidogyne incognita*, in India. *Tropical Pest Management* 27(1):29-31.

Singh, R. S. and Sitaramaiah, K. 1994. *Plant Pathogens: the Nematode*, Oxford and IBH for rural areas publishing co. Pvt. Ltd, 4-136.

Colour Plates

Chapter 2. Root Rot and Web Blight

(a) Symptoms on leaves

(b) Symptoms on pods

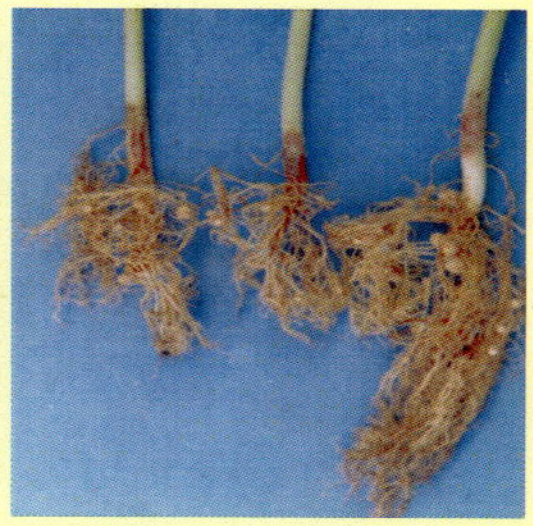

(c) Symptoms on stems and roots

(b) Completely infected plant

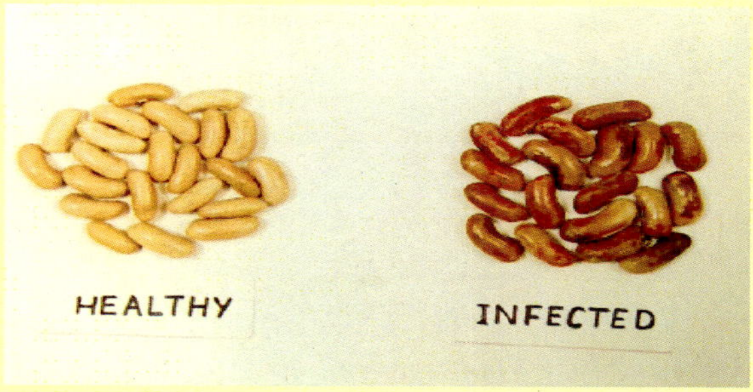

(e) Symptoms on seeds

Plate 1. Symptoms of root rot and web blight of French bean

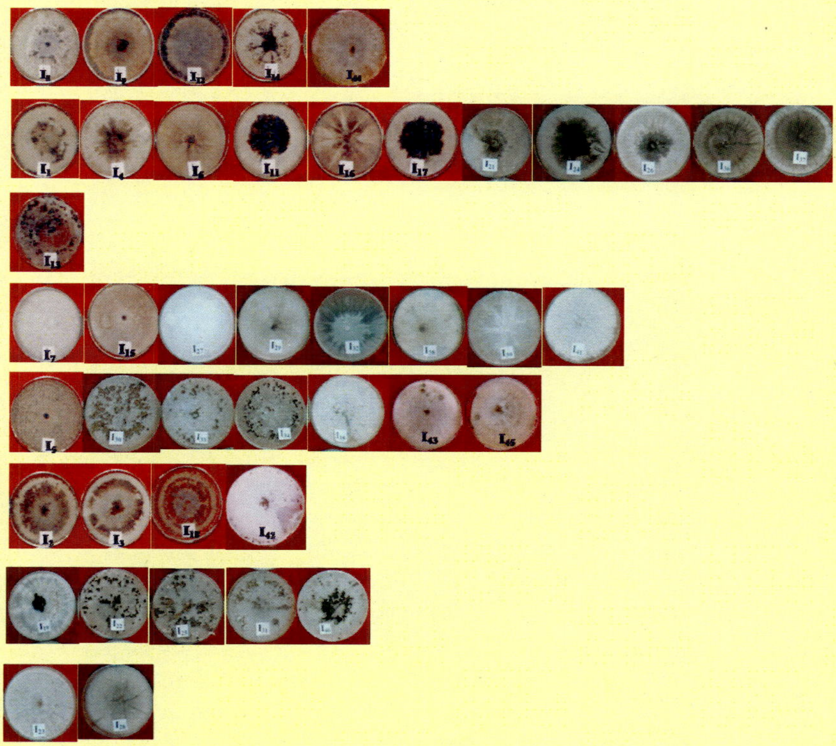

Plate 2a. Cultural and sclerotial variability of *R. solani*

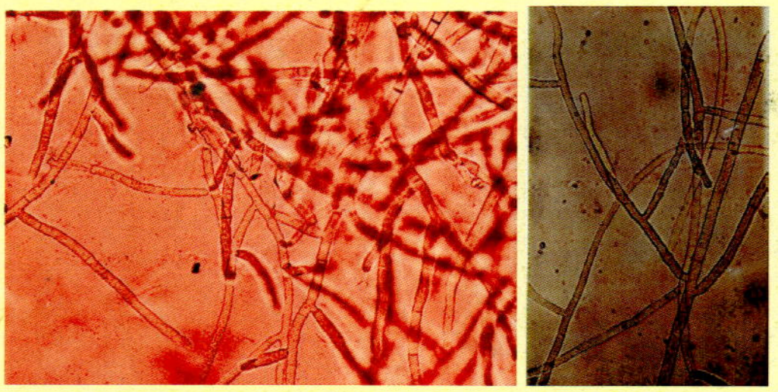

Plate 2b. Mycelium of *R. solani*

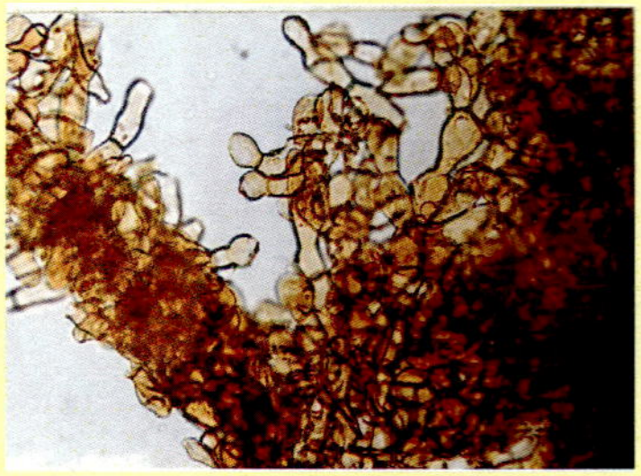

Plate 2c. Moniloid cells of Sclerotia of *R. solani*

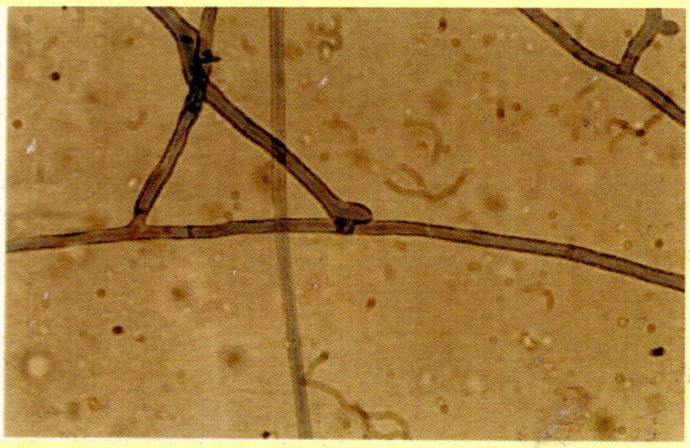

Plate 2d. Type 2 reaction (AG-4 × I15)

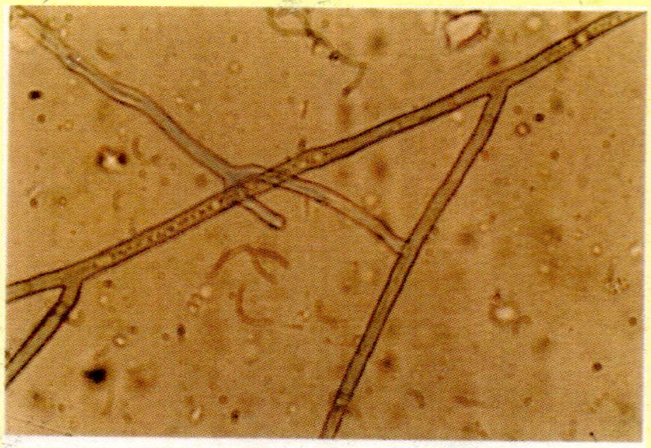

Plate 2e. Type 3 reaction (AG-1-IB × I4)

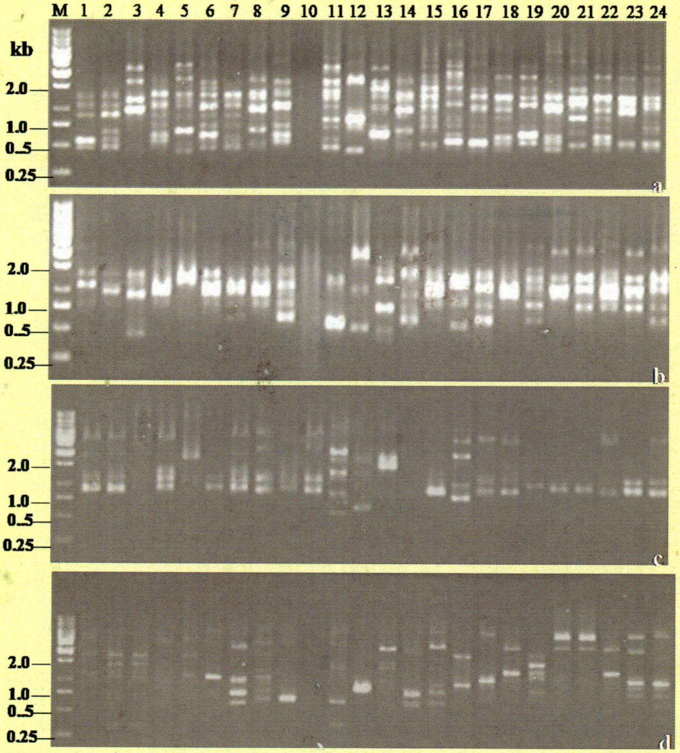

Plate 3a. DNA fingerprint patterns of isolates of *Rhizoctonia solani* generated by RAPD with primers (a) OPA-04 (b) OPA-10 (c) OPA-13 (d) OPA-18

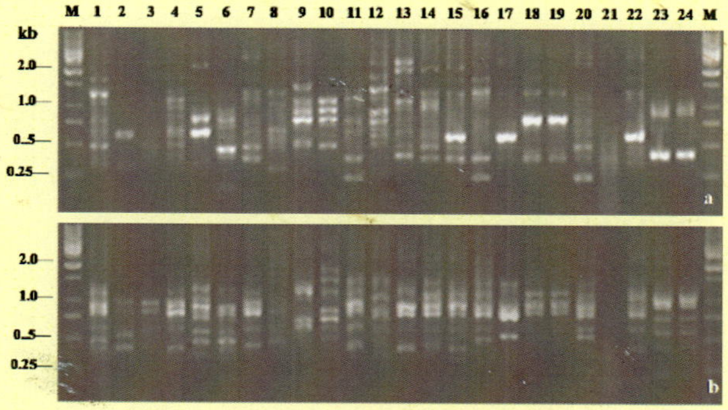

Plate 3b. DNA fingerprint patterns of isolates of *Rhizoctonia solani* generated by URP with primers (a) URP13R (b) URP17R

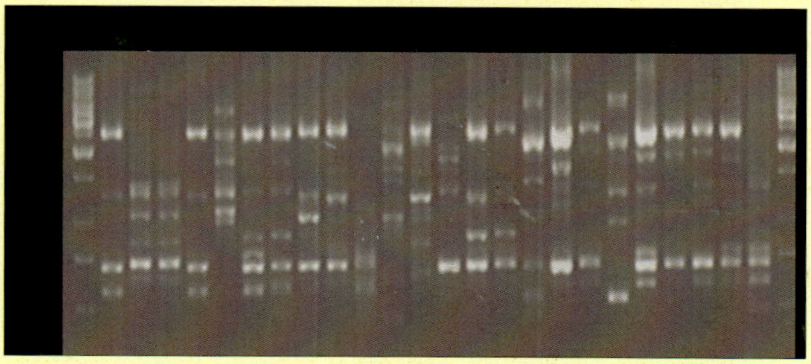

Plate 3c. DNA fingerprint patterns of isolates of *Rhizoctonia solani* generated by ISSR with primers SSR-02

Chapter 3. Anthracnose

Plate 1. Symptoms of anthracnose on pods of French bean

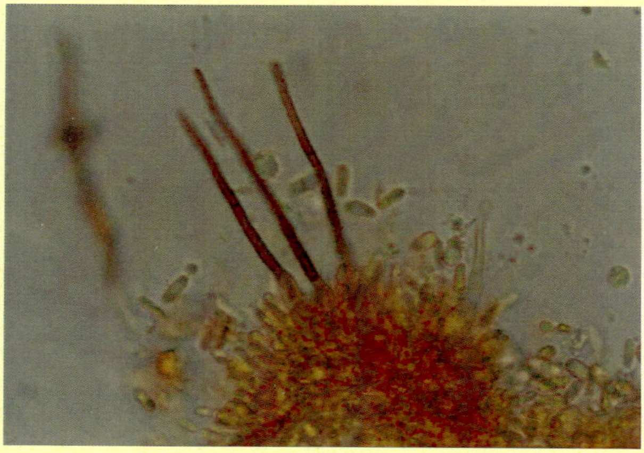

Plate 2. Acervulus of *Colletotrichum lindemuthianum*

Chapter 4. Powdery Mildew

Plate 1. Leaves of French bean showing the symptoms of powdery mildew, caused by *Erysiphe polygoni*

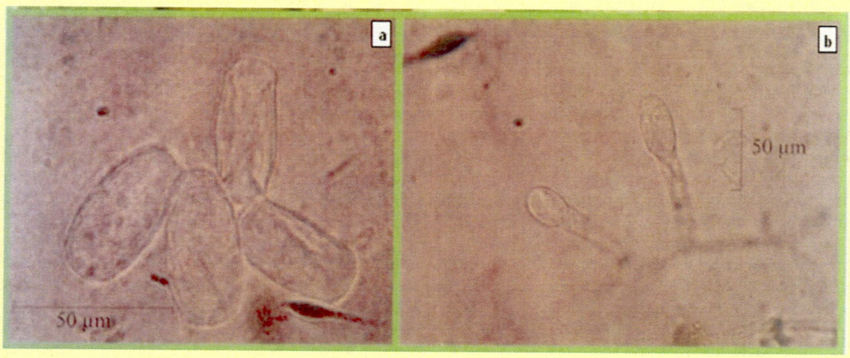

Plate 2. (a) Conidia and (b) conidiophores of *Erysiphe polygoni*

Chapter 5. Rust

Plate 1a. Symptoms of rust on French bean leaves

Plate 1b. Symptoms of rust on stem and petiole of French bean

Plate 1c. Symptoms of rust on pods of French bean

Chapter 6. Angular Leaf Spot

Plate 1a. Symptoms on upper and lower surface of leaves

Plate 1b. Symptoms on cotyledonary leaves

Plate 1c. Infected pods **Plate 1d. Completely infected plant**

Plate 1. Symptoms of angular leaf spot of French bean

Plate 2. Coremia of *Pseudocercospora griseola* (syn: *Phaeoisariopsis griseola*)

Chapter 7. Phytophthora Blight

Plate 1. Symptoms of collar rot (*Phytophthora nicotianae* var. *nicotianae*) on French bean plants

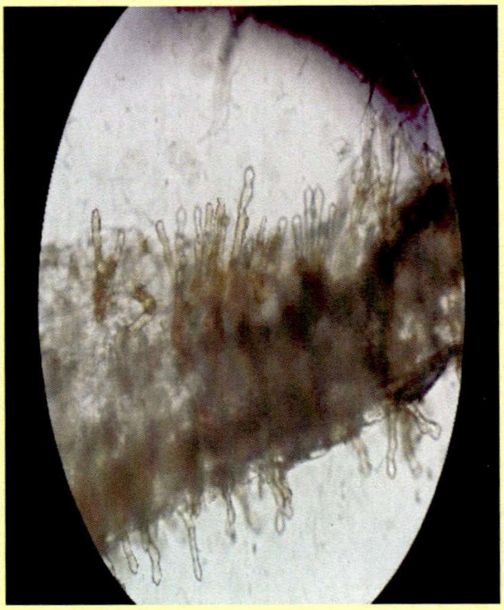

Plate 2. Sympodial growth of sporangiophore and papillate sporangia of *Phytophthora nicotianae* var. *nicotianae* causing infection on French bean

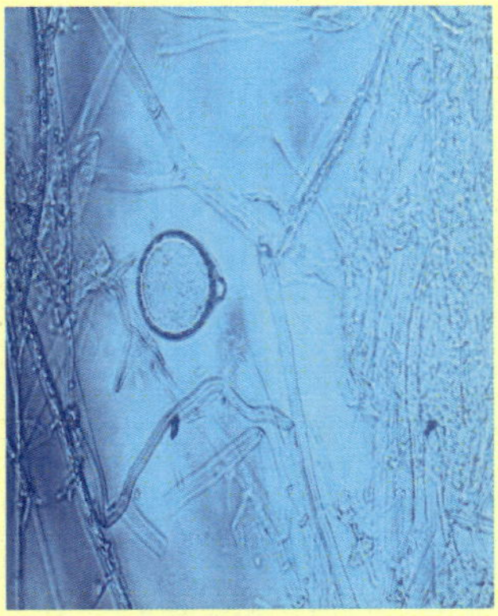

Plate 3. Sporangium of *Phytophthora nicotianae* var. *nicotianae*

Chapter 8. Floury Leaf Spot

Plate 1. Symptoms of Floury leaf spot of French bean on lower leaf

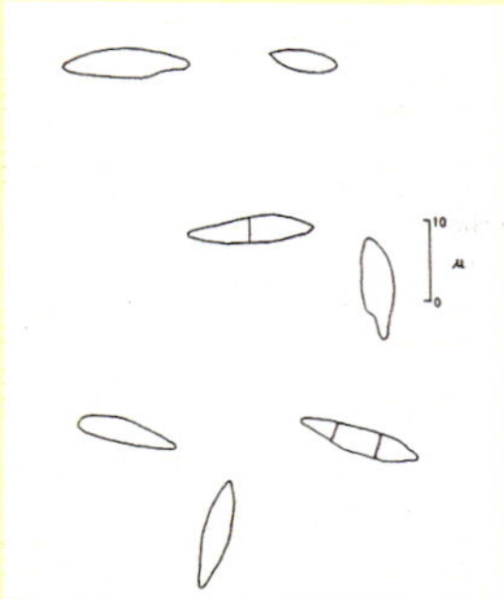

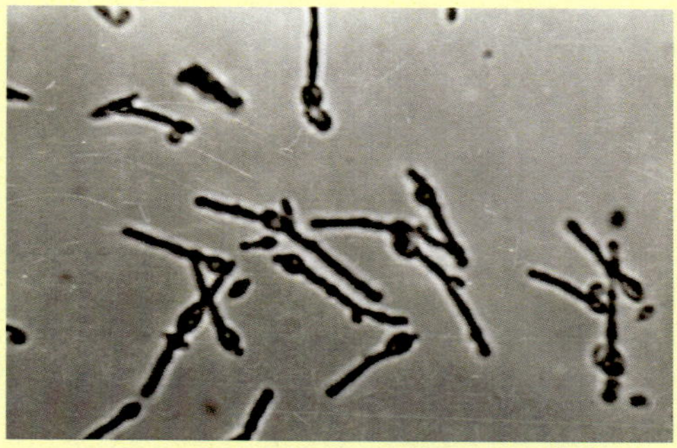

Plate 2a. Conidia of *Ramularia phaseoli* **Plate 2b. Germinating conidia of *R. phaseoli***

(Courtesy: Dr. K.D. Sharma)

Chapter 9. Charcoal Rot

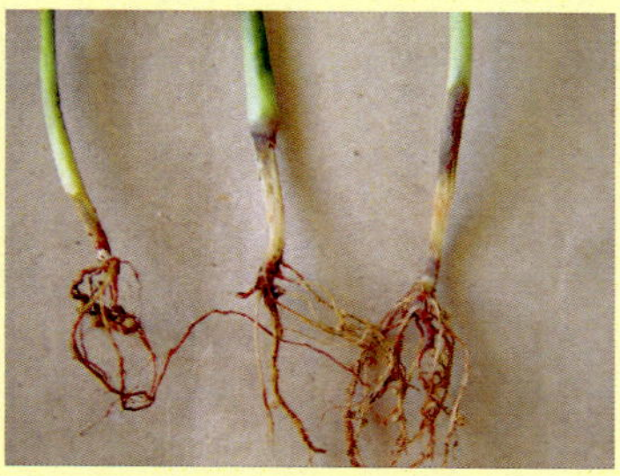

Plate 1 a, b. Symptoms of root rot caused due to infection of *Macrophomina phaseolina*

Plate 1c. Symptoms of charcoal rot on stem

Plate 1d. Pycnidia on stem surface

Plate 1e. Culture of *Macrophomina phaseolina*

Courtesy: Dr. Ritu Mawar

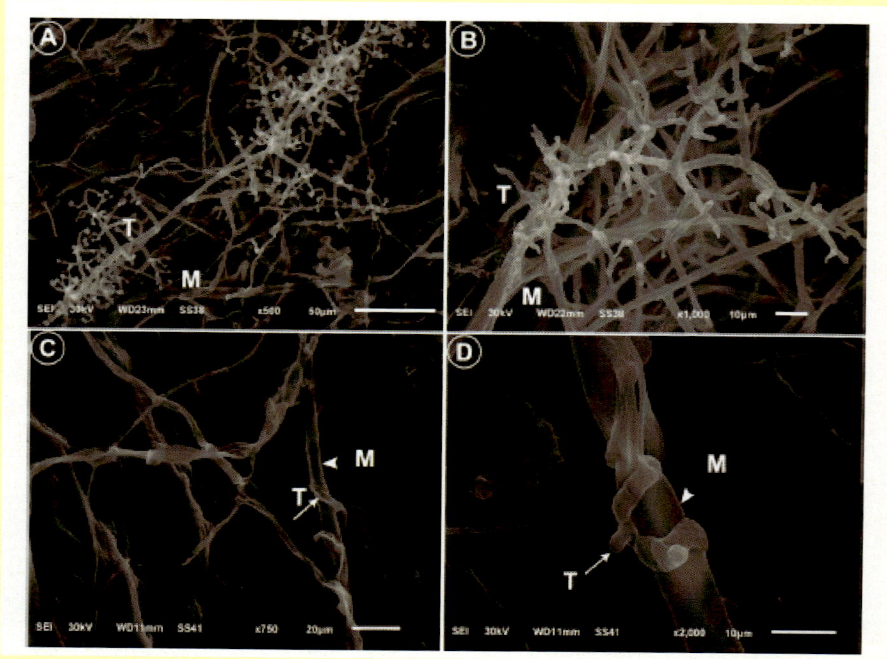

Plate. 2. SEM micrograph showing the parasitic action of *Trichoderma atroviride* T22 (T) against *Macrophomina phaseolina* (M). (A) Growth of T hyphae interacting with those of M. (B) A magnified part showing the growth of T over the hyphae of M. (C) Mycoparasitism process as parallel growth and coiling formation of T hyphae (arrow) over the hyphae of M (arrowhead). (D) A magnified part showing coiling of T hyphae (arrow) over the hyphae of M (arrowhead). Scale bar = 10 μm.

Chapter 12. Crown Rot

Plate 1a. Symptoms on seedling

Plate 1b. Symptoms on grown up plants

Plate 1c. Symptoms on leaf-dorsal view

Plate 1d. Symptoms on leaf-ventral view

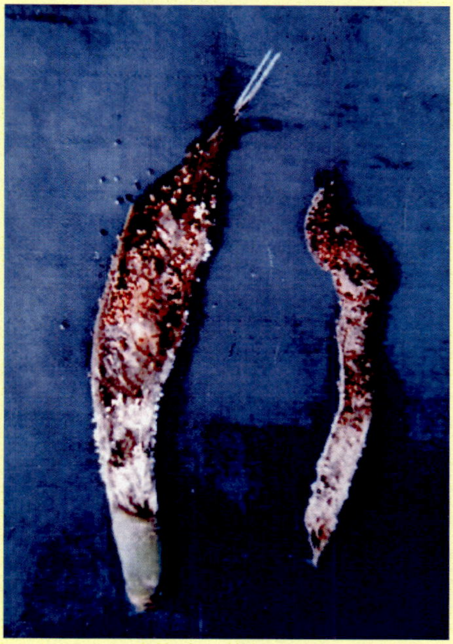

Plate 1e. Symptoms on pods

Plate 1. Symptoms of crwon rot of French bean

Plate 2. ***In vitro*** **inhibition of** ***Sclerotium rolfsii*** **by** ***Gliocladium virens***

Chapter 13. Fusarium Root Rot

Plate 1. Symptoms of Fusarium root rot

Chapter 14. Fusarium Yellows

Plate 1. (a) Symptoms of Fusarium yellows

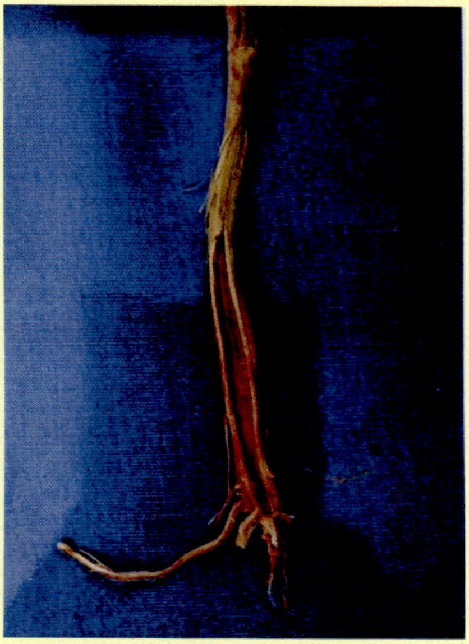

(b) Browning of vascular bundles in *Fusarium* infected stem

Chapter 15. Cercospora Leaf Spot

Plate 1. Symptoms of Cercospora leaf spot on French bean leaves

Chapter 19. Common Blight

Plate 1. Symptoms of Bacterial blight of bean

(*Courtesy: Dr. S. Kansal*)

Chapter 21. Bacterial Wilt

Plate. 1. Symptoms of bacterial wilt of dry beans caused by *Curtobacterium flaccumfaciens* pv. *flaccumfaciens* on common bean. Symptoms first appear as interveinal chlorotic areas (a,b), leading to leaf wilting (c) and flabby or flaccid areas on the leaves (d,e). Subsequently, tissue necrosis on leaves surrounded by a chlorotic margin are observed (f–h). Severe infections on common bean lead to overall wilting and defoliation as well as plant mortality (i)

(Source: Osdaghi *et al*., 2020)

Chapter 22. Bean Common Mosaic

Plate 1. Symptoms of Bean common mosaic

Chapter 23. Bean Yellow Mosaic

Plate 1. Symptoms of bean yellow mosaic

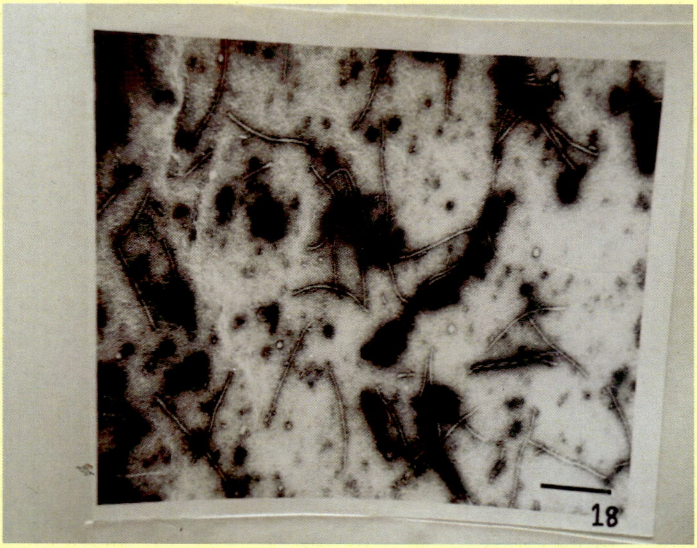

Plate 2. Flexuous, filamentous particles of *Bean Yellow Mosaic Virus*

Related Books

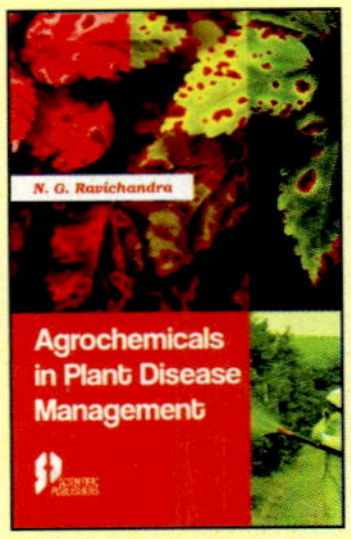

ISBN : 978-93-88043-56-4

ISBN : 978-81-72337-43-8

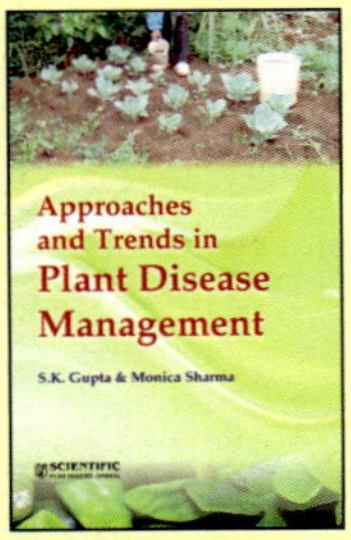

ISBN : 978-81-72338-87-9

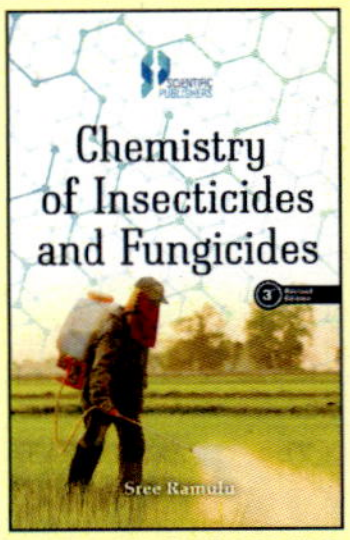

ISBN : 978-93-89832-02-0

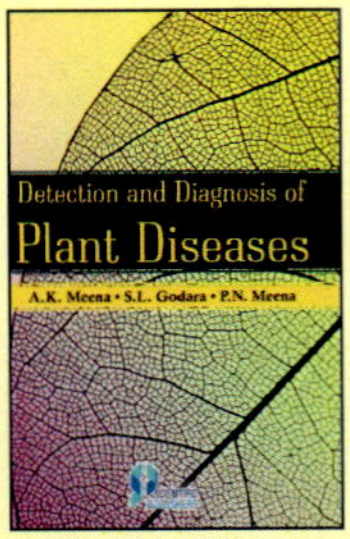

ISBN : 978-93-89184-40-2

ISBN : 978-93-86652-23-2

ISBN : 978-81-72335-43-4

ISBN : 978-93-86102-74-4

ISBN : 978-93-88043-63-2

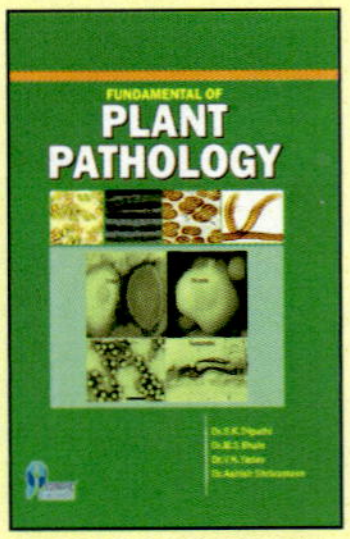

ISBN : 978-93-92590-41-2

ISBN : 978-81-72337-01-8

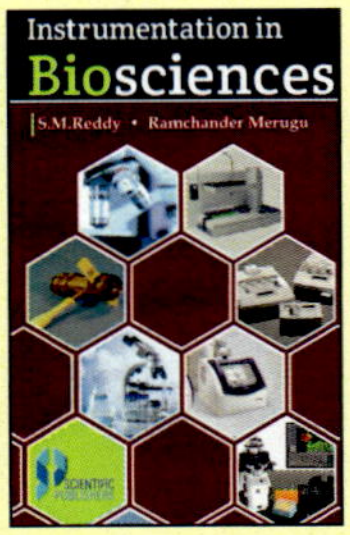

ISBN : 978-93-89412-54-3

ISBN : 978-81-72337-13-1

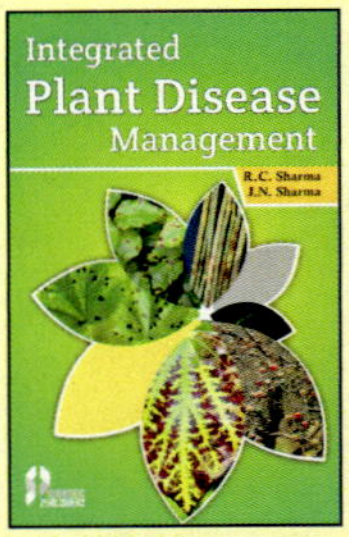

ISBN : 978-81-72334-00-0

ISBN : 978-81-72335-60-1

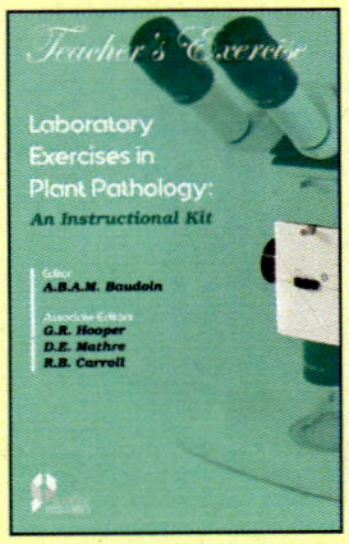

ISBN : 978-81-72337-18-6

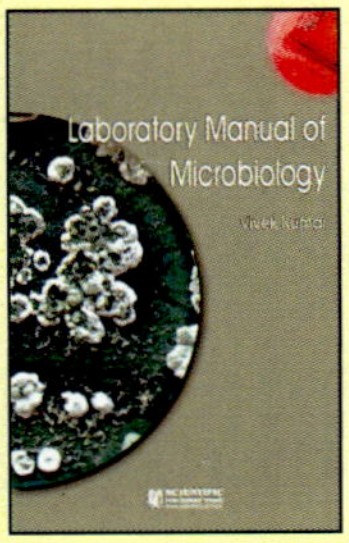

ISBN : 978-81-72337-06-3

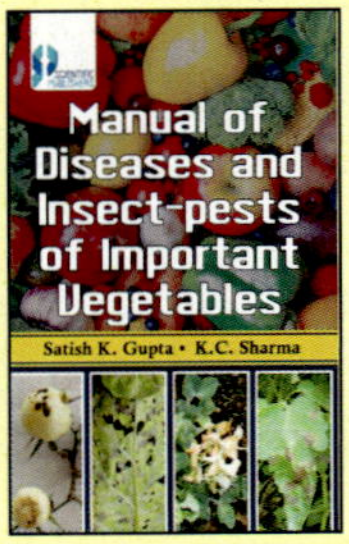

ISBN : 978-93-89412-28-4

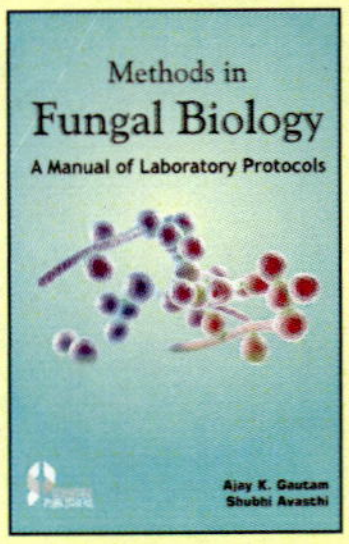

ISBN : 978-93-88043-67-0

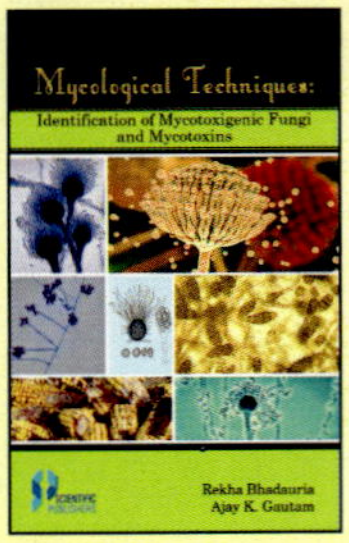

ISBN : 978-93-88043-68-7

ISBN : 978-93-89412-71-0

ISBN : 978-93-87741-73-7

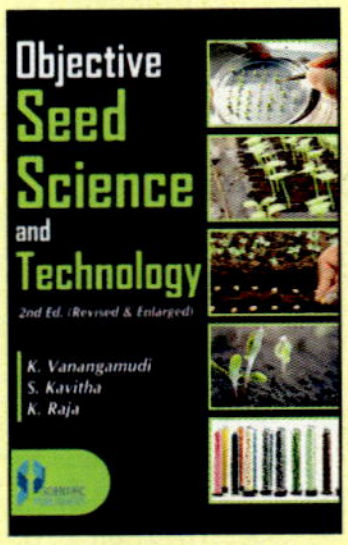

ISBN : 978-93-89412-60-4

ISBN : 978-93-86347-26-8

ISBN : 978-93-92590-62-7

For More Details Visit Website